Shirley Dobbie

RHS GOOD PLANT GUIDE

RHS GOOD PLANT GUIDE

DORLING KINDERSLEY
LONDON • NEW YORK • SYDNEY
www.dk.com

A DORLING KINDERSLEY BOOK
www.dk.com

PROJECT EDITOR **Simon Maughan**
EDITOR **Tracie Lee**
ART EDITOR **Ursula Dawson**
MANAGING EDITOR **Louise Abbott**
MANAGING ART EDITOR **Lee Griffiths**
DTP DESIGN **Sonia Charbonnier**
PRODUCTION **Ruth Charlton**
PICTURE RESEARCH **Sam Ruston, Neale Chamberlain**

First edition published in Great Britain in 1998 by Dorling Kindersley
Publishers Limited, 9 Henrietta Street, London WC2E 8PS
This revised and expanded edition published 2001

4 6 8 10 9 7 5

A CIP catalogue record for this book is available from
the British Library.

ISBN 0 7513 0812 9

Colour reproduction by GRB Editrice, Italy.
Printed and bound in Italy by Lego.

CONTENTS

INTRODUCTION

GARDENERS TODAY have plenty of choice when buying plants. Not only are plant breeders constantly producing new and exciting cultivars, but plants are now more widely available, in garden centres and nurseries and also in homecare outlets and superstores – and in less traditional plant-buying situations, information and advice may not be readily available. It is perhaps no wonder that gardeners sometimes find choosing the right plant a bewildering business.

The *RHS Good Plant Guide* was conceived to help gardeners select outstanding and reliable plants for their garden, whatever their level of expertise and experience. Because all of the plants it features are recommended by the Royal Horticultural Society – the majority holding its coveted Award of Garden Merit (see pages 10–12) – even the novice gardener can choose plants from its pages with confidence.

However, these recommendations refer only to the plants themselves, and not to those who grow them. It is impossible to guarantee the quality of plants offered for sale by any nursery, garden centre or store, and here gardeners must exercise a degree of common sense and good judgment, in order to ensure that the individual plants they select and take home will thrive. The following pages include guidance on recognising healthy, well-grown plants and bulbs, and also show how to get them home safely and get them off to a good start.

Ever-widening plant choice
*Exotic flowers, like this blue poppy (*Meconopsis grandis*), may tempt the gardener, but check the conditions they require before buying.*

CHOOSING FOR YOUR GARDEN

Selecting plants appropriate to your site and soil is essential, and each plant in the *A–Z of AGM Plants* has its preferences indicated. Well-prepared soil, feeding and, in dry conditions, watering in the early stages are also important for the well-being of most plants (although in the *Planting Guide* you will find plants to grow in poor soil or dry sites). Entries in the *A–Z* include basic care for the plant concerned, along with hints and tips on pruning and winter protection where hardiness is borderline. By following this advice, the plants you choose using the *RHS Good Plant Guide* should perform well, either as garden decoration or in cropping, fulfilling all the expectations conferred on them by the RHS recommendation.

USING THE GUIDE

THE *RHS GOOD PLANT GUIDE* is designed to help you choose plants in two different ways.

THE A–Z OF AGM PLANTS

Here, over 1,000 plants have full entries and are illustrated with photographs. This is the section to consult to find details about a plant, whether you are reading its name on a label or have noted it down from a magazine article or a television or radio broadcast. It will tell you what type of plant it is, how it grows, what its ornamental features are, where it grows and looks best and how to care for it. For quick reference, symbols (right) summarise its main requirements.

THE PLANTING GUIDE

This section provides "shopping lists" of plants for every purpose – whether practical, such as a group of plants for a damp, shady site, or for themed plantings – a range of plants to attract birds into your garden, for example. Fruit and vegetable cultivars are also recommended. Page references are given to plants with entries and portraits elsewhere.

Show stoppers (facing page)
Flower shows such as Chelsea provide opportunities to see new plants.

SYMBOLS USED IN THE GUIDE

♈ The RHS Award of Garden Merit

SOIL MOISTURE PREFERENCES/TOLERANCES

◌ Well-drained soil
◑ Moist soil
● Wet soil

SUN/SHADE PREFERENCES/TOLERANCES

☼ Full sun
◐ Partial shade: either dappled shade or shade for part of the day
● Full shade

NB Where two symbols appear from the same category, the plant is suitable for a range of conditions.

HARDINESS RATINGS

(given in conjunction with the AGM and preceded by the AGM symbol)

♈H1 Plants requiring a warm, permanently frost-free climate, which must be grown in heated glasshouses or as house plants in the UK. In practice some can be grown outside in summer as bedding or in containers.
♈H2 Plants requiring unheated glass (or the protection of an unheated glasshouse): some may be placed outdoors in summer.
♈H3 Plants hardy outside in some regions or in favoured sites, or those that, in temperate countries, are usually grown outside in summer but need frost-free cover over winter (e.g. dahlias).
♈H4 Fully hardy plants.
♈H1+3 In frost-prone climates, conservatory plants that may be moved outdoors in summer.

THE PLANTS IN THE GUIDE

THE RANGE OF PLANTS – well over 2,000 – featured in the *RHS Good Plant Guide* has been carefully chosen to provide the best selection across all the different types of plant, and all the situations for which plants may be required. All should be readily available from sources in the United Kingdom, although gardeners in other countries may have to do a little more research to track down some specimens. The majority hold the Royal Horticultural Society's Award of Garden Merit, the trophy symbol for which is now a recognised "kitemark" of quality throughout the UK and even further afield, seen in nursery catalogues, on plant labels and in a variety of gardening publications.

ABOUT THE AGM

The RHS Award of Garden Merit, re-instituted in 1992, recognises plants of outstanding excellence, whether grown in the open or under glass. Its remit is broad, encompassing all ornamental plants as well as fruit and vegetables. Any type of plant can be entered for the award, be it the most minuscule alpine plant or a majestic tree.

Besides being the highest accolade the Society can give a plant, the AGM system is primarily designed to be of practical use to ordinary gardeners, helping them in making a choice from the many thousands of plants currently available. When choosing an AGM plant, the gardener can be certain that it fulfils the following criteria:

G. clarkei 'Kashmir White'
Plants that are versatile and easily grown earn their AGMs for sheer garden value.

Camellia ×williamsii 'Brigadoon'
Choice forms of flowering plants are recognised by the award scheme.

Hydrangea macrophylla 'Altona'
The best selections of many garden stalwarts are recommended.

- It should be excellent for ornamental use, either in the open garden or under glass
- It should be of good constitution
- It should be available in the horticultural trade, or at least be available for propagation
- It should not be particularly susceptible to any pest or disease
- It should not require any highly specialised care, other than providing the appropriate conditions for the type of plant or individual plant concerned (for example, lime-free soil if required)
- It should not be subject to an unreasonable degree of reversion in its vegetative or floral characteristics.

To explain this point simply, many plants with unusual characteristics differing from the species, such as double flowers or variegated leaves, have often been propagated from a single plant – a natural mutation or "sport" – that has appeared spontaneously. Plants bred from these "sports" – especially when raised from seed – are liable to have only the normal leaf colour or flower form. It takes several generations of careful and controlled propagation for the special feature to be stable enough for plants to be recognised and registered with a distinct cultivar name (which is usually chosen by the breeder) and allowed to be offered for sale. Many never breed reliably enough from seed with the special feature, and can only be reproduced by vegetative methods such as taking cuttings.

HOW PLANTS GAIN AWARDS

An AGM award is made either following plant trials, usually conducted at the Society's display

Prunus laurocerasus *This evergreen is handsome year-round both as a free-standing shrub and as a hedge.*

Anemone blanda **'White Splendour'** *Floriferous cultivars for every site and season gain the RHS award.*

Acer negundo **'Flamingo'** *AGM plants that have variegated foliage should not be prone to excessive reversion.*

garden at Wisley in Surrey, but also at other locations where expertise in growing a particular genus is to be found, or on the recommendation of a panel of experts. The AGM is not graded, so there is no attempt to differentiate the good from the better or best, but in those groups in which there are many cultivars, standards are necessarily high so that the resulting selection offers helpful guidance to the gardener.

OTHER PLANTS IN THE GUIDE

In some categories of plants, notably annuals (particularly those for summer bedding) and vegetables, new cultivars appear so rapidly, often superseding others offered for sale, that the trialling process needed for AGM assessment is hard-pressed to keep up with developments. In order to give a wider choice in these somewhat under-represented categories, the *RHS Good Plant Guide* includes, in its *Planting Guide* section, other selected cultivars which have proved their reliability over the years. Some have been grown at the RHS gardens at Wisley and around the UK; some are recommended by other bodies, such as the respected Fleuroselect organisation, an association based in the Netherlands that specialises in recommending cultivars and seed series of annuals and biennials from top breeders, and whose awards are recognised throughout Europe.

Putting plants to the test
Viola *trials at the RHS gardens at Wisley, helping to select the best forms for gardeners.*

FINDING PLANTS BY NAME

The botanical or "Latin" names for plants are used throughout the Guide, simply because these are the names that gardeners will find on plant tags and in nursery lists: these names are also international, transcending any language barriers.

To the uninitiated, plant nomenclature may seem confusing, and sometimes plant names appear to be very similar. It is important to recognise that the AGM is given only to specific plants within a genus – and not, for example, to another cultivar within the same species, however similar that plant might look to its awarded relative. Some plants have synonyms – older or alternative names – by which they may sometimes be referred to: the *RHS Good Plant Guide* gives popular recognised synonyms for a number of plants. A brief guide to plant nomenclature can be found on pages 24–25. Many common names for plants or plant types that feature in the *Guide* are also given, both in the *A–Z of AGM Plants* and in the Index.

SHOPPING FOR GOOD PLANTS

THE FIRST STEP in ensuring that your garden will be full of healthy plants is to choose and buy carefully.

WHERE TO BUY

Plants can be found for sale in such a variety of situations today that there are no hard and fast rules to be applied. Generally, only buy plants where you feel reasonably sure that the plant offered is actually what it says it is, that it has been well-grown, and is not going to bring any pests and diseases into your garden.

PLANTS BY POST

Buying plants by mail order is one of the easiest (and most addictive!) ways of obtaining particular plants that you are keen to acquire. It gives you a huge choice, far greater than in most garden centres, and also gives you access to specialist nurseries that concentrate on certain plant groups or genera, which may not be open to visitors. Armed with a buyer's guide such as *The RHS Plant Finder*, you can track down almost any specimen from the comfort of your own home, and provided that someone will be at home on the day it arrives, it should be delivered in perfect health. Mail-order nurseries are usually happy to replace any plant damaged in transit. The plants will arrive, usually, with their roots wrapped or "balled" in a little compost, and are best planted out immediately.

WHEN TO BUY

Trees, shrubs and roses that are available as bare-rooted specimens are invariably deciduous. Woody plants with roots balled and wrapped may be deciduous or evergreen. All are available only in the dormant season; they are ideally purchased in late autumn or early spring, and should be planted as soon as possible.

Container-grown plants are now much more popular, and are offered for sale all year round. They can be planted year-round too, but the traditional planting times of spring and autumn, when conditions are not too extreme, are still the best. No plants should be put into freezing or baked, dry soil, but kept in their containers until conditions are more favourable.

In frost-prone climates, only fully hardy plants should be planted in winter. Spring is the time to plant plants of borderline hardiness, and to buy bedding. Beware buying summer bedding too early, however tempting it looks after the dark, bare days of winter. Bedding sold in early and mid-spring is intended to be bought by people with greenhouses and cold frames, who can grow the plants on under cover until all danger of frost has passed. Plant out bedding too early in the garden, and you risk losing it all to late frost.

YOUR SHOPPING LIST

Whether you are stocking a new garden from scratch, replacing a casualty or simply wanting to add extra touches with some plants for a shady corner, say, or for a patio tub, informed choice is the key to buying the right plants, and to getting the very best and healthiest specimens.

There are many factors to consider about your garden (see pages 16–17) and, equally, buying tips that will help you make the best choice from the selection of plants on offer (see pages 18–19). Many gardeners prefer to set out with specific plants in mind. However, no one could deny that impulse buys are one of gardening's great pleasures – and with the *RHS Good Plant Guide* to hand, you will be able to snap up just the right plants for your garden, and avoid expensive mistakes.

Well-grown plants for sale
Healthy, clearly labelled plants in neat, well-kept surroundings are a good indication of excellent nursery care.

CHOOSING THE RIGHT PLANTS

ALTHOUGH THERE ARE ways to get around many of plants' climatic and soil requirements, you will avoid extra work and expense by choosing plants well-suited by the conditions your garden offers.

HARDINESS

Most of us have gardens that experience some degree of frost in winter. Optimism is no substitute for action if plants susceptible to frost damage are to survive. Preferably, move them into a greenhouse for the winter, but a frost-free shed or cool utility room can also be used. In cold gardens, make full use of the ever-widening choice of plants offered as annual summer bedding. Reserve warm spots for shrubs and perennials of borderline hardiness, and protect them in winter with a

Extending the range
Container growing may be the answer for gardeners who covet plants with special soil needs, like these acid-loving azaleas, that their own gardens cannot meet.

thick mulch such as straw around the base, kept dry with plastic sheeting or bubble polythene.

SUN AND SHADE

It is well worth "mapping" your garden to see how much sun areas receive at different times of the day and year. For good growth and the best display of features such as coloured foliage, always match plants' sun and shade requirements.

YOUR SOIL

Most plants tolerate soil that falls short of perfect, and many survive in conditions that are very far from ideal. It is always preferable – and far more labour-saving – to choose plants that suit your soil, rather than manipulate your soil to suit plants, by adding, say, quantities of peat or lime. The effect never lasts, and is now also considered environmentally inadvisable – local insects, birds and other fauna may be unable or unwilling to feed on plants to which they are unaccustomed. It is, however, important to add nutrients to the soil, and two birds can be killed with one stone if you add these not as powder or pellets, but as a mulch that will also improve soil texture – compost or rotted manure.

Determining your soil acidity or its reverse, alkalinity, measured by units known as pH values, is important. Most plants tolerate a broad range of pH values around neutral, but some groups have specifically evolved to be suited by soil that is either definitely acid or definitely alkaline. There is no point ignoring these soil preferences and trying to grow, for example, azaleas in a chalky soil; they will not thrive. Choose plants that will enjoy your soil or, if you really covet some specialised plant groups, grow them in containers, in a potting mix that meets their needs.

DRAINAGE

The expression "well-drained yet moisture-retentive" is one of the most widely used yet seemingly baffling expressions in gardening. However, it is not such a contradiction in terms as it seems. What it applies to is a soil that in composition is not dominated by pebbles or grains of sand, which cannot hold water, nor clay which binds up and retains water in a solid gluey mass. It is the presence of a dark, spongy organic material known as humus that enables soil to hold moisture *and* air simultaneously, so that they are available to plants. Working in organic matter rich in humus (see page 21) is the most useful way to improve soil texture, either across the whole garden or as plants are planted.

CHOOSING A HEALTHY PLANT

TRY TO RESIST buying plants that are in poor condition, even if it is the sole example on offer of a plant you really crave. Nursing a pathetic specimen could take a whole growing season, during which time the plant will scarcely reward you, and with container-grown plants now so widely available it is likely that, later in the season, you will find a healthier specimen that you can plant up straight away. In practice, however, there are few gardeners that have never taken pity on a neglected or undernourished plant – but you must always harden your heart to those showing signs of pest infestation or disease. You risk importing problems that could spread to other plants.

WHAT TO BUY

Plants offered for sale should be clearly labelled, healthy, undamaged, and free from pests and diseases. Inspect the plant thoroughly for signs of neglect. With root-balled plants, check that the root ball is firm and evenly moist, with the netting or hessian wrapping intact. Containerised plants should always have been raised in containers. Do not buy plants that appear to have been hastily uprooted from a nursery bed or field and potted up. There should be no sign of roots at the surface.

Look under the pot for protruding roots, a sign that the plant is pot-bound (has been in its container for too long). Always lift the pot to make sure roots have not penetrated the standing area, another sign of a root-bound specimen. Crowded roots do not penetrate the surrounding soil readily after planting, and the plant will not establish well.

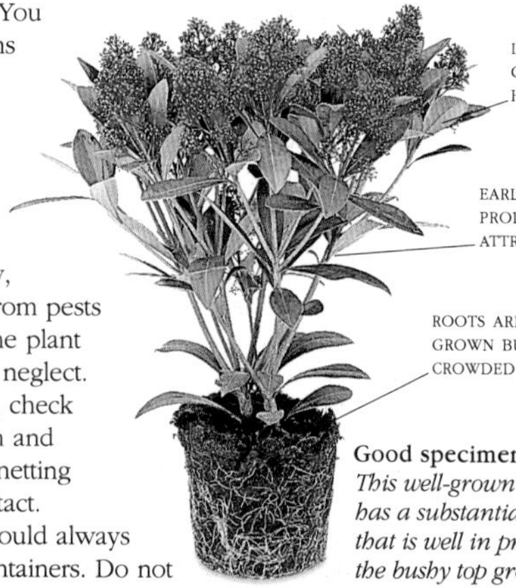

LEAVES ARE GLOSSY AND HEALTHY

EARLY PRUNING HAS PRODUCED AN ATTRACTIVE SHAPE

ROOTS ARE WELL-GROWN BUT NOT CROWDED

Good specimen
This well-grown skimmia has a substantial root ball that is well in proportion to the bushy top growth.

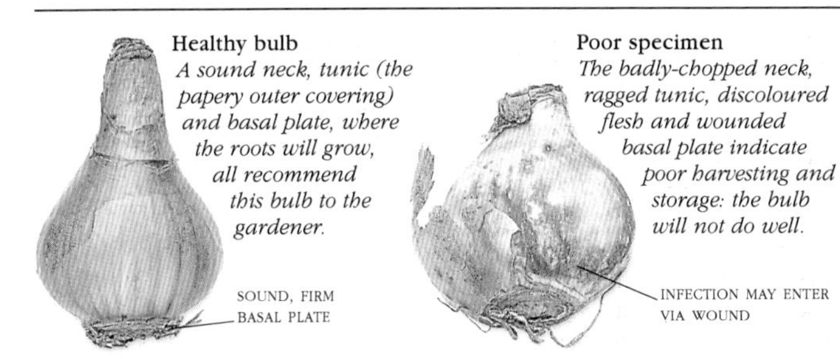

Healthy bulb
A sound neck, tunic (the papery outer covering) and basal plate, where the roots will grow, all recommend this bulb to the gardener.

Poor specimen
The badly-chopped neck, ragged tunic, discoloured flesh and wounded basal plate indicate poor harvesting and storage: the bulb will not do well.

HEALTHY GROWTH

Strong top growth is another key factor. Avoid plants with pale or yellowing foliage, damaged shoot tips or etiolated growth (soft, pale, over-extended and weak-looking shoots). Look at the compost surface: it should not be covered in weeds, moss or liverworts.

With herbaceous plants, small healthy specimens are cheaper than large ones and will soon grow away once planted. Groups of three or four plants often look better than single specimens. With all types of herbaceous plants, including young annuals and bedding, pick out the stockiest, bushiest specimens, with (if appropriate) plenty of flower buds, rather than open flowers.

PLANT SHAPE AND FORM

When buying trees and shrubs, which are to make a long-lasting contribution to your garden, consider their shape as well as their good health. Choose plants with a well-balanced branch framework, and with top growth that is not far too large for the rootball or pot. A stocky, multi-stemmed plant usually makes the best well-rounded shrub, while young trees with a single trunk should have just that – and a good, straight, sound one too – right from the start. A one-sided shrub may of course be acceptable for wall-training, but will put on more growth from the base, and thus cover space better, if its shoot tips are pruned after planting.

BUYING BULBS

Buy bulbs (see above) as if they were onions you were intending to cook with and eat – reject any that are soft, discoloured, diseased or damaged. While it is not ideal, there is no harm in buying bulbs that have begun to sprout a little.

PREPARING THE SOIL

WHILE THE BEST plants for your garden are those suited to the growing conditions available, most soils can be improved to extend the range of plants that can be grown.

IDENTIFYING YOUR SOIL TYPE

Investigating your soil to discover just what its qualities are is one of the most useful things you can do to ensure thriving plants and successful gardening. Generally, the soil will either be sandy or have a clay texture, or somewhere between the two. If your soil has a light, loose texture and drains rapidly, it is probably sandy. Sandy soil has a rough, gritty feel when rubbed and makes a characteristic rasping, scraping sound against the blade of a spade. Although easy to dig, it is low in fertility. Clay soil is heavy, sticky, cold, easy to mould when wet, and can become waterlogged. It is difficult to work, but is often very fertile. A good mix of the two – a "medium loam" – is ideal.

To further establish which plants will be best suited by the soil in your garden, you can discover its pH value (whether and to what degree it is acid or alkaline) and the levels of various nutrients it contains using very simple testing kits that are available at most garden centres.

CLEARING WEEDS

In a new garden, or if you are planting in a previously uncultivated area, your first task will be to clear away any debris and weeds. Pre-planting clearance of all weeds is essential, since they will compete with your plants for light, moisture and nutrients. Where the ground is infested with perennial weeds, an alternative to laborious hand-clearing is to spray it over in the season before planting with a systemic herbicide that will kill their roots. This is most effective when the weeds are growing strongly, usually in early summer. Although this may mean delaying planting, patience at this stage will be amply rewarded later. Repeated applications may be needed.

Isolated perennial weeds can be forked out carefully, removing all root fragments. Annual weeds may be hoed or sprayed off immediately before planting.

WORKING THE SOIL

Digging or forking over the soil helps to break down compacted areas and increase aeration, which encourages good plant growth. The deeper you dig the better, but never bring poor-quality subsoil up to the surface: plants need all the nourishment that the darker, more

Digging with ease
The correct tools and technique make digging more comfortable, and safer for your back. Test the weight and height of a spade before buying, and always keep a straight back when using it.

Forking over
Drive the fork into the ground, then lift and turn the fork over to break up and aerate soil. Spread a layer of organic matter over the soil first so that it is incorporated into the soil as you work.

nutritious topsoil can give them. If you need to dig deep, remove the topsoil, fork over the subsoil, then replace the topsoil.

Dig heavy, clay soils in late autumn. The weathering effects of frost and cold over the winter will help improve the soil's texture by breaking it down into smaller pieces. Avoid digging when the soil is wet, as this will damage the soil structure. It will save work if you incorporate soil conditioners as you go.

ADDING SOIL CONDITIONERS

Digging and forking will improve soil texture to some extent, but to really bring a soil to life, the addition of a soil conditioner is invaluable. The drainage, aeration, fertility and moisture-holding properties of most soil types can be improved simply by adding well-rotted organic matter, rich in humus. This is best done in the season before planting, to allow the soil to settle. Garden compost and leafmould enhance the moisture retention of a sandy soil and improve nutrient levels. Applied regularly, they improve the structure of clay soil. Clay soils can be further opened up by the addition of horticultural gravel to a depth of at least 30cm (12in). Organic mulches of bark, cocoa shells or chippings are also eventually broken down into the soil.

PLANTING OUT

WHATEVER YOU ARE planting, make sure that you give your plants a really promising start. Careful planting saves both time and money, and well-chosen plants positioned in optimum conditions will perform well and should resist attack from pests and diseases.

PRE-PLANTING PLANNING

Before planting, check the potential heights and spreads of plants to ensure that you leave the correct distances between them. When designing plant groups, consider different plants' season of interest, and their appearance in winter.

WHEN TO PLANT

The best seasons for planting are autumn and spring. Autumn planting allows plants to establish quickly before the onset of winter, since the soil is still warm and moist enough to permit root growth. Spring planting is better in cold areas for plants that are not fully hardy, or dislike wet winter conditions.

Perennials can be planted at any time of the year, except during extreme conditions. They should grow rapidly and usually perform well within their first year. To reduce stress on a perennial when planting during dry or hot weather, prune off

Basic planting
Soak plants well in a bowl of water. Position the plant so that the root ball surface is flush with soil level, then fill round the sides.

Settling the soil
Backfill the hole, gently firming the soil to ensure good contact with the roots. Water in well, then add a layer of mulch over the root area.

its flowers and the largest leaves before planting. In full, hot sun, shade the plant with garden netting.

PLANTING TECHNIQUES

For container-grown or rootballed plants, dig a hole about twice the size of the rootball. If necessary, water the hole well to ensure that the surrounding soil is thoroughly moist. Carefully remove the plant from its pot and gently tease out the roots with your fingers. Check that the plant is at the correct depth in its hole, then backfill the hole with a mix of compost, fertiliser and soil. Firm the soil, and water thoroughly to settle it around the roots. Apply a mulch around, but not touching, the plant base to aid moisture retention and suppress weeds.

For bare-rooted plants, the planting technique is essentially the same, but it is vital that the roots never dry out before replanting. Dig holes in advance, and if there is any delay in planting, heel the plants in or store in moist sand or garden compost until conditions are more suitable. Bare-rooted plants need a planting hole wide enough to accommodate their roots when fully spread. The depth must be sufficient to allow the final soil level to be the same as it was in the pot or nursery. After planting, tread the soil gently to firm.

WALL SHRUBS AND CLIMBERS

Always erect supports before planting. Plant wall shrubs and climbers at least 25cm (10in) from walls and fences, so that the roots are not in a rain shadow, and lean the plant slightly inward toward the wall. Fan out the main shoots and attach them firmly to their support. Shrubs and non-clinging climbers will need further tying-in as they grow; the shoots of twining climbers may also need gentle guidance.

ANNUALS AND BEDDING

In order to ensure that these plants look their best for the little time that they are in flower, it is essential that they are well-planted. A moist soil and regular deadheading are keys to success. Bedding plants need regular feeding throughout the growing season, especially in containers, but many hardy annuals flower best in soil that is not enriched. Many annuals are available as seedling "plugs", which should be grown on under cover. When planted, the well-developed root system suffers little damage, ensuring rapid growth.

PLANTING BULBS

In general, bulbs can be planted at between one and three times their own depth. Plant small bulbs quite shallowly; those of bigger plants like large tulips and lilies more deeply.

UNDERSTANDING PLANT NAMES

THROUGHOUT THE *RHS Good Plant Guide*, all plants are listed by their current botanical names. The basic unit of plant classification is the species, with a two-part name correctly given in italic text: the first part is the genus, and the second part is the species name or "epithet".

GENUS

A group of one or more plants that share a wide range of characteristics, such as *Chrysanthemum* or *Rosa*, is known as a genus. A genus name is quite like a family surname, because it is shared by a group of individuals that are all closely related. Hybrid genera (crosses between plants from two genera, such as x *Halimiocistus*), are denoted by a multiplication sign before the genus name.

SPECIES

A group of plants capable of breeding together to produce similar offspring are known as a species. In a two-part botanical name, the species epithet distinguishes a species from other plants in the same genus; rather like a Christian or given name. A species epithet usually refers to a particular feature of that species, like *alpinum* or *tricolor*, or it may refer to the person who first discovered the plant.

SUBSPECIES, VARIETY AND FORMA

Naturally occurring variants of a species – subspecies, variety or forma – are given an additional name in italics, prefixed by the abbreviations "subsp.", "var." or "f.".

Genus/Species
Malus floribunda *is of the same genus as apples (*Malus domestica*); its species name means "mass of flowers".*

Variety
Dictamnus albus *var.* purpureus *has purplish flowers instead of the pink-white of the species.*

Hybrid
The multiplication sign after the genus name in Osmanthus x burkwoodii *denotes a species hybrid.*

These are minor subdivisions of a species, differing slightly in their botanical structure or appearance.

HYBRIDS

If different species within the same genus are cultivated together, they may cross-breed, giving rise to hybrids sharing attributes of both parents. This process is exploited by gardeners who wish to combine the valued characteristics of two distinct plants. The new hybrid is then increased by propagation. An example is *Camellia* x *williamsii*, which has the parents *C. japonica* and *C. saluensis*.

CULTIVARS

Variations of a species that are selected or artificially raised are given a vernacular name. This appears in single quotation marks after the species name. Some cultivars are also registered with trademark names, often used commercially instead of the valid cultivar name. If the parentage is obscure or complex, the cultivar name may directly follow the generic name – *Dahlia* 'Conway'. In a few cases, particularly the roses, the plant is known by a popular selling name, which is not the correct cultivar name; here, the popular name comes before the cultivar name, as in *Rosa* BONICA 'Meidomonac'.

GROUPS AND SERIES

Several very similar cultivars may, for convenience, be classified in named Groups or Series that denote their similarities. Sometimes, they can be a deliberate mixture of cultivars, of the same character but with flowers in different colours.

Cultivar
Species parentage may not be given if it is complex or unknown, as with Osteospermum *'Buttermilk'.*

Cultivar of species
Ophiopogon planiscapus *'Nigrescens' is an unusual form cultivated for its black leaves.*

Seed series
Antirrhinum *Sonnet Series is a mixture of bright-coloured cultivars for summer bedding.*

THE A–Z OF AGM PLANTS

ATTRACTIVE AND RELIABLE plants for every garden, and for every part of the garden, can be found in this section, each carrying the recommendation of the RHS Award of Garden Merit. However much their ornamental features and season of interest appeal, always check their hardiness, eventual size, and site and soil requirements before you buy.

ABELIA 'EDWARD GOUCHER'

This semi-evergreen shrub with arching branches bears glossy, dark green leaves which are bronze when young. Trumpet-shaped, lilac-pink flowers appear from summer to autumn. Like most abelias, it is suitable for a sunny border.

CULTIVATION *Grow in well-drained, fertile soil, in sun with shelter from cold winds. Remove dead or damaged growth in spring, cutting some of the older stems back to the ground after flowering to promote new growth.*

☼ ◊ ♈H4 ↕1.5m (5ft) ↔2m (6ft)

ABELIA FLORIBUNDA

An evergreen shrub with arching shoots, from which tubular, bright pink-red flowers hang in profuse clusters in early summer. The leaves are oval and glossy dark green. Ideal for a sunny border, but may not survive cold winters except in a very sheltered position.

CULTIVATION *Grow in well-drained, fertile soil, in full sun with shelter from cold, drying winds. Prune older growth back after flowering, removing any dead or damaged growth in spring.*

☼ ◊ ♈H3 ↕3m (10ft) ↔4m (12ft)

ABELIA X *GRANDIFLORA*

A rounded, semi-evergreen shrub bearing arching branches. Cultivated for its attractive, glossy dark green leaves and profusion of fragrant, pink-tinged white flowers which are borne from mid-summer to autumn. Suitable for a sunny border.

CULTIVATION *Grow in well-drained, fertile soil, in full sun with shelter from cold, drying winds. Prune back older growth back after flowering, and remove damaged growth in spring.*

☼ ◊ ♀H4 ↕3m (10ft) ↔4m (12ft)

ABUTILON 'KENTISH BELLE'

A semi-evergreen shrub bearing dark purple-brown shoots and slender, arching branches. The large, bell-shaped flowers, which hang from the branches during summer and autumn, are apricot-yellow and red. The leaves are shallowly lobed and dark green. In cold areas, provide shelter, warmth and support, by training against a warm wall.

CULTIVATION *Grow in well-drained, fertile soil, in sun. Prune annually in late winter to preserve a well-spaced, healthy framework.*

☼ ◊ ♀H2–3 ↕↔ to 2.5m (8ft)

ABUTILON MEGAPOTAMICUM

The trailing abutilon is a semi-evergreen shrub bearing pendulous, bell-shaped, red and yellow flowers from summer to autumn. The oval leaves are bright green and heart-shaped at the base. In cold climates, train against a warm wall or grow in a conservatory.

CULTIVATION *Best in well-drained, moderately fertile soil, in full sun. Remove any wayward shoots during late winter or early spring.*

☼ ◊ ♀H3 ↕↔ 2m (6ft)

ABUTILON VITIFOLIUM 'VERONICA TENNANT'

A fast-growing, upright, deciduous shrub that may attain the stature of a small, bushy tree. Masses of large, bowl-shaped flowers hang from the stout, grey-felted shoots in early summer. The softly grey-hairy leaves are sharply toothed. May not survive in cold areas. 'Tennant's White', with pure white flowers, is also recommended.

CULTIVATION *Grow in well-drained, moderately fertile soil, in full sun. Prune young plants after flowering to encourage a good shape; do not prune established plants.*

☼ ◊ ♀H3 ↕5m (15ft) ↔2.5m (8ft)

ACAENA MICROPHYLLA

This summer-flowering, mat-forming perennial bears heads of small, dull red flowers with spiny bracts which develop into decorative burrs. The finely divided, mid-green leaves are bronze-tinged when young and evergreen through all but the hardest winters. Good for a rock garden, trough or raised bed.

CULTIVATION *Grow in well-drained soil, in full sun or partial shade. Pull out rooted stems around the main plant to restrict spread.*

☼☀ ◇ ♈H4 ↕5cm (2in) ↔15cm (6in)

ACANTHUS SPINOSUS

Bear's breeches is a striking, architectural perennial bearing long, arching, dark green leaves which have deeply cut and spiny edges. From late spring to mid-summer, pure white, two-lipped flowers with purple bracts are borne on tall, sturdy stems; they are good for cutting and drying. Grow in a spacious border.

CULTIVATION *Best in deep, well-drained, fertile soil, in full sun or partial shade. Provide plenty of space to display its architectural merits.*

☼☀ ◇ ♈H4 ↕1.5m (5ft) ↔60cm (24in)

ACER GRISEUM

The paper-bark maple is a slow-growing, spreading, deciduous tree valued for its peeling orange-brown bark. The dark green leaves, divided into three leaflets, turn orange to red and scarlet in autumn. Tiny yellow flowers are carried in hanging clusters during early or mid-spring, followed by brown, winged fruits.

CULTIVATION *Grow in moist but well-drained, fertile soil, in sun or partial shade. In winter only, remove shoots that obscure the bark on the trunk and lower parts of the main branches.*

☼ ◐ ♁ ♀H4 ↕↔ 10m (30ft)

ACER GROSSERI VAR. *HERSII*

This variety of the snake-bark maple with boldly green- and white-streaked bark is a spreading to upright, deciduous tree. The three-lobed, triangular, bright green leaves turn orange or yellow in autumn. Hanging clusters of tiny, pale yellow flowers appear in spring, followed by pink-brown, winged fruits.

CULTIVATION *Grow in moist but well-drained, fertile soil, in full sun or partial shade. Shelter from cold winds. Remove shoots that obscure the bark on the trunk and main branches in winter.*

☼ ◐ ♁ ♀H4 ↕↔ 15m (50ft)

ACER JAPONICUM 'ACONITIFOLIUM'

A deciduous, bushy tree or large shrub bearing deeply lobed, mid-green leaves which turn brilliant dark red in autumn. It is very free-flowering, producing upright clusters of conspicuous, reddish-purple flowers in mid-spring, followed by brown, winged fruits. 'Vitifolium' is similar, not so free-flowering but with fine autumn colour.

CULTIVATION *Grow in moist but well-drained, fertile soil, in partial shade. In areas with severe winters, mulch around the base in autumn. Remove badly placed shoots in winter only.*

◐ ◊ 🏆H4 ↕5m (15ft) ↔ 6m (20ft)

ACER NEGUNDO 'FLAMINGO'

Round-headed, deciduous tree with pink-margined, oval leaflets which turn white in summer. With regular pruning, it can be grown as a shrub; this also produces larger leaves with an intensified colour. Flowers are tiny and inconspicuous. Plain-leaved var. *violaceum* has glaucous shoots and long tassels of violet flowers.

CULTIVATION *Grow in any moist but well-drained, fertile soil, in full sun or partial shade. For larger leaves and a shrubby habit, cut back to a framework every 1 or 2 years in winter. Remove any branches with all-green leaves.*

☼ ◐ ◊ 🏆H4 ↕15m (50ft) ↔ 10m (30ft)

Japanese Maples (*Acer palmatum*)

Cultivars of *Acer palmatum*, the Japanese maple, are mostly small, round-headed, deciduous shrubs, although some, such as 'Sango-kaku', will grow into small trees. They are valued for their delicate and colourful foliage, which often gives a beautiful display in autumn. The leaves of 'Butterfly', for example, are variegated grey-green, white and pink, and those of 'Osakazuki' turn a brilliant red before they fall. In mid-spring, hanging clusters of small, reddish-purple flowers are produced, followed by winged fruits later in the season. Japanese maples are excellent for gardens of any size.

CULTIVATION *Grow in moist but well-drained, fertile soil, in sun or partial shade. Restrict pruning and training to young plants only; remove badly placed or crossing shoots in winter to develop a well-spaced network of branches. Keep pruning to a minimum on established plants.*

☼ ◑ ◊ ♀H4

1 ↕↔ 5m (15ft)

2 ↕3m (10ft) ↔ 1.5m (5ft)

3 ↕2m (6ft) ↔ 3m (10ft)

4 ↕2m (6ft) ↔ 3m (10ft)

1 *Acer palmatum* 'Bloodgood' **2** *A. palmatum* 'Butterfly' **3** *A. palmatum* 'Chitoseyama' **4** *A. palmatum* 'Garnet'

5 ↕ 5m (15ft) ↔ 4m (12ft)

MORE CHOICES

'Burgundy Lace' Very deeply cut red-purple leaves, 4m (12ft) tall and widely spreading

var. *coreanum* 'Korean Gem' Green leaves turning crimson/scarlet in autumn.

var. *dissectum* 'Crimson Queen' (see p.556)

var. *dissectum* (see p.630)

var. *dissectum* 'Inabe-Shidare' Deeply divided purple-red leaves.

'Seiryû' (see p.630)

6 ↕↔ 6m (30ft)

7 ↕↔ 1.5m (5ft)

8 ↕ 6m (20ft) ↔ 5m (15ft)

5 *A. palmatum* 'Linearilobum' **6** *A. palmatum* 'Osakazuki' **7** *A. palmatum* 'Red Pygmy'
8 *A. palmatum* 'Sango-kaku' (*syn.* 'Senkaki')

ACER PENSYLVANICUM 'ERYTHROCLADUM'

This striped maple is an upright, deciduous tree. Its leaves are bright green, turning clear yellow in autumn. In winter, brilliant pink or red young shoots make a fiery display; they become orange-red with white stripes as they mature. Hanging clusters of small, greenish-yellow flowers in spring are followed by winged fruits. Best grown as a specimen tree.

CULTIVATION *Grow in fertile, moist but well-drained soil in sun or part shade. Remove crossing or damaged shoots from late autumn to mid-winter only.*

☼☀ ◊ ♀H4 ↕12m (40ft) ↔10m (30ft)

ACER PLATANOIDES 'CRIMSON KING'

Like *Acer platanoides*, the Norway maple, this cultivar is a large, spreading, deciduous tree. It is grown for its dark red-purple leaves, which deepen to dark purple as they mature. The foliage is preceded in spring by clusters of small, red-tinged yellow flowers. These develop into winged fruits. A colourful specimen tree.

CULTIVATION *Grow in fertile, moist but well-drained soil, in sun or pártial shade. Prune in late autumn to mid-winter only, to remove any crossing, crowded, or unhealthy growth.*

☼☀ ◊ ♀H4 ↕25m (80ft) ↔15m (50ft)

ACER PLATANOIDES 'DRUMMONDII'

This is a much smaller, more spreading deciduous tree than the Norway maple, and is as broad as it is tall. Its leaves have a wide, pale green to cream margin, and colour well in autumn. In spring, clusters of yellow flowers are seen; from these later develop winged fruits. A bright specimen tree for a medium-sized garden.

CULTIVATION *Grow in fertile, moist but well-drained soil, in sun or partial shade. Prune in late autumn to mid-winter only, to remove any crossing, crowded, or unhealthy growth.*

☼☀ ◊ ♈H4 ↕↔10–12m (30–40ft)

ACER PSEUDOPLATANUS 'BRILLIANTISSIMUM'

This small, slow-growing cultivar of sycamore is a spreading, deciduous tree with a dense head. It bears colourful, five-lobed leaves, which turn from salmon-pink to yellow then dark green as they mature. Hanging clusters of tiny, yellow-green flowers appear in spring, followed by winged fruit. An attractive maple for smaller gardens.

CULTIVATION *Grow in any soil, in sun or partial shade. Tolerates exposed sites. Prune young plants in winter to develop well-spaced branches and a clear trunk.*

☼☀ ◊ ♈H4 ↕6m (20ft) ↔8m (25ft)

ACER RUBRUM 'OCTOBER GLORY'

This cultivar of the red, or swamp, maple is a round-headed to open-crowned, deciduous tree with glossy dark green foliage, turning bright red in early autumn. The upright clusters of tiny red flowers in spring are followed by winged fruits. A fine specimen tree for a large garden.

CULTIVATION *Grow in any fertile, moist but well-drained soil, although best autumn colour is seen in acid soil. Choose a site with full sun or in partial shade, and restrict pruning to between late autumn and mid-winter to remove any crossing or congested branches*

☼ ◑ ♢ ♀H4 ↕20m (70ft) ↔10m (30ft)

ACER TATARICUM SUBSP. *GINNALA*

The Amur maple is a rounded, bushy, deciduous tree with slender, arching branches, and glossy bright green leaves, very deeply lobed; in autumn, these become a deep, rich bronze-red. In spring, it bears upright clusters of cream flowers, from which develop red winged fruits. Best as a specimen tree.

CULTIVATION *Grow in any fertile, moist but well-drained soil, in full sun or in partial shade. Restrict pruning to between late autumn and mid-winter to remove any crossing or congested branches, if necessary.*

☼ ♢ ♀H4 ↕10m (30ft) ↔8m (25ft)

ACHILLEA AGERATIFOLIA

This small yarrow is a fast-growing, creeping perennial, forming mats of silvery, hairy leaf rosettes above which, in summer, small white flowerheads stand on upright stems. Grow at the front of a sunny border or in a rock or scree garden, or in planting spaces in a paved area.

CULTIVATION *Grow in any moderately fertile, free-draining soil, in sun. Divide plants that have spread too widely or become straggly in spring or autumn. Deadhead to encourage further flowers.*

☼ ◊ ♀H4 ↕5–8cm (2–3in) ↔to 45cm (18in)

ACHILLEA 'CORONATION GOLD'

This cultivar of yarrow is a clump-forming perennial that bears large, flat heads of golden-yellow flowers from mid-summer to early autumn. The luxuriant, evergreen, fern-like leaves are silver-grey, complementing the flower colour (contact may aggravate skin allergies). Excellent for a mixed or herbaceous border, and for cutting and drying.

CULTIVATION *Grow in moist but well-drained soil in an open site in full sun. Divide large or congested clumps to maintain vigour.*

☼ ◊♦ ♀H4 ↕75–90cm (30–36in) ↔45cm (18in)

ACHILLEA FILIPENDULINA 'GOLD PLATE'

This strong-growing, clump-forming and upright, evergreen perennial is similar to *A.* 'Coronation Gold' (p.39), but taller, with grey-green leaves. The flat-headed clusters of bright golden-yellow flowers are borne on strong stems from early summer to early autumn; they make good cut flowers. Grow in a mixed or herbaceous border. Contact with the foliage may aggravate skin allergies.

CULTIVATION *Grow in moist but well-drained soil in an open, sunny site. To maintain good performance, divide clumps when large and congested.*

☼ ◊◊ ♈H4 ↕1.2m (4ft) ↔45cm (18in)

ACHILLEA x *LEWISII* 'KING EDWARD'

This woody-based, low mound-forming perennial bears pale yellow flowerheads in dense, flat-headed clusters in early summer. They fade as they age, but attractively so. The serrated, semi-evergreen leaves are narrow, fern-like, and soft grey-green. It looks well in a wild flower or rock garden. Contact with the foliage may aggravate skin allergies.

CULTIVATION *Grow in moist but well-drained soil in an open site in full sun. Divide clumps every 3 years or so to maintain vigour.*

☼ ◊◊ ♈H4 ↕8–12cm (3–5in) ↔23cm (9in or more)

ACHILLEA 'MOONSHINE'

A clump-forming, evergreen perennial with narrow, feathery, grey-green leaves. Light yellow flowerheads with slightly darker centres appear from early summer to early autumn in flattish clusters; they dry well for arrangements. Excellent for mixed borders and for informal, wild or cottage-style plantings. May not survive over winter in cold climates without protection.

CULTIVATION *Grow in well-drained soil in an open, sunny site. Divide every 2 or 3 years in spring to maintain vigour. Protect from cold with a winter mulch.*

☼ ◊ 🏆H3 ↕ ↔ 60cm (24in)

ACHILLEA TOMENTOSA

The woolly yarrow is a mat-forming perennial that bears dense, flat-headed clusters of lemon-yellow flowerheads from early summer to early autumn. The woolly, grey-green leaves are narrow and fern-like. Tolerant of dry conditions, it suits a wild flower or rock garden, or grow near the front of a border. The flowers dry well. Contact with foliage may aggravate skin allergies.

CULTIVATION *Grow in moist but well-drained soil in an open, sunny site. To maintain good flowering, divide clumps when they become large and congested.*

☼ ◊◆ 🏆H4 ↕ to 35cm (14in) ↔ 45cm (18in)

ACONITUM 'BRESSINGHAM SPIRE'

A compact perennial producing very upright spikes of hooded, deep violet flowers from mid-summer to early autumn. The leaves are deeply divided and glossy dark green. Ideal for woodland or borders in partial or dappled shade. For lavender flowers on a slightly taller plant, look for *A. carmichaelii* 'Kelmscott'. All parts of these plants are poisonous.

CULTIVATION *Best in cool, moist, fertile soil, in partial shade, but will tolerate most soils and full sun. The tallest stems may need staking.*

☼◑ ♦ ♈H4 ↕90–100cm (36–39in) ↔30cm (12in)

ACONITUM 'SPARK'S VARIETY'

This upright perennial is taller than 'Bressingham Spire' (above) bearing spikes of hooded, deep violet flowers which are clustered together on branched stems; these are borne in mid- and late summer. The rich green leaves are deeply divided. Like all aconites, it is ideal for woodland gardens or shaded sites. All parts of the plant are poisonous.

CULTIVATION *Grow in cool, moist soil, in partial or deep shade. Taller stems may need staking. Divide and replant every third year in autumn or late winter to maintain vigour.*

◑● ♦ ♈H4 ↕1.5m (5ft) ↔ 45cm (18in)

ACTINIDIA KOLOMIKTA

A vigorous, deciduous climber with large, deep green leaves that are purple-tinged when young and develop vivid splashes of white and pink as they mature. Small, fragrant white flowers appear in early summer. Female plants produce small, egg-shaped, yellow-green fruits, but only if a male plant is grown nearby. Train against a wall or up into a tree.

CULTIVATION *Best in well-drained, fertile soil. For best fruiting, grow in full sun with protection from strong winds. Tie in new shoots as they develop, and remove badly placed shoots in summer.*

☼ ◊ ♈H4 ↕5m (15ft)

ADIANTUM PEDATUM

A deciduous relative of the maidenhair fern bearing long, mid-green fronds up to 35cm (14in) tall. These have glossy dark brown or black stalks which emerge from creeping rhizomes. There are no flowers. *A. aleuticum* is very similar and also recommended; var. *subpumilum* is a dwarf form of this fern, only 15cm (6in) tall. Grow in a shady border or light woodland.

CULTIVATION *Best in cool, moist soil, in deep or partial shade. Remove old or damaged fronds in early spring. Divide and replant rhizomes every few years.*

◐● ♦ ♈H4 ↕↔ 30–40cm (12–16in)

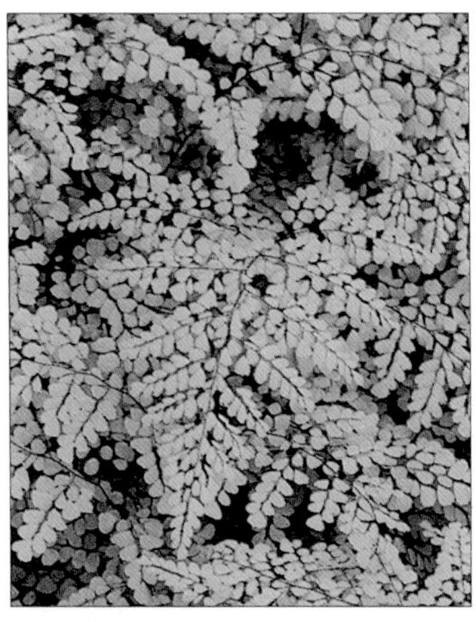

ADIANTUM VENUSTUM

The Himalayan maidenhair fern has black-stalked, triangular, mid-green fronds, beautifully divided into many small leaflets. The new foliage is bright bronze-pink when it emerges in late winter to early spring from creeping rhizomes. It is evergreen above –10°C (14°F). There are no flowers. Decorative ground cover for a woodland garden or shady border.

CULTIVATION *Grow in moderately fertile, moist but well-drained soil in partial shade. Remove old or damaged fronds in spring, and divide the rhizomes every few years in early spring.*

☀ ◊◆ ♀H4 ↕15cm (6in) ↔indefinite

AEONIUM HAWORTHII

A succulent subshrub with slender branches, each crowned by a neat rosette of bluish-green, fleshy leaves with red margins. Clusters of pale yellow to pinkish-white flowers are seen in spring. It is a popular pot plant for conservatories and porches.

CULTIVATION *Under glass, grow in standard cactus compost in filtered light, and allow the compost to dry out between waterings. Outdoors, grow in moderately fertile, well-drained soil in partial shade. Minimum temperature 10°C (50°F).*

☀ ◊ ♀H1 ↕↔60cm (24in)

AEONIUM 'ZWARTKOP'

An upright, succulent subshrub with few branches, each tipped by a rosette of black-purple leaves. Large, pyramid-shaped clusters of bright yellow flowers appear in late spring. Makes an unusually coloured pot plant. *A. arboreum* 'Atropurpureum' is similarly striking.

CULTIVATION *Grow in standard cactus compost in filtered light under glass, allowing the compost to dry out between waterings. Outdoors, grow in reasonably fertile, well-drained soil in partial shade. Minimum temperature 10°C (50°F).*

◑ ◊ 🏆H1 ↕↔to 2m (6ft)

AESCULUS CARNEA 'BRIOTII'

This cultivar of red horse chestnut is a spreading tree, admired in early summer for its large, upright cones of dark rose-red flowers. The leaves, dark green, are divided into 5–7 leaflets. The flowers are followed by spiny fruits. For large gardens; for a smaller tree, look for the sunrise horse chestnut, *A.* x *neglecta* 'Erythroblastos', to 10m (30ft), or the red buckeye, *A. pavia*, to 5m (15ft).

CULTIVATION *Grow in deep, fertile, moist but well-drained soil in full sun or partial shade. Remove dead, diseased or crossing branches during winter.*

☼◑ ◊♦ 🏆H4 ↕20m (70ft) ↔15m (50ft)

AESCULUS PARVIFLORA

A large, thicket-forming, deciduous shrub, closely related to the horse chestnut, that bears large-lobed, dark green leaves. The foliage is bronze when young, turning yellow in autumn. Upright white flowerheads, up to 30cm (12in) tall, appear in mid-summer, followed by smooth-skinned fruits.

CULTIVATION *Grow in moist but well-drained, fertile soil, in sun or partial shade; it will not grow in wet ground. If necessary, restrict spread by pruning stems to the ground after leaf fall.*

☼◑ ◊♦ ♀H4 ↕3m (10ft) ↔5m (15ft)

AETHIONEMA 'WARLEY ROSE'

A short-lived, evergreen or semi-evergreen, compact shrub bearing clusters of bright pink, cross-shaped flowers in late spring and early summer. The small, narrow leaves are blue-grey. For flowers of a much paler pink, look for *A. grandiflorum*, very similar if a little taller. Aethionemas are ideal for a rock garden or on a wall.

CULTIVATION *Best in well-drained, fertile, alkaline soil, but tolerates poor, acid soils. Choose a site in full sun.*

☼ ◊ ♀H4 ↕↔ 15–20cm (6–8in)

AGAPANTHUS CAMPANULATUS SUBSP. *PATENS*

A vigorous, clump-forming perennial bearing round heads of bell-shaped, light blue flowers on strong, upright stems during late summer and early autumn. The narrow, strap-shaped, greyish-green leaves are deciduous. Useful in borders or large containers as a late-flowering perennial. This and 'Loch Hope', with deep blue flowers, are the hardiest of the AGM-recommended agapanthus.

CULTIVATION *Grow in moist but well-drained, fertile soil or compost, in full sun. Water freely when in growth, and sparingly in winter.*

H3 ↕to 45cm (18in) ↔30cm (12in)

AGAPANTHUS CAULESCENS

A clump-forming perennial with leek-like stems bearing large, rounded, open flowerheads of bell-shaped, violet-blue flowers. These appear from mid-summer to early autumn above the strap-shaped, mid-green, deciduous leaves. Very useful as a late-flowering perennial that can also be container-grown in climates with cold winters.

CULTIVATION *Grow in moist but well-drained, fertile soil, in full sun. In cold areas, take under cover during winter. Minimum temperature 2°C (35°F).*

H1 ↕1–1.2m (3–4ft) ↔60cm (24in)

AGAVE VICTORIAE-REGINAE

A frost-tender, succulent perennial bearing basal rosettes of triangular, dark green leaves with white marks. The central leaves curve inward, each tipped with a brown spine. Upright spikes of creamy-white flowers appear in summer. A good specimen plant: in frost-prone areas, grow in containers for summer display, taking it under cover for winter shelter.

CULTIVATION *Best in sharply drained, moderately fertile, slightly acid soil, or standard cactus compost. Site in full sun. Minimum temperature 2°C (35°F).*

☼ ◊ H1 ↕↔ to 50cm (20in)

AJUGA REPTANS 'ATROPURPUREA'

An excellent evergreen perennial for ground cover, spreading freely over the soil surface by means of rooting stems. Dark blue flowers are borne in whorls along the upright stems during late spring and early summer. The glossy leaves are deep bronze-purple. *A. reptans* 'Catlin's Giant' has similarly coloured leaves. Invaluable for border edging under shrubs and robust perennials.

CULTIVATION *Best in moist but well-drained, fertile soil, but tolerates most soils. Site in sun or partial shade.*

☼ ◐ ◊ ♦ H4 ↕15cm (6in) ↔1m (3ft)

AJUGA REPTANS 'BURGUNDY GLOW'

A low-growing, creeping, evergreen ground cover perennial with partly hairy stems carrying attractive, silvery-green leaves which are suffused deep wine-red. Dark blue flowers are borne in tall, spike-like whorls in late spring and early summer.

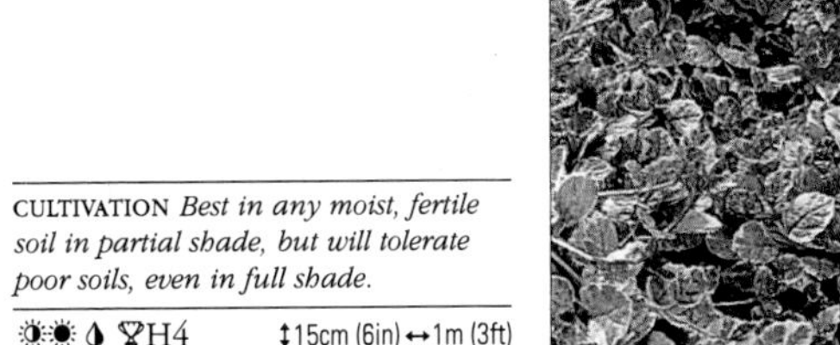

CULTIVATION *Best in any moist, fertile soil in partial shade, but will tolerate poor soils, even in full shade.*

H4 ↕15cm (6in) ↔1m (3ft)

ALCHEMILLA MOLLIS

Lady's mantle is a drought-tolerant, clump-forming, tallish ground cover perennial that produces sprays of tiny, bright greenish-yellow flowers from early summer to early autumn; these are ideal for cutting, and dry well for winter arrangements. The pale green leaves are rounded with crinkled edges. It looks well in a wildflower garden; for a rock garden, *A. erythropoda* is similar but smaller, with blue-tinged leaves.

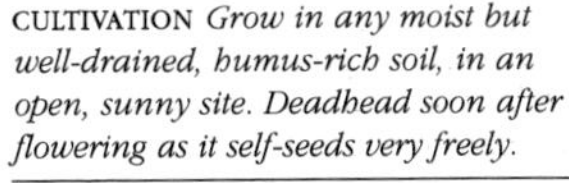

CULTIVATION *Grow in any moist but well-drained, humus-rich soil, in an open, sunny site. Deadhead soon after flowering as it self-seeds very freely.*

H4 ↕60cm (24in) ↔75cm (30in)

TALL ORNAMENTAL ONIONS (*ALLIUM*)

Tall onions grown for garden display are bulbous perennials from the *Allium* genus; their attractive flowerheads look excellent in a mixed border, especially grouped together. The tiny summer flowers are usually massed into dense, rounded or hemispherical heads – like those of *A. giganteum* – or they may hang loosely, like the yellow flowers of *A. flavum*. When crushed, the strap-shaped leaves release a pungent aroma; they are often withered by flowering time. The seedheads tend to dry out intact, standing well into autumn and continuing to look attractive. Some alliums self-seed and will naturalize.

CULTIVATION *Grow in fertile, well-drained soil in full sun to simulate their dry native habitats. Plant bulbs 5–10cm (2–4in) deep in autumn; divide and replant older clumps at the same time or in spring. In climates with cold winters, provide a thick winter mulch for* A. cristophii *and* A. caeruleum.

☼ ◊ ♈H3, H4

1 ↕1m (3ft) ↔ 15cm (6in)

MORE CHOICES

A carinatum subsp. *pulchellum* Purple flowers, 30–45cm (12–18in) tall.

A. cernuum 'Hidcote' Nodding pink flowers.

A. hollandicum Purplish-pink flowers, 1m (3ft) tall, very similar to 'Purple Sensation'.

2 ↕60cm (24in) ↔ 2.5cm (1in)

1 *Allium* 'Beau Regard' **2** *A. caeruleum*

3 ↕30–60cm (12–24in) ↔ 18cm (7in)

5 ↕1.5m–2m (5–6ft) ↔ 15cm (6in)

4 ↕ to 35cm (14in) ↔ 5cm (2in)

6 ↕1.2m (4ft) ↔ 15cm (6in)

7 ↕80cm (32in) ↔ 20in (8in)

8 ↕1m (3ft) ↔ 7cm (3in)

3 *A. cristophii* 4 *A. flavum* 5 *A. giganteum* 6 *A.* 'Gladiator'
7 *A.* 'Globemaster' 8 *A. hollandicum* 'Purple Sensation'

SMALL ORNAMENTAL ONIONS (*ALLIUM*)

Shorter alliums are summer-flowering, bulbous perennials for the front of a border or rock garden. They form clumps as they establish, some, such as *A. moly*, will self-seed. The flowers are borne in clustered heads which may be large or small; those of *A. karataviense* can be 8cm (3in) across despite its small stature. Flower colours range from bright gold to purple, blue, and pale pink. The seedheads are attractive too, lasting well into winter. The strap-shaped leaves are often withered by flowering time. Those of ornamental chives, such as *A. schoenopraesum* 'Pink Perfection' and the very similar 'Black Isle Blush', are edible.

CULTIVATION *Best in well-drained, humus-rich soil in full sun. Plant bulbs 5–10cm (2–4in) deep in autumn; divide and replant old or crowded clumps in autumn or spring. Provide a thick, dry winter mulch for* A. karataviense *where winters are cold.*

☼ ◊ ♈H4

1 ↕10–25cm (4–10cm) ↔ 10cm (4in)

2 ↕15–25cm (6–10in) ↔ 5cm (2in)

3 ↕ 5–20cm (2–8in) ↔ 3cm (1¼in)

4 ↕30–60cm (12–24in) ↔ 5cm (2in)

1 *Allium karataviense* **2** *A. moly* **3** *A. oreophilum* **4** *A. schoenoprasum* 'Pink Perfection'

ALNUS GLUTINOSA 'IMPERIALIS'

This attractive cultivar of the common alder is a broadly conical tree with deeply dissected, lobed, mid-green leaves. Groups of yellow-brown catkins are seen in late winter, followed by small oval cones in summer. This is a beautiful foliage tree, particularly good close to water as it tolerates poor, wet soil.

CULTIVATION *Thrives in any moderately fertile, moist but not waterlogged soil, in full sun. Prune after leaf fall, if necessary, to remove any damaged or crossing branches.*

☼ ♦ ♈H4 ↕25m (80ft) ↔5m (15ft)

ALOE VARIEGATA

The partridge-breasted aloe, a stemless succulent, forms clumps of narrow, fleshy leaves, dark green with white edges and horizontal white bands. Clusters of hanging pink or scarlet flowers may form in summer. For a desert garden in areas with a min. temp. of 10°C (50°F), or grow as a house or container plant; smaller than the plain-leaved *Aloe vera*, it is more manageable in a pot.

CULTIVATION *Under glass, grow in well-drained potting compost in a sunny, well-ventilated position. Water sparingly during winter. Outdoors, grow in fertile, well-drained soil in full sun.*

☼ ◊ ♈H1 ↕20cm (8in) ↔indefinite

ALONSOA WARSCEWICZII

This species of mask flower is a compact, bushy perennial, grown for its bright scarlet, sometimes white, flowers. These are on display from summer to autumn amid the dark green leaves. Useful as summer bedding or in a mixed border, it also provides good cut flowers.

CULTIVATION *Outdoors, grow in any fertile, well-drained soil in full sun, or in loam-based potting compost if grown in a container. Water moderately.*

☼ ◊ ♈H3 ↕45–60cm (18–24in) ↔30cm (12in)

ALOYSIA TRIPHYLLA

The lemon verbena is cultivated for its strongly lemon-scented, bright green foliage, used for culinary purposes or in pot-pourri. Pick and dry the leaves in summer before flowering. Tiny, pale lilac to white flowers are borne in slender clusters in late summer on this deciduous shrub. In mild areas, grow in a sunny border; elsewhere, grow in a cool greenhouse.

CULTIVATION *Grow in well-drained, poor, dry soil in full sun. Mulch well in cold areas to give winter protection. Cut back to a low framework each spring to maintain a good shape.*

☼ ◊ ♈H2 ↕↔2m (6ft)

ALSTROEMERIA LIGTU HYBRIDS

These summer-flowering, tuberous perennials produces heads of widely flared flowers which are considerably varied in colour from white to shades of pink, yellow or orange, often spotted or streaked with contrasting colours. The mid-green leaves are narrow and twisted. Ideal for a sunny, mixed or herbaceous border, the cut flowers are good for indoor arrangements.

CULTIVATION *Grow in moist but well-drained, fertile soil, in full sun. Mulch thickly in areas with very cold winters. Leave undisturbed to form clumps.*

☼ ◊ 🏆H4 ↕50cm (20in) ↔75cm (30in)

AMELANCHIER X *GRANDIFLORA* 'BALLERINA'

A spreading, deciduous tree grown for its profusion of white spring flowers and colourful autumn foliage. When young, the glossy leaves are tinted bronze, becoming mid-green in summer, then red and purple in autumn. The sweet, juicy fruits are red at first, ripening to purplish-black in summer. They can be eaten if cooked, and are attractive to birds.

CULTIVATION *Grow in moist but well-drained, fertile, neutral to acid soil, in full sun or partial shade. Allow shape to develop naturally; only minimal pruning is necessary, in winter.*

☼☀ ◊♦ 🏆H4 ↕6m (20ft) ↔8m (25ft)

AMELANCHIER LAMARCKII

A many-stemmed, upright, deciduous shrub bearing leaves that are bronze when young, maturing to dark green in summer, then brilliant red and orange in autumn. Hanging clusters of white flowers are produced in spring. The ripe, purple-black fruits that follow are edible when cooked, and are attractive to birds. Also known as *A. canadensis.*

CULTIVATION *Grow in moist but well-drained, humus-rich, neutral to acid soil, in sun or partial shade. Develops its shape naturally with only minimal pruning when dormant in winter.*

☼ ◑ ◊ ◆ ♀H4 ↕10m (30ft) ↔12m (40ft)

ANAPHALIS TRIPLINERVIS 'SOMMERSCHNEE'

A clump-forming perennial carrying pale grey-green, white-woolly leaves. The tiny yellow flowerheads, surrounded by brilliant white bracts, appear during mid- and late summer in dense clusters; they are excellent for cutting and drying. Provides good foliage contrast in borders that are too moist for the majority of other grey-leaved plants.

CULTIVATION *Grow in any reasonably well-drained, moderately fertile soil that does not dry out in summer. Choose a position in full sun or partial shade.*

☼ ◑ ◊ ♀H4 ↕80–90cm (32–36in) ↔45–60cm (18–24in)

ANCHUSA AZUREA 'LODDON ROYALIST'

An upright, clump-forming perennial that is much-valued in herbaceous borders for its spikes of intensely dark blue flowers. These are borne on branching stems in early summer, above the lance-shaped and hairy, mid-green leaves which are arranged at the base of the stems. For a rock garden, *A. cespitosa* looks similar in miniature, only 5–10cm (2–4in) tall.

CULTIVATION *Grow in deep, moist but well-drained, fertile soil, in sun. Often short-lived, but easily propagated by root cuttings. If growth is vigorous, staking may be necessary.*

☼ ♢ H4 ↕90cm (36in) ↔60m (24in)

ANDROSACE CARNEA SUBSP. *LAGGERI*

An evergreen, cushion-forming perennial that bears small clusters of tiny, cup-shaped, deep pink flowers with yellow eyes, in late spring. The pointed, mid-green leaves are arranged in small, tight rosettes. Rock jasmines grow wild in alpine turf and rock crevices, making them ideal for rock gardens or troughs. *A. sempervivoides* has scented flowers, ideal for a raised bed.

CULTIVATION *Grow in moist but sharply drained, gritty soil, in full sun. Provide a top-dressing of grit or gravel to keep the stems and leaves dry.*

☼ ♢ H4 ↕5cm (2in) ↔8–15cm (3–6in)

ANDROSACE LANUGINOSA

A mat-forming, evergreen perennial producing compact heads of flat, small pink flowers with dark pink or greenish-yellow eyes, in mid- and late summer. The deep grey-green leaves are borne on trailing, reddish-green stems which are covered in silky hairs. Thrives in scree gardens, raised beds or troughs.

CULTIVATION *Grow in gritty, moist but well-drained soil. Choose a site in full sun. In areas with wet winters, provide protection under a cloche.*

☼ ◊ ♀H4 ↕to 10cm (4in) ↔to 30cm (12in)

ANEMONE BLANDA 'WHITE SPLENDOUR'

A spreading, spring-flowering perennial that soon forms clumps of stems growing from knobbly tubers. The solitary, upright, flattish white flowers, with pink-tinged undersides, are borne above oval, dark green leaves divided into delicately lobed leaflets. Excellent for naturalizing in sunny or shaded sites with good drainage. Mix it with 'Radar', with white-centred magenta flowers, or 'Ingramii', with deep blue flowers.

CULTIVATION *Grow in well-drained soil that is rich in humus. Choose a position in full sun or partial shade.*

☼ ◐ ◊ ♀H4 ↕↔ 15cm (6in)

ANEMONE HUPEHENSIS 'HADSPEN ABUNDANCE'

Upright, woody-based, late-flowering border perennial that spreads by shoots growing from the roots. Reddish-pink flowers, with petal margins that gradually fade to white, are borne on branched stems during mid- and late summer. The deeply divided, long-stalked, dark green leaves are oval and sharply toothed. *A. hupehensis* 'Prinz Heinrich' is similar, but spreads more vigorously.

CULTIVATION *Grow in moist, fertile, humus-rich soil, in sun or partial shade. Provide a mulch in cold areas.*

☼◐ ◊◆ ♈H4 ↕60–90cm (24–36in) ↔40cm (16in)

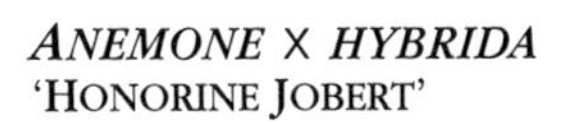

ANEMONE X *HYBRIDA* 'HONORINE JOBERT'

This upright, woody-based perennial with branched, wiry stems is an invaluable long-flowering choice for late summer to mid-autumn, when single, cupped white flowers, with pink-tinged undersides and golden-yellow stamens, are borne above divided, mid-green leaves. (For pure white flowers without a hint of pink, look for 'Géante des Blanches'.) It can be invasive.

CULTIVATION *Grow in moist but well-drained, moderately fertile, humus-rich soil, in sun or partial shade.*

☼◐ ◊◆ ♈H4 ↕1.2–1.5m (4–5ft) ↔indefinite

ANEMONE NEMOROSA 'ROBINSONIANA'

A vigorous, carpeting perennial that produces masses of large, star-shaped, pale lavender-blue flowers on maroon stems from spring to early summer, above deeply-divided, mid-green leaves which die down in midsummer. Excellent for underplanting or a woodland garden; it naturalizes with ease. 'Allenii' has deeper blue flowers; choose 'Vestal' for white flowers.

CULTIVATION *Grow in loose, moist but well-drained soil that is rich in humus, in light dappled shade.*

☀ 💧 ♀H4 ↕8–15cm (3–6in) ↔30cm (12in) or more

ANEMONE RANUNCULOIDES

This spring-flowering, spreading perennial is excellent for naturalizing in damp woodland gardens. The large, solitary, buttercup-like yellow flowers are borne above the "ruffs" of short-stalked, rounded, deeply lobed, fresh green leaves.

CULTIVATION *Grow in moist but well-drained, humus-rich soil, in semi-shade or dappled sunlight. Tolerates drier conditions when dormant in summer.*

☀ 💧💧 ♀H4 ↕5–10cm (2–4in) ↔to 45cm (18in)

ANTENNARIA MICROPHYLLA

A mat-forming, semi-evergreen perennial carrying densely white-hairy, spoon-shaped, grey-green leaves. In late spring and early summer, heads of small, fluffy, rose-pink flowers are borne on short stems. Use in a rock garden, as low ground cover at the front of a border or in crevices in walls or paving. The flowerheads dry well for decoration.

CULTIVATION *Best in well-drained soil that is no more than moderately fertile. Choose a position in full sun.*

☼ ◊ H4 ↕5cm (2in) ↔ to 45cm (18in)

ANTHEMIS PUNCTATA SUBSP. *CUPANIANA*

A mat-forming, evergreen perennial that produces a flush of small but long-lasting, daisy-like flowerheads in early summer, and a few blooms later on. The white flowers with yellow centres are borne singly on short stems, amid dense, finely cut, silvery-grey foliage which turns dull grey-green in winter. Excellent for border edges.

CULTIVATION *Grow in well-drained soil, in a sheltered, sunny position. Cut back after flowering to maintain vigour.*

☼ ◊ H3–4 ↕30cm (12in) ↔45cm (18in)

ANTIRRHINUM SONNET SERIES

Snapdragons are short-lived perennials best grown as annuals. The Sonnet Series produces upright spikes of fragrant, two-lipped flowers in a broad range of colours, available mixed or in single shades, from early summer into autumn. Planted in groups, the deep green leaves on woody-based stems are barely visible. Excellent for cut flowers or summer bedding.

CULTIVATION *Grow in sharply drained, fertile soil, in full sun. Deadhead whole flower spikes to prolong flowering.*

☼ ◊ ♀H3 ↕30–60cm (12–24in) ↔30cm (12in)

AQUILEGIA VULGARIS 'NIVEA'

An upright, vigorous, clump-forming perennial, sometimes sold as 'Munstead's White', bearing leafy clusters of nodding, short-spurred, pure white flowers in late spring and early summer. Each greyish-green leaf is deeply divided into lobed leaflets. Attractive in light woodland or in a herbaceous border; plant with soft blue *Aquilegia* 'Hensol Harebell' for a luminous mix in light shade.

CULTIVATION *Best in moist but well-drained, fertile soil. Choose a position in full sun or partial shade.*

☼ ◐ ◊ ♀H4 ↕90cm (36in) ↔45cm (18in)

AQUILEGIA VULGARIS 'NORA BARLOW'

This upright, vigorous perennial is much-valued for its leafy clusters of funnel-shaped, double-pompon flowers. These are pink and white with pale green petal tips, and appear from late spring to early summer. The greyish-green leaves are deeply divided into narrow lobes. Suits herbaceous borders and cottage garden-style plantings.

CULTIVATION *Grow in moist but well-drained, fertile soil. Position in an open, sunny site.*

☼ ◊ 🏆H4 ↕90cm (36in) ↔45cm (18in)

ARABIS PROCURRENS 'VARIEGATA'

A mat-forming, evergreen or semi-evergreen perennial bearing loose clusters of cross-shaped white flowers on tall, slender stems during late spring. The narrow, mid-green leaves, arranged into flattened rosettes, have creamy-white margins and are sometimes pink-tinged. Useful in a rock garden. May not survive in areas with cold winters.

CULTIVATION *Grow in any well-drained soil, in full sun. Remove completely any stems with plain green leaves.*

☼ ◊ 🏆H3 ↕5–8cm (2–3in) ↔30–40cm (12–16in)

ARALIA ELATA 'VARIEGATA'

The variegated Japanese angelica tree is deciduous, with a beautiful, exotic appearance. The large leaves are divided into 80 or more leaflets, irregularly edged with creamy white. Flat clusters of small white flowers appear in late summer and early autumn, followed by round black fruits. Suitable for a shady border or wooded streambank in a large garden, and less prone to suckering than the plain-leaved *Aralia elata.*

CULTIVATION *Grow in fertile, humus-rich, moist soil, in sun or part shade. Remove any branches with all-green foliage in late winter. Needs a sheltered site; strong winds can damage leaves.*

☀◐ 💧 🏆H4 ↕↔ 5m (15ft)

ARAUCARIA HETEROPHYLLA

The Norfolk Island pine is a grand, cone-shaped conifer, valued for its geometrical shape and unusual branches of whorled foliage. The leaves are tough, light green, and scale-like. There are no flowers. An excellent, fast-growing, gale-tolerant tree for coastal sites; in cold climates, it can be grown as a conservatory plant.

CULTIVATION *Grow in moderately fertile, moist but well-drained soil in an open site with shelter from cold drying winds. Tolerates partial shade when young.*

☀ 💧💧 🏆H1 ↕25–45m (80–150ft) ↔6–8m (20–25ft)

ARBUTUS X *ANDRACHNOIDES*

This strawberry tree is a broad, sometimes shrubby tree, with peeling, red-brown bark. The mid-green leaves are finely toothed and glossy. Clusters of small white flowers are borne from autumn to spring, only rarely followed by fruits. Excellent for a large shrub border, or as a specimen tree.

CULTIVATION *Grow in fertile, well-drained soil rich in organic matter, in a sheltered but sunny site. Protect from cold winds, even when mature, and keep pruning to a minimum, in winter if necessary. It tolerates alkaline soil.*

☼ ◊ 🏆H4 ↕↔8m (25ft)

ARBUTUS MENZIESII

The madrone is a spreading tree with beautiful, peeling, red-brown bark. Its dark green leaves are glossy and toothed. In early summer, the tree is covered with upright clusters of pure white flowers, to be followed by warty, orange-red, small spherical fruits, which ripen in the next growing season. A good tree for a large shrub border or woodland where there is acid soil.

CULTIVATION *Grow in fertile, humus-rich, well-drained, acid soil in a sheltered site in full sun. Protect from cold winds. Keep pruning to a minimum, in winter if necessary.*

☼ ◊ 🏆H4 ↕↔15m (50ft)

ARBUTUS UNEDO

The strawberry tree is a spreading, evergreen tree with attractive, rough, shredding, red-brown bark. Hanging clusters of small, urn-shaped white flowers, which are sometimes pink-tinged, open during autumn as the previous season's strawberry-like red fruits ripen. The glossy deep green leaves are shallowly toothed. Excellent for a large shrub border, with shelter from wind.

CULTIVATION *Best in well-drained, fertile, humus-rich, acid soil. Tolerates slightly alkaline conditions. Choose a sheltered site in sun. Prune low branches in spring, but keep to a minimum.*

☼ ◊ ♈H4 ↕↔ 8m (25ft)

ARENARIA MONTANA

This sandwort is a low-growing, spreading, vigorous, evergreen perennial freely bearing shallowly cup-shaped white flowers in early summer. The small, narrowly lance-shaped, greyish-green leaves on wiry stems form loose mats. Easily grown in wall or paving crevices, or in a rock garden.

CULTIVATION *Grow in sandy, moist but sharply drained, poor soil, in full sun. Must have adequate moisture.*

☼ ◊ ♈H4 ↕2–5cm (¾–2in) ↔30cm (12in)

ARGYRANTHEMUM 'JAMAICA PRIMROSE'

A bushy, evergreen perennial that bears daisy-like, primrose-yellow flowerheads with darker yellow centres throughout summer above fern-like, greyish-green leaves. In frost-prone areas, grow as summer bedding or in containers, bringing under cover for the winter. 'Cornish Gold' – shorter, with flowers of a deep yellow – is also recommended.

CULTIVATION *Grow in well-drained, fairly fertile soil or compost, in a warm, sunny site. Mulch outdoor plants well. Pinch out shoot tips to encourage bushiness. Min. temp. 2°C (35°F).*

☼ ◊ 🏆H1+3 ↕1.1m (3½ft) ↔1m (3ft)

ARGYRANTHEMUM 'VANCOUVER'

This compact, summer-flowering, evergreen subshrub is valued for its double, daisy-like pink flowerheads with rose-pink centres and fern-like, grey-green leaves. Suits a mixed or herbaceous border; in frost-prone areas, grow as summer bedding or in containers that can be sheltered in frost-free conditions over winter.

CULTIVATION *Grow in well-drained, fairly fertile soil, in sun. Apply a deep, dry mulch to outdoor plants. Pinch out growing tips to encourage bushiness. Minimum temperature 2°C (35°F).*

☼ ◊ 🏆H1+3 ↕90cm (36in) ↔80cm (32in)

ARMERIA JUNIPERIFOLIA

This tiny, hummock-forming, evergreen subshrub bears small, purplish-pink to white flowers which are carried in short-stemmed, spherical clusters during late spring. The small, linear, grey-green leaves are hairy and spine-tipped, and are arranged in loose rosettes. Native to mountain pastures and rock crevices, it is ideal for a rock garden or trough. Also known as *A. caespitosa*.

CULTIVATION *Grow in well-drained, poor to moderately fertile soil, in an open position in full sun.*

☼ ◊ ♈H4 ↕5–8cm (2–3in) ↔to 15cm (6in)

ARMERIA JUNIPERIFOLIA 'BEVAN'S VARIETY'

A compact, cushion-forming, evergreen subshrub that bears small, deep rose-pink flowers. These are carried in short-stemmed, rounded clusters during late spring, over the loose rosettes of small and narrow, pointed, grey-green leaves. Suits a rock garden or trough; for the front of the border, look for *Armeria* 'Bee's Ruby', a similar perennial plant growing to 30cm (12in) tall, with chive-like flowerheads.

CULTIVATION *Grow in well-drained, poor to moderately fertile soil. Choose an open site in full sun.*

☼ ◊ ♈H4 ↕to 5cm (2in) ↔to 15cm (6in)

ARTEMISIA ABSINTHIUM 'LAMBROOK SILVER'

A clump-forming, woody-based, evergreen perennial cultivated for its mass of ferny, aromatic, silvery-grey foliage. The closely related 'Lambrook Mist' is very similar, if slightly less hardy. The greyish-yellow flowerheads in late summer are of little ornamental value. Suitable for a rock garden or border, but shelter is needed if grown in an exposed site.

CULTIVATION *Grow in well-drained, fertile soil, in full sun. Short-lived on poorly drained soils. Cut to the base in autumn to maintain a compact habit.*

☼ ◊ 🏆H4 ↕to 75cm (30in) ↔60cm (24in)

ARTEMISIA LUDOVICIANA 'SILVER QUEEN'

An upright, bushy, clump-forming, semi-evergreen perennial bearing narrow, downy leaves, sometimes jaggedly toothed; silvery-white when young, they become greener with age. White-woolly plumes of brown-yellow flowers are borne from mid-summer to autumn. Indispensable in a silver-themed border. 'Valerie Finnis' is also recommended; its leaves have more deeply cut edges.

CULTIVATION *Grow in well-drained soil, in an open, sunny site. Cut back in spring for best foliage effect.*

☼ ◊ 🏆H4 ↕75cm (30in) ↔60cm (24in) or more

ARTEMISIA 'POWIS CASTLE'

A vigorous, shrubby, woody-based perennial forming a dense, billowing clump of finely cut, aromatic, silver-grey leaves. Sprays of insignificant, yellow-tinged silver flowerheads are borne in late summer. Excellent in a rock garden or border. May not survive in areas with cold winters.

CULTIVATION *Grow in well-drained, fertile soil, in full sun. Will die back in heavy, poorly drained soils and may be short-lived. Cut to the base in autumn to maintain a compact habit.*

☼ ◊ ♈H3 ↕60cm (24in) ↔90cm (36in)

ASPARAGUS DENSIFLORUS 'MYERSII'

The asparagus fern is an arching and trailing, evergreen perennial forming spires of narrow, feathery, leaf-like, light green stems. In warm climates, it bears clusters of small, pink-tinged white flowers in summer, followed by bright red berries. In cold areas, it is better grown under cover and makes an impressive conservatory specimen, or a much smaller house- or hanging basket plant.

CULTIVATION *Grow in moist but well-drained, fertile soil, in partial shade. Minimum temperature 2°C (35°F).*

◑ ◊ ♈H1 ↕60–90cm (24–36in) ↔1–1.2m (3–4ft)

ASPLENIUM SCOLOPENDRIUM

The hart's tongue fern has irregular crowns of shuttlecock-like, tongue-shaped, leathery, bright green fronds, to 40cm (16in) long. Heart-shaped at the bases, they often have wavy margins, markedly so in the cultivar 'Crispum Bolton's Nobile'. On the undersides of mature fronds, rust-coloured spore cases are arranged in a herringbone pattern. There are no flowers. Good in alkaline soil.

CULTIVATION *Grow in moist but well-drained, humus-rich, preferably alkaline soil with added grit, in partial shade.*

◑ 💧💧 🏆H4 ↕45–70cm (18–28in) ↔60cm (24in)

ASTER ALPINUS

This spreading, clump-forming perennial is grown for its mass of daisy-like, purplish-blue or pinkish-purple flowerheads with deep yellow centres. These are borne on upright stems in early and mid-summer, above short-stalked, narrow, mid-green leaves. A low-growing aster, it is suitable for the front of a border or in a rock garden. Several outstanding cultivars are available.

CULTIVATION *Grow in well-drained, moderately fertile soil, in sun. Mulch annually after cutting back in autumn.*

☼ 💧 🏆H4 ↕to 25cm (10in) ↔45cm (18in)

ASTER AMELLUS 'KING GEORGE'

A clump-forming, bushy perennial bearing loose clusters of large, daisy-like, violet-blue flowerheads with yellow centres which open from late summer to autumn. The rough, mid-green leaves are hairy and lance-shaped. An invaluable late-flowering border plant; other recommended cultivars include 'Framfieldii' (lavender-blue flowers), 'Jacqueline Genebrier' (bright red-purple) and 'Veilchenkönigen' (deep purple).

CULTIVATION *Grow in open, well-drained, moderately fertile soil, in full sun. Thrives in alkaline conditions.*

☼ ◊ ♀H4 ↕↔ 45cm (18in)

ASTER 'ANDENKEN AN ALMA PÖTSCHKE'

This vigorous, upright, clump-forming perennial carries sprays of large, daisy-like, bright salmon-pink flowerheads with yellow centres from late summer to mid-autumn, on stiff stems above rough, stem-clasping, mid-green leaves. Good for cutting, or in late-flowering displays. *A. novae-angliae* 'Harrington's Pink' is very similar, with paler flowers.

CULTIVATION *Grow in moist but well-drained, fertile, well-cultivated soil, in sun or semi-shade. Divide and replant every third year to maintain vigour and flower quality. May need staking.*

☼ ◐ ◊ ♀H4 ↕1.2m (4ft) ↔60cm (24in)

ASTER X *FRIKARTII* 'MÖNCH'

This upright, bushy perennial provides a continuous show of long-lasting, daisy-like, lavender-blue flowerheads with orange centres during late summer and early autumn. The dark green leaves are rough-textured and oblong. A useful plant for adding cool tones to a late summer or autumn display, as is the very similar 'Wunder von Stäfa'.

CULTIVATION *Best in well-drained, moderately fertile soil. Position in an open, sunny site. Mulch annually after cutting back in late autumn.*

☼ ◊ 🏆H4 ↕70cm (28in) ↔35–40cm (14–16in)

ASTER LATERIFOLIUS 'HORIZONTALIS'

A clump-forming, freely branching perennial bearing clusters of daisy-like, sometimes pink-tinged white flowerheads with darker pink centres, from mid-summer to mid-autumn. The slender, hairy stems bear small, lance-shaped, mid-green leaves. An invaluable late-flowerer for a mixed border.

CULTIVATION *Grow in moist but well-drained, moderately fertile soil, in partial shade. Keep moist in summer.*

◑ ◊ 🏆H4 ↕60cm (24in) ↔30cm (12in)

ASTER 'LITTLE CARLOW'

A clump-forming, upright perennial that produces large clusters of daisy-like, violet-blue flowers with yellow centres, in early and mid-autumn. The dark green leaves are oval to heart-shaped and toothed. Valuable for autumn displays; the flowers cut and dry very well. Closely related to *A. cordifolius* 'Chieftain' (mauve flowers) and 'Sweet Lavender', also recommended.

CULTIVATION *Best in moist, moderately fertile soil, in partial shade, but tolerates well-drained soil, in full sun. Mulch annually after cutting back in late autumn. May need staking.*

☼☀ ◊♦ ♀H4 ↕90cm (36in) ↔45cm (18in)

ASTILBE × *ARENDSII* 'FANAL'

A leafy, clump-forming perennial grown for its long-lasting, tapering, feathery heads of tiny, dark crimson flowers in early summer; they later turn brown, keeping their shape well into winter. The dark green leaves, borne on strong stems, are divided into several leaflets. Grow in a damp border or woodland garden, or use for waterside plantings. 'Brautschleier' is similar, with creamy-white flower plumes.

CULTIVATION *Grow in moist, fertile, preferably humus-rich soil. Choose a position in full sun or partial shade.*

☼☀ ♦ ♀H4 ↕60cm (24in) ↔45cm (18in)

ASTILBE X *CRISPA* 'PERKEO'

A summer-flowering, clump-forming perennial, low-growing compared with other astilbes, that bears small, upright plumes of tiny, star-shaped, deep pink flowers. The stiff, finely cut, crinkled, dark green leaves are bronze-tinted when young. Suitable for a border or rock garden; the flowers colour best in light shade. 'Bronce Elegans' is another good, compact astilbe where space is limited.

CULTIVATION *Grow in reasonably moist, fertile soil that is rich in organic matter. Choose a position in partial shade.*

☀ 💧 🏆H4 ↕15–20cm (6–8in) ↔15cm (6in)

ASTILBE 'SPRITE'

A summer-flowering, leafy, clump-forming dwarf perennial that is suitable for waterside plantings. The feathery, tapering plumes of tiny, star-shaped, shell-pink flowers arch elegantly over a mass of broad, mid-green leaves composed of many narrow leaflets. *Astilbe* 'Deutschland' has the same arching, as opposed to upright, flower plumes, in cream.

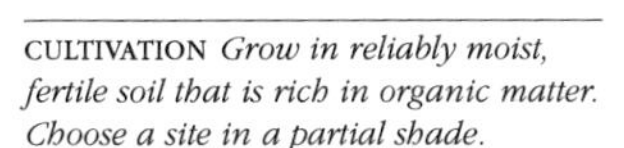

CULTIVATION *Grow in reliably moist, fertile soil that is rich in organic matter. Choose a site in a partial shade.*

☀ 💧 🏆H4 ↕50cm (20in) ↔to 1m (3ft)

ASTILBE 'STRAUSSENFEDER'

A vigorous, clump-forming perennial that bears loose, arching sprays of rich coral-pink flowers in late summer and early autumn; the flowerheads turn brown when dry, and persist into winter. The dark green leaves, divided into oval leaflets, are bronze-tinted when young. Ideal for damp borders, bog gardens or waterside plantings. Sometimes called 'Ostrich Plume'.

CULTIVATION *Grow in moist, fertile, rich soil, in sun or partial shade. Requires ample moisture in the growing season. Apply an annual mulch in spring of organic matter to hold water in the soil.*

☼☀ 💧 ♈H4 ↕90cm (36in) ↔60cm (24in)

ASTRANTIA MAXIMA

Sometimes known as Hattie's pincushion, this mat-forming perennial produces domed, rose-pink flowerheads with star-shaped collars of papery, greenish-pink bracts, on tall stems during summer and autumn. The mid-green leaves are divided into three toothed lobes. Flowers are good for cutting and drying, for use in cottage-style arrangements.

CULTIVATION *Grow in any moist, fertile, preferably humus-rich soil, in sun or semi-shade. Tolerates drier conditions.*

☼☀ 💧💧 ♈H4 ↕60cm (24in) ↔30cm (12in)

ASTRANTIA MAJOR 'SUNNINGDALE VARIEGATED'

A clump-forming perennial bearing attractive, deeply lobed, basal leaves which have unevenly variegated, creamy-yellow margins. From early summer, domes of tiny, green or pink, often deep purple-red flowers with star-shaped collars of pale-pink bracts, are carried on wiry stems. Thrives in a moist border, woodland garden or on a stream bank.

CULTIVATION *Grow in any moist but well-drained, fertile soil. Needs full sun to obtain the best leaf colouring.*

☼ ◊ 🏆H4 ↕30–90cm (12–36in) ↔45cm (18in)

ATHYRIUM FILIX-FEMINA

The lady fern has much divided, light green, deciduous fronds which are borne like upright shuttlecocks, about 1m (3ft) long, arching outward with age. Frond dissection is very varied, and the stalks are sometimes red-brown. There are no flowers. Useful for shaded sites, like a woodland garden. Its cultivars 'Frizelliae' and 'Vernoniae' have unusual, distinctive fronds.

CULTIVATION *Grow in moist, fertile, neutral to acid soil enriched with leaf mould or garden compost. Choose a shaded, sheltered site.*

☀ ♦ 🏆H4 ↕to 1.2m (4ft) ↔60–90cm (24–36in)

AUCUBA JAPONICA 'CROTONIFOLIA' (FEMALE)

This variegated form of spotted laurel is a rounded, evergreen shrub with large, glossy dark green leaves boldly speckled with golden-yellow. Upright clusters of small purplish flowers are borne in mid-spring, followed by red berries in autumn. Ideal for dense, semi-formal hedging.

CULTIVATION *Grow in any but water-logged soil, in full sun for best foliage colour, or in shade. Provide shelter in cold areas. Plant with male cultivars to ensure good fruiting. Tolerates light pruning at any time; cut back in spring to promote bushiness.*

☼ ◑ ◊ ◆ ♀H4 ↕↔ 3m (10ft)

AURINIA SAXATILIS

An evergreen perennial that forms dense clusters of bright yellow flowers in late spring which give rise to its common name, gold dust. The flowers of its cultivar 'Citrinus', also recommended, are a more lemony yellow. The oval, hairy, grey-green leaves are arranged in clumps. Ideal for rock gardens, walls and banks. Also sold as *Alyssum saxatilis.*

CULTIVATION *Grow in moderately fertile soil that is reliably well-drained, in a sunny site. Cut back after flowering to maintain compactness.*

☼ ◊ ♀H4 ↕20cm (8in) ↔to 30cm (12in)

BALLOTA PSEUDODICTAMNUS

An evergreen subshrub that forms mounds of rounded, yellow-grey-green leaves on upright, white-woolly stems. Whorls of small, white or pinkish-white flowers, each enclosed by a pale green funnel, are produced in late spring and early summer. May not survive in areas with cold, wet winters.

CULTIVATION *Grow in poor, very well-drained soil, in full sun with protection from excessive winter wet. Cut back in early spring to keep plants compact.*

☼ ◊ 🏆H3–4 ↕45cm (18in) ↔60cm (24in)

BAPTISIA AUSTRALIS

Blue false indigo is a gently spreading, upright perennial with a long season of interest. The bright blue-green leaves, on grey-green stems, are divided into three oval leaflets. Spikes of indigo-blue flowers, often flecked white or cream, open throughout early summer. The dark grey seed pods can be dried for winter decoration.

CULTIVATION *Grow in deep, moist but well-drained, fertile, preferably neutral to acid soil, in full sun. Once planted, it is best left undisturbed.*

☼ ◊ 🏆H4 ↕1.5m (5ft) ↔60cm (24in)

BEGONIAS WITH DECORATIVE FOLIAGE

These perennial begonias are typically grown as annuals for their large, usually asymmetrical, ornamental leaves which are available in a variety of colours. For example, there are lively leaves of 'Merry Christmas' outlined with emerald green, or there is the more subtle, dark green, metallic foliage of *B. metallica.* Some leaves are valued for their unusual patterns; *B. masoniana* is appropriately known as the iron-cross begonia. Under the right conditions, 'Thurstonii' may reach shrub-like proportions, but most, like 'Munchkin', are more compact. Grow as summer bedding, in a conservatory or as house plants.

CULTIVATION *Grow in fertile, well-drained, neutral to acid soil or compost, in light dappled shade. Promote compact, leafy growth by pinching out shoot tips during the growing season. When in growth, feed regularly with a nitrogen-rich fertiliser. Minimum temperature 15°C (59°F).*

☀ ◊ ♀H1

1 ↕↔ 60cm (2ft)

2 ↕50cm (20in) ↔ 45cm (18in)

1 *Begonia listada* **2** *B. masoniana*

3 ↕25cm (8in) ↔ 30cm (12in)

4 ↕90cm (36in) ↔ 60cm (24in)

5 ↕20cm (8in) ↔ 25cm (10in)

6 ↕30cm (12in) ↔ 45cm (18in)

7 ↕2m (6ft) ↔ 45cm (18in)

8 ↕20cm (8in) ↔ 25cm (10in)

3 *B.* 'Merry Christmas **4** *B. metallica* **5** *B.* 'Munchkin'
6 *B.* 'Silver Queen' **7** *B.* 'Thurstonii' **8** *B.* 'Tiger Paws'

FLOWERING BEGONIAS

Usually grown outdoors as annuals, these bold-flowered begonias are very variable in size and shape, offering a range of summer uses to the gardener: for specific information on growth habit, check the label or ask advice when buying. Upright or compact begonias, such as 'Pin Up' or the Olympia series, are ideal for summer bedding; for containers and hanging baskets, there are pendulous or trailing varieties like 'Illumination Orange'. Begonias can also be grown as house plants. The flowers also come in a wide variety of sizes and colours; they are either single or double, and appear in loose clusters throughout summer.

CULTIVATION *Fertile, humus-rich, neutral to acid soil or compost with good drainage. Flowers are best in partial shade; they suffer in direct sun. When in growth, give a balanced fertilizer. Those of H1 hardiness will not survive below 15°C (59°F).*

☀ ◊ ♈H1, H2–3

2 ↕↔ to 40cm (16in)

1 ↕ to 45cm (18in) ↔ to 35cm (14in)

3 ↕ 20cm (8in) ↔ 20–22cm (8–9in)

4 ↕ 60cm (24in) ↔ 30cm (12in)

1 *B.* 'Alfa Pink' (H2–3) **2** *Begonia* 'All Round Dark Rose Green Leaf' (H2–3)
3 *B.* 'Expresso Scarlet' (H2–3) **4** *B.* 'Illumination Orange' (H2–3)

5 ↕ to 35cm (14in) ↔ to 30cm (12in)

6 ↕ 75cm (30in) ↔ 60cm (24in)

7 ↕↔ 30cm (12in)

8 ↕↔ to 20cm (8in)

9 ↕ 60cm (24in) ↔ 45cm (18in)

10 ↕ 25cm (10in) ↔ 20cm (8in)

11 ↕ 75cm (30in) ↔ 45cm (18in)

5 *B.* 'Inferno Apple Blossom' (H2–3) **6** *B.* 'Irene Nuss' (H1) **7** *B.* 'Nonstop' (H2–3) **8** *B.* 'Olympia White' (H2–3) **9** *B.* 'Orange Rubra' (H1) **10** *B.* 'Pin Up' (H2–3) **11** *B. sutherlandii* (H1)

BELLIS PERENNIS 'POMPONETTE'

This double-flowered form of the common daisy is usually grown as a biennial for spring bedding. Pink, red or white flowerheads with quill-shaped petals appear from late winter to spring, above the dense clumps of spoon-shaped, bright green leaves. 'Dresden China' and 'Rob Roy' are other recommended bellis.

CULTIVATION *Grow in well-drained, moderately fertile soil, in full sun or partial shade. Deadhead to prolong flowering and to prevent self-seeding.*

☼☀ ◊ ♀H4 ↕↔ 10–20cm (4–8in)

BERBERIS DARWINII

The Darwin barberry is a vigorous, arching, evergreen shrub that carries masses of small, deep golden-orange flowers on spiny stems from mid- to late spring; these are followed by blue berries in autumn. The leaves are glossy dark green and spiny. Use as a vandal-resistant or barrier hedge.

CULTIVATION *Grow in any but water-logged soil, in full sun or partial shade with shelter from cold, drying winds. Trim after flowering, if necessary.*

☼☀ ◊♦ ♀H4 ↕3m (10ft) or more ↔3m (10ft)

BERBERIS × *OTTAWENSIS* 'SUPERBA'

This spiny, rounded, deciduous, spring-flowering shrub bears clusters of small, pale yellow, red-tinged flowers which are followed by red berries in autumn. The red-purple leaves turn crimson before they fall. Effective as a specimen shrub or in a mixed border.

CULTIVATION *Grow in almost any well-drained soil, preferably in full sun. Thin out dense growth in mid-winter.*

☼ ◊♦ ♀H4 ↕↔ 2.5m (8ft)

BERBERIS × *STENOPHYLLA* 'CORALLINA COMPACTA'

While *Berberis stenophylla* is a large, arching shrub, ideal for informal hedging, this cultivar of it is tiny; a small, evergreen shrub bearing spine-tipped, deep green leaves on arching, spiny stems. Quantities of tiny, light orange flowers appear from mid-spring, followed by small, blue-black berries.

CULTIVATION *Best in fertile, humus-rich soil that is reliably drained, in full sun. Cut back hard after flowering.*

☼ ◊ ♀H4 ↕↔ to 30cm (12in)

BERBERIS THUNBERGII 'BAGATELLE'

A very compact, spiny, spring-flowering, deciduous shrub with deep red-purple leaves which turn orange and red in autumn. The pale yellow flowers are followed by glossy red fruits. Good for a rock garden, but may not survive in areas with cold winters. *B. thunbergii* 'Atropurpurea Nana', or 'Crimson Pygmy', is another small purple-leaved berberis, to 60cm (24in) tall.

CULTIVATION *Grow in well-drained soil, in full sun for best flower and foliage colour. Thin out dense, overcrowded growth in mid- to late winter.*

☼ ◊ ♈H3 ↕30cm (12in) ↔40cm (16in)

BERBERIS THUNBERGII 'ROSE GLOW'

A compact, spiny, deciduous shrub with reddish-purple leaves that gradually become flecked with white as the season progresses. Tiny, pale yellow flowers appear in mid-spring, followed by small red berries. Good as a barrier hedge.

CULTIVATION *Grow in any but water-logged soil, in full sun or partial shade. Cut out any dead wood in summer.*

☼ ◐ ◊ ♦ ♈H4 ↕2m (6ft) or more ↔2m (6ft)

BERBERIS VERRUCULOSA

A slow-growing, compact, spring-flowering barberry that makes a fine evergreen specimen shrub. The cup-shaped, golden-yellow flowers are carried amid the spine-tipped, glossy dark green leaves on spiny, arching stems. Oval to pear-shaped black berries develop in autumn.

CULTIVATION *Best in well-drained, humus-rich, fertile soil, in full sun. Keep pruning to a minimum.*

☼ ◊ 🏆H4 ↕↔ 1.5m (5ft)

BERBERIS WILSONIAE

A very spiny, semi-evergreen, arching shrub forming dense mounds of grey-green foliage which turns red and orange in autumn. Clusters of pale yellow flowers in summer are followed by coral-pink to pinkish-red berries. Makes a good barrier hedge. Avoid seed-grown plants as they may be inferior hybrids.

CULTIVATION *Grow in any well-drained soil, in sun or partial shade. Flowering and fruiting are best in full sun. Thin out dense growth in mid-winter.*

☼ ◐ ◊ 🏆H4 ↕1m (3ft) ↔2m (6ft)

BERGENIA 'BALLAWLEY'

This clump-forming, evergreen perennial, one of the first to flower in spring, bears bright crimson flowers which are carried on sturdy red stems. The leathery, oval leaves turn bronze-red in winter. Suits a woodland garden, or plant in groups to edge a mixed border. *B. cordifolia* 'Purpurea' has similarly coloured leaves, with deep magenta flowers.

CULTIVATION *Grow in any well-drained soil, in full sun or light shade. Shelter from cold winds. Mulch in autumn.*

☼◑ ◊◊ ♈H4 ↕to 60cm (24in) ↔60cm (24in)

BERGENIA 'SILBERLICHT'

An early-flowering, clump-forming, evergreen perennial bearing clusters of cup-shaped white flowers, often flushed pink, in spring. (For pure white flowers on a similar plant, look for 'Bressingham White', or for deep pink, 'Morgenröte'.) The mid-green leaves are leathery, with toothed margins. Good under-planting for shrubs, which give it some winter shelter.

CULTIVATION *Grow in any well-drained soil, in full sun or partial shade. Shelter from cold winds to avoid foliage scorch. Provide a mulch in autumn.*

☼◑ ◊◊ ♈H4 ↕30cm (12in) ↔50cm (20in)

BETULA NIGRA

The black birch is a tall, conical to spreading, deciduous tree with glossy, mid- to dark green, diamond-shaped leaves. It has shaggy, red-brown bark that peels in layers on young trees; on older specimens, the bark becomes blackish or grey-white and develops cracks. Yellow-brown male catkins are conspicuous in spring. Makes a fine specimen tree, but only for a large garden.

CULTIVATION *Grow in moist but well-drained, moderately fertile soil, in full sun. Remove any damaged, diseased or dead wood in late autumn.*

☼ ◊ 🏆H4 ↕18m (60ft) ↔12m (40ft)

BETULA PENDULA 'YOUNGII'

Young's weeping birch is a deciduous tree with an elegant, weeping habit. The yellow-brown male catkins appear in early spring before the triangular leaves; the foliage turns golden-yellow in autumn. An attractive tree for a small garden, more dome-shaped than 'Tristis' or 'Laciniata', other popular weeping birches, growing wider than it is tall.

CULTIVATION *Any moist but well-drained soil, in an open, sunny site. Keep pruning to a minimum; remove any shoots on the trunk in late autumn.*

 🏆H4 ↕8m (25ft) ↔10m (30ft)

BETULA UTILIS VAR. *JACQUEMONTII*

The West Himalayan birch is an open, broadly conical, deciduous tree with smooth, peeling white bark. Catkins are a feature in early spring, and the dark green leaves turn rich golden-yellow in autumn. Plant where winter sun will light up the bark, particularly brilliantly white in the cultivars 'Silver Shadow', 'Jermyns' and 'Grayswood Ghost'.

CULTIVATION *Grow in any moist but well-drained soil, in sun. Remove any damaged or dead wood from young trees in late autumn; once established, keep pruning to a minimum.*

☼ ◊ ♈H4 ↕15m (50ft) ↔7.5m (23ft)

BIDENS FERULIFOLIA

This clump-forming, spreading, short-lived perennial tends to be grown as an annual. A profusion of star-shaped, bright golden-yellow flowers are borne over a long period from mid-spring until the first frosts. The leaves are fresh green and finely divided. Ideal for trailing over the edges of hanging baskets and other containers.

CULTIVATION *Grow in moist but well-drained, fairly fertile soil or compost, in sun. Short-lived, but easily propagated by stem cuttings in autumn. Minimum temperature 2°C (35°F).*

☼ ◊◆ ♈H1+3 ↕to 30cm (12in) ↔indefinite

BRACHYGLOTTIS 'SUNSHINE'

A bushy, mound-forming, evergreen shrub bearing oval leaves which are silvery-grey when young, becoming dark green with white-felted undersides as they develop. Daisy-like yellow flowers appear from early to mid-summer. Some gardeners prefer it as a foliage plant, pinching or snipping off the flower buds before they open. Thrives in coastal sites.

CULTIVATION *Grow in any well-drained soil, in a sunny, sheltered site. Trim back after flowering. Responds well to hard pruning in spring.*

☼ ◊ 🏆H4 ↕1–1.5m (3–5ft) ↔2m (6ft) or more

BRACTEANTHA BRIGHT BIKINI SERIES

These strawflowers are upright annuals or short-lived perennials with papery, double flowers in red, pink, orange, yellow and white from late spring to autumn. The leaves are grey-green. Use to edge a border, or grow in a window box; flowers are long-lasting and cut and dry well. For single colours rather than a mixture, try 'Frosted Sulphur' (lemon-yellow), 'Silvery Rose' and 'Reeves Purple'.

CULTIVATION *Grow in moist but well-drained, moderately fertile soil. Choose a position in full sun.*

☼ ◊ 🏆H3 ↕to 30cm (12in) ↔30cm (12in)

BRUNNERA MACROPHYLLA 'HADSPEN CREAM'

This clump-forming perennial with attractive foliage is ideal for ground cover in borders and among deciduous trees. In mid- and late spring, upright clusters of small, bright blue flowers appear above heart-shaped leaves, plain green in *Brunnera macrophylla*, but with irregular, creamy-white margins in this attractive cultivar.

CULTIVATION *Grow in moist but well-drained, humus-rich soil. Choose a position that is cool and lightly shaded.*

◑ ◊◊ ♀H4 ↕45cm (18in) ↔60cm (24in)

BUDDLEJA ALTERNIFOLIA

A dense, deciduous shrub carrying slender, arching branches. Fragrant, lilac-purple flowers are produced in neat clusters during early summer among the narrow, grey-green leaves. Makes a good wall shrub, or can be trained with a single, clear trunk as a striking specimen tree. Attractive to beneficial insects.

CULTIVATION *Best in chalky soil, but can be grown in any soil that is well-drained, in full sun. Cut stems back to strong buds after flowering; responds well to hard pruning in spring.*

☼ ◊ ♀H4 ↕↔4m (12ft)

BUDDLEJA DAVIDII

All cultivars of *B. davidii*, the butterfly bush, are fast-growing, deciduous shrubs with a wide range of flower colours. As the popular name suggests, the flowers attract butterflies and other beneficial garden insects in profusion. The long, arching shoots carry lance-shaped, mid- to grey-green leaves, up to 25cm (10in) long. Conical clusters of bright, fragrant flowers, usually about 30cm (12in) long, are borne at the end of arching stems from summer to autumn; those of 'Royal Red' are the largest, up to 50cm (20in) long. These shrubs respond well to hard pruning in spring, which keeps them a compact size for a small garden.

CULTIVATION *Grow in well-drained, fertile soil, in sun. Restrict size and encourage better flowers by pruning back hard to a low framework each spring. To prevent self-seeding, cut spent flowerheads back to a pair of leaves or sideshoots; this may also result in a second blooming.*

☼ ◊ 𝕐H4

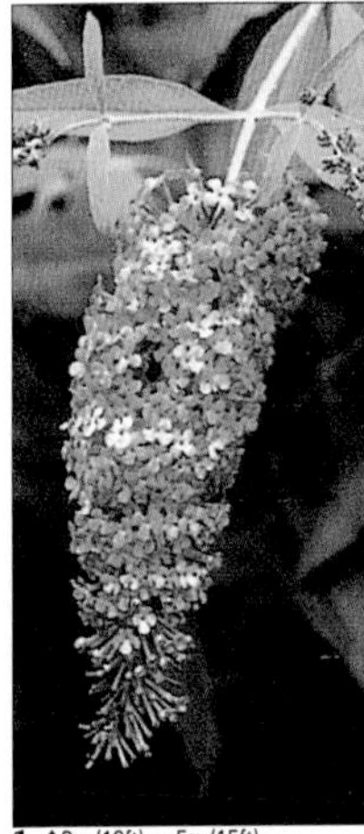

1 ↕3m (10ft) ↔ 5m (15ft)

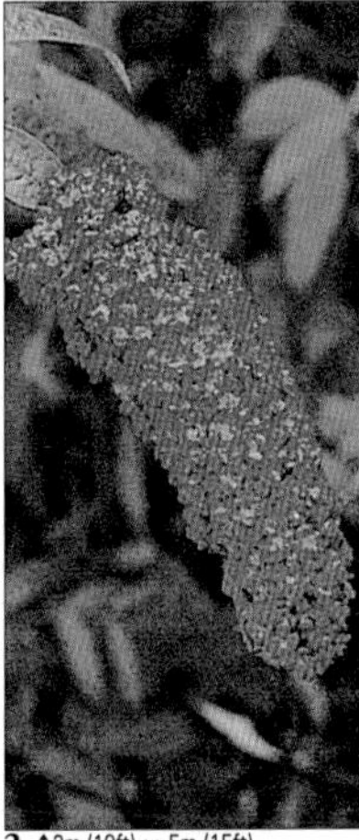

2 ↕3m (10ft) ↔ 5m (15ft)

3 ↕3m (10ft) ↔ 5m (15ft)

1 *B. davidii* 'Empire Blue' **2** *B. davidii* 'Royal Red' **3** *B. davidii* 'White Profusion'

BUDDLEJA GLOBOSA

The orange ball tree is a deciduous or semi-evergreen shrub bearing, unusually for a buddleja, round clusters of tiny, orange-yellow flowers which appear in early summer. The lance-shaped leaves are dark green with woolly undersides. A large shrub, it is prone to becoming bare at the base and does not respond well to pruning, so grow towards the back of a mixed border.

CULTIVATION *Best on well-drained, chalky soil, in a sunny position with shelter from cold winds. Pruning should be kept to a minimum or the next year's flowers will be lost.*

☼ ◊ ♈H4 ↕↔ 5m (15ft)

BUDDLEJA 'LOCHINCH'

A compact, deciduous shrub, very similar to a *Buddleja davidii* (see p. 93), bearing long spikes of lilac-blue flowers from late summer to autumn. The leaves are downy and grey-green when young, becoming smooth and developing white-felted undersides as they mature. Very attractive to butterflies.

CULTIVATION *Grow in any well-drained, moderately fertile soil, in sun. Cut back all stems close to the base each year, as the buds begin to swell in spring.*

☼ ◊ ♈H3–4 ↕ 2.5m (8ft) ↔ 3m (10ft)

BUXUS SEMPERVIRENS 'ELEGANTISSIMA'

This variegated form of the common box is a rounded, dense, evergreen shrub bearing small and narrow, glossy bright green leaves edged with cream. The flowers are of little significance. Responding well to trimming, it is very good as an edging plant or for use as a low hedge. 'Latifolia Maculata' is also variegated, with yellow leaf markings.

CULTIVATION *Grow in any well-drained soil, in sun or light shade. Trim in spring and summer; overgrown shrubs respond well to hard pruning in late spring.*

☼ ◐ ◊ 🏆H4 ↕↔ 1.5m (5ft)

BUXUS SEMPERVIRENS 'SUFFRUTICOSA'

A very dense, slow-growing box carrying small, evergreen, glossy bright green leaves. Widely used as an edging plant or for clipping into precise shapes. During late spring or early summer, inconspicuous flowers are produced. Excellent as a hedge.

CULTIVATION *Grow in any well-drained, fertile soil, in sun or semi-shade. The combination of dry soil and full sun can cause scorching. Trim hedges in summer; overgrown specimens can be hard-pruned in late spring.*

☼ ◐ ◊ 🏆H4 ↕1m (3ft) ↔1.5m (5ft)

CALENDULA 'FIESTA GITANA'

This dwarf pot marigold is a bushy, fast-growing annual which produces masses of usually double flowerheads in pastel orange or yellow, including bi-colours, from summer to autumn. The leaves are hairy and aromatic. Excellent for cutting, bedding and containers.

CULTIVATION *Grow in well-drained, poor to moderately fertile soil, in full sun or partial shade. Deadhead regularly to prolong flowering.*

☼☀ ◊ ♀H4 ↕to 30cm (12in) ↔30–45cm (12–18in)

CALLICARPA BODINIERI VAR. *GIRALDII* 'PROFUSION'

An upright, deciduous shrub grown mainly for its long-lasting autumn display of shiny, bead-like, deep violet berries. The large, pale green, tapering leaves are bronze when they emerge in spring, and pale pink flowers appear in summer. Brings a long season of interest to a shrub border; for maximum impact, plant in groups. May not survive in areas with cold winters.

CULTIVATION *Grow in any well-drained, fertile soil, in full sun or dappled shade. Cut back about 1 in 5 stems to the base each year in early spring.*

☼☀ ◊ ♀H3 ↕3m (10ft) ↔ 2.5m (8ft)

CALLISTEMON CITRINUS 'SPLENDENS'

This attractive cultivar of the crimson bottlebrush is an evergreen shrub usually with arching branches. Dense spikes of brilliant red flowers appear in spring and summer, amid the grey-green, lemon-scented leaves which are bronze-red when young. Grow at the base of a sunny wall to give protection from winter cold.

CULTIVATION *Best in well-drained, fertile, neutral to acid soil, in full sun. Pinch out tips young of young plants to promote bushiness. Tolerates hard pruning in spring.*

☼ ◊ 🏆H3 ↕2–8m (6–25ft) ↔1.5–6m (5–20ft)

CALLISTEPHUS MILADY SUPER MIXED

This sturdy, partially wilt-resistant mixture of variably coloured, fast-growing annuals is ideal for use in bedding and containers. The double, rounded flowerheads, borne from late summer to autumn, are pink, red, scarlet, blue, or white. The leaves are mid-green and toothed.

CULTIVATION *Grow in moist but well-drained, fertile, neutral to alkaline soil, in a sheltered, sunny site. Deadhead regularly to prolong flowering.*

☼ ◊♦ 🏆H3 ↕to 30cm (12in) ↔25cm (10in)

CALLUNA VULGARIS

Cultivars of *C. vulgaris* are upright to spreading, fine-leaved heathers. They make excellent evergreen ground cover plants if weeds are suppressed before planting. Dense spikes of bell-shaped flowers appear from mid-summer to late autumn, in shades of red, purple, pink or white; 'Kinlochruel' is quite distinctive with its double white flowers in long clusters. Seasonal interest is extended into winter by cultivars with coloured foliage, such as 'Robert Chapman' and 'Beoley Gold'. Heathers are very attractive to bees and other beneficial insects, and make good companions to dwarf conifers.

CULTIVATION *Best in well-drained, humus-rich, acid soil, in an open, sunny site, to recreate their native moorland habitats. Trim off flowered shoots in early spring with shears; remove overlong shoots wherever possible, cutting back to their point of origin below the flower cluster.*

☼ ◊ 🏆H4

1 ↕25cm (10in) ↔ 35cm (14in)

2 ↕35cm (14in) ↔ to 75cm (30in)

3 ↕25cm (10in) ↔ 40cm (16in)

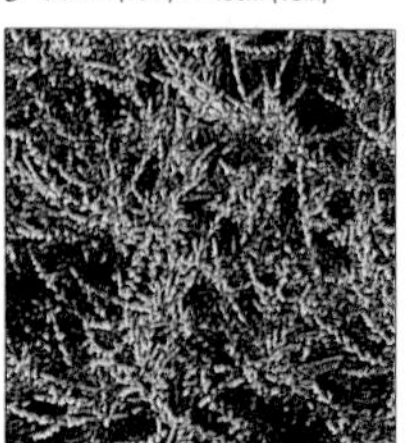

4 ↕25cm (10in) ↔ 65cm (26in)

1 *C. vulgaris* 'Beoley Gold' **2** *C. vulgaris* 'Darkness' **3** *C. vulgaris* 'Kinlochruel' **4** *C. vulgaris* 'Robert Chapman'

CALTHA PALUSTRIS

The marsh marigold is a clump-forming, aquatic perennial which thrives in a bog garden or at the margins of a stream or pond. Cup-shaped, waxy, bright golden-yellow flowers appear on tall stems in spring, above the kidney-shaped, glossy green leaves. For double flowers, look for the cultivar 'Flore Pleno'.

CULTIVATION *Best in boggy, rich soil, in an open, sunny site. Tolerates root restriction in aquatic planting baskets in water no deeper than 23cm (9in), but prefers shallower conditions.*

☼ ● ♀H4 ↕10–40cm (4–16in) ↔45cm (18in)

CAMELLIA 'INSPIRATION'

A dense, upright, evergreen shrub or small tree bearing masses of saucer-shaped, semi-double, deep pink flowers from mid-winter to late spring. The dark green leaves are oval and leathery. Good for the back of a border or as a specimen shrub.

CULTIVATION *Best in moist but well-drained, fertile, neutral to acid soil, in partial shade with shelter from cold, drying winds. Mulch around the base with shredded bark. After flowering, prune back young plants to encourage a bushy habit and a balanced shape.*

◑ ◊● ♀H4 ↕4m (12ft) ↔2m (6ft)

CAMELLIA JAPONICA

These long-lived and elegant, evergreen shrubs or small trees for gardens with acid soil are very popular in shrub borders, woodland gardens, or standing on their own in the open ground or in containers. The oval leaves are glossy and dark green; they serve to heighten the brilliance of the single to fully double flowers in spring. Single and semi-double flowers have a prominent central boss of yellow stamens. Most flowers are suitable for cutting, but blooms may be spoiled by late frosts.

CULTIVATION *Grow in moist but well-drained, humus-rich, acid soil in partial shade. Choose a site sheltered from early morning sun, cold winds, and late-season frosts. Do not plant too deeply; the top of the root ball must be level with the firmed soil. Maintain a mulch 5–7cm (2–3in) deep of leaf mould or shredded bark. Little pruning is necessary, although moderate trimming of young plants can help to create a balanced shape.*

☀ ◊◊ ♀H4

1 ↕5m (15ft) ↔ 8m (25ft) **2** ↕9m (28ft) ↔ 8m (25ft) **3** ↕9m (28ft) ↔ 8m (25ft)

4 ↕2m (6ft) ↔ 1m (3ft) **5** ↕9m (28ft) ↔ 8m (25ft) **6** ↕9m (28ft) ↔ 8m (25ft)

1 *Camellia japonica* 'Adolphe Audusson' **2** 'Alexander Hunter' **3** 'Berenice Boddy' **4** 'Bob's Tinsie' **5** 'Coquettii' **6** 'Elegans'

MORE CHOICES

'Akashigata' Deep pink flowers.

'Bob Hope' Dark red.

'C. M. Hovey' Crimson/scarlet

'Doctor Tinsley' Pinky-white.

'Grand Prix' Bright red with yellow stamens

'Hagoromo' Pale pink

'Masayoshi' White with red marbling.

'Miss Charleston' Ruby-red.

'Nuccio's Gem' White.

10 ↕10m (30ft)

11 ↕10m (30ft)

13 ↕10m (30ft)

14 ↕10m (30ft)

7 *Camellia japonica* 'Gloire de Nantes' **8** 'Guilio Nuccio' **9** 'Jupiter' (*syn.*'Paul's Jupiter') **10** 'Lavinia Maggi' **11** 'Mrs D. W. Davis' **12** 'R. L. Wheeler' **13** 'Rubescens Major' **14** 'Tricolor'

CAMELLIA 'LASCA BEAUTY'

An open, upright shrub greatly valued for its very large, semi-double, pale pink flowers, which appear in mid-spring. They stand out against the dark green foliage. Grow in a cool greenhouse, and move outdoors to a partially shaded site in early summer.

CULTIVATION *Grow in lime-free potting compost in bright filtered light. Water freely with soft water when in growth; more sparingly in winter. Apply a balanced fertilizer once in mid-spring and again in early summer.*

☀ 💧 ♀H2-3 ↕2–5m (6–15ft) ↔1.5–3m (5–10ft)

CAMELLIA 'LEONARD MESSEL'

This spreading, evergreen shrub with oval, leathery, dark green leaves is one of the hardiest camellias available. It produces an abundance of large, flattish to cup-shaped, semi-double, clear pink flowers from early to late spring. Suits a shrub border.

CULTIVATION *Best in moist but well-drained, fertile, neutral to acid soil. Position in semi-shade with shelter from cold, drying winds. Maintain a mulch of shredded bark or leafmould around the base. Pruning is not necessary.*

☀ 💧 ♀H4 ↕4m (12ft) ↔3m (10ft)

CAMELLIA 'MANDALAY QUEEN'

This large, widely branching shrub bears deep rose-pink, semi-double flowers in spring. The broad, leathery leaves are dark green. Best grown in a cool greenhouse, but move it outdoors in summer.

CULTIVATION *Best in lime-free potting compost with bright filtered light. Water freely with soft water when in growth, more sparingly in winter. Apply a balanced fertilizer in mid-spring and again in early summer.*

Partial shade · Moist · AGM H2 ↕ to 15m (50ft) ↔5m (15ft)

CAMELLIA SASANQUA 'NARUMIGATA'

An upright shrub or small tree valued for its late display of fragrant, single-petalled white flowers in mid- to late autumn. The foliage is dark green. Makes a good hedge, though in very cold areas it may be better against a warm, sunny wall. *C. sasanqua* 'Crimson King' is a very similar shrub, with red flowers.

CULTIVATION *Grow in moist but well-drained, humus-rich, acid soil, and maintain a thick mulch. Choose a site in full sun or partial shade, with shelter from cold winds. Tolerates hard pruning after flowering.*

Full sun / Partial shade · Moist · AGM H3 ↕to 6m (20ft)↔to 3m (10ft)

CAMELLIA x *WILLIAMSII*

Cultivars of *C.* x *williamsii* are strong-growing, evergreen shrubs much-valued for their bright, lustrous foliage and the unsurpassed elegance of their rose-like flowers which range from pure white to crimson. Most flower in mid- and late spring, although 'Anticipation' and 'Mary Christian' begin to flower from late winter. Many are fully hardy and can be grown in any climate, although the flowers are susceptible to damage in hard frosts. They make handsome specimens for a cool conservatory or shrub border, and 'J.C. Williams', for example, will train against a cold wall. Avoid sites exposed to morning sun.

CULTIVATION *Best in moist but well-drained, acid to neutral soil, in partial shade. Shelter from frost and cold winds, and mulch with shredded bark. Prune young plants after flowering to promote bushiness; wall-trained shrubs should be allowed to develop a strong central stem.*

◑ ◊◊ ♆H4

1 ↕4m (12ft) ↔ 2m (6ft)

2 ↕3m (10ft) ↔ 2.5m (8ft)

3 ↕5m (15ft) ↔ 2.5m (8ft)

1 *C.* x *williamsii* 'Anticipation' **2** *C.* x *williamsii* 'Brigadoon' **3** *C.* x *williamsii* 'Donation'

4 ↕↔ 4m (12ft)

5 ↕4m (12ft) ↔ 2.5m (8ft)

6 ↕4m (12ft) ↔ 2.5m (8ft)

7 ↕4m (12ft) ↔ 2.5m (8ft)

8 ↕4m (12ft) ↔ 2.5m (8ft)

9 ↕↔ 3m (10ft)

4 *C.* x *williamsii* 'George Blandford' **5** *C.* x *williamsii* 'Joan Trehane' **6** *C.* x *williamsii* 'J.C. Williams' **7** *C.* x *williamsii* 'Mary Christian' **8** *C.* x *williamsii* 'Saint Ewe' **9** *C.* x *williamsii* 'Water Lily'

CAMPANULA COCHLEARIIFOLIA

Fairies' thimbles is a low-growing, rosette-forming perennial bearing, in mid-summer, abundant clusters of open, bell-shaped, mauve-blue or white flowers. The bright green leaves are heart-shaped. It spreads freely by means of creeping stems, and can be invasive. Particularly effective if allowed to colonize areas of gravel, paving crevices or the tops of dry walls.

CULTIVATION *Prefers moist but well-drained soil, in sun or partial shade. To restrict spread, pull up unwanted plants.*

☼☀ ◊ ♀H4 ↕to 8cm (3in) ↔to 50cm (20in) or more

CAMPANULA GLOMERATA 'SUPERBA'

A fast-growing, clump-forming perennial carrying dense heads of large, bell-shaped, purple-violet flowers in summer. The lance-shaped to oval, mid-green leaves are arranged in rosettes at the base of the plant and along the stems. Excellent in herbaceous borders or informal, cottage-style gardens.

CULTIVATION *Best in moist but well-drained, neutral to alkaline soil, in sun or semi-shade. Cut back after flowering to encourage a second flush of flowers.*

☼☀ ◊♦ ♀H4 ↕75cm (30in) ↔1m (3ft) or more

CAMPANULA LACTIFLORA 'LODDON ANNA'

An upright, branching perennial producing sprays of large, nodding, bell-shaped, soft lilac-pink flowers from mid-summer above mid-green leaves. Makes an excellent border perennial, but may need staking in an exposed site. Mixes well with the white-flowered 'Alba', or the deep purple 'Prichard's Variety'.

CULTIVATION *Best in moist but well-drained, fertile soil, in full sun or partial shade. Trim after flowering to encourage a second, although less profuse, flush of flowers.*

☼☀ 💧💧 🏆H4 ↕1.2–1.5m (4–5ft) ↔60cm (24in)

CAMPANULA PORTENSCHLAGIANA

The Dalmatian bellflower is a robust, mound-forming, evergreen perennial with long, bell-shaped, deep purple flowers from mid- to late summer. The leaves are toothed and mid-green. Good in a rock garden or on a sunny bank. May become invasive. *Campanula poschkarskyana* can be used in a similar way; look for the recommended cultivar 'Stella'.

CULTIVATION *Best in moist but well-drained soil, in sun or partial shade. Very vigorous, so plant away from smaller, less robust plants.*

☼☀ 💧 🏆H4 ↕to 15cm (6in) ↔50cm (20in) or more

CAMPSIS X *TAGLIABUANA* 'MADAME GALEN'

Woody-stemmed climber that will cling with aerial roots against a wall, fence or pillar, or up into a tree. From late summer to autumn, clusters of trumpet-shaped, orange-red flowers open among narrow, toothed leaves. For yellow flowers, choose *C. radicans* f. *flava*.

CULTIVATION *Prefers moist but well-drained, fertile soil, in a sunny, sheltered site. Tie in new growth until the allotted space is covered by a strong framework. Prune back hard each winter to promote bushiness.*

☼ ◊◆ 🏆H4 ↕10m (30ft) or more

CARDIOCRINUM GIGANTEUM

The giant lily is a spectacular, summer-flowering, bulbous perennial with trumpet-shaped white flowers which are flushed with maroon-purple at the throats. The stems are stout, and the leaves broadly oval and glossy green. It needs careful siting and can take up to seven years to flower. Grow in woodland, or a sheltered border in shade.

CULTIVATION *Best in deep, moist but well-drained, reliably cool, humus-rich soil, in semi-shade. Intolerant of hot or dry conditions. Slugs can be a problem.*

◐ ◆ 🏆H4 ↕1.5–4m (5–12ft) ↔45cm (18in)

CAREX ELATA 'AUREA'

Bowles' golden sedge is a colourful, tussock-forming, deciduous perennial for a moist border, bog garden or the margins of a pond or stream. The bright leaves are narrow and golden-yellow. In spring and early summer, small spikes of relatively inconspicuous, dark brown flowers are carried above the leaves. Often sold as *C.* 'Bowles' Golden'.

CULTIVATION *Grow in moist or wet, reasonably fertile soil. Position in full sun or partial shade.*

☼☀ 💧💧 🏆H4 ↕to 70cm (28in) ↔45cm (18in)

CAREX OSHIMENSIS 'EVERGOLD'

A very popular, evergreen, variegated sedge, bright and densely tufted with narrow, dark green, yellow-striped leaves. Spikes of tiny, dark brown flowers are borne in mid- and late spring. Tolerates freer drainage than many sedges and is suitable for a mixed border.

CULTIVATION *Needs moist but well drained, fertile soil, in sun or partial shade. Remove dead leaves in summer.*

☼☀ 💧 🏆H4 ↕30cm (12in) ↔35cm (14in)

CARPENTERIA CALIFORNICA

The tree anemone is a summer-flowering, evergreen shrub bearing large, fragrant white flowers with showy yellow stamens. The glossy, dark green leaves are narrowly oval. It is suitable for wall-training, which overcomes its sometimes sprawling habit. In cold areas, it must be sited against a warm, sheltered wall.

CULTIVATION *Grow in well-drained soil, in full sun with shelter from cold winds. In spring, remove branches that have become exhausted by flowering, cutting them back to their bases.*

☼ ◊ ♈H3 ↕2m (6ft) or more ↔2m (6ft)

CARYOPTERIS X *CLANDONENSIS* 'HEAVENLY BLUE'

A compact, upright, deciduous shrub grown for its clusters of intensely dark blue flowers which appear in late summer and early autumn. The irregularly toothed leaves are grey-green. In cold areas, position against a warm wall.

CULTIVATION *Grow in well-drained, moderately fertile, light soil, in full sun. Prune all stems back hard to low buds in late spring. A woody framework will develop, which should not be cut into.*

☼ ◊ ♈H3 ↕↔ 1m (3ft)

CASSIOPE 'EDINBURGH'

A heather-like, upright, evergreen shrub producing nodding, bell-shaped flowers in spring; these are white with small, greenish-brown outer petals. The scale-like, dark green leaves closely overlap along the stems. Suits a rock garden (not among limestone) or a peat bed. *C. lycopoides* has very similar flowers, but is mat-forming, almost prostrate, only 8cm (3in) tall.

CULTIVATION *Grow in reliably moist, humus-rich, acid soil, in partial shade. Trim after flowering.*

H4 ↕↔ to 25cm (10in)

CEANOTHUS ARBOREUS 'TREWITHIN BLUE'

A vigorous, spreading, evergreen shrub valued for its profusion of fragrant, mid-blue flowers in spring and early summer. The leaves are dark green and rounded. In cold climates, grow against a warm wall; otherwise, suitable for a large, sheltered border.

CULTIVATION *Grow in well-drained, fertile soil, in full sun with shelter from cold, drying winds. Tip-prune young plants in spring. Once established, prune only to shape, after flowering.*

H3 ↕6m (20ft) ↔8m (25ft)

CEANOTHUS 'AUTUMNAL BLUE'

A vigorous, evergreen shrub that produces a profusion of tiny but vivid, rich sky-blue flowers from late summer to autumn. The leaves are broadly oval and glossy dark green. One of the hardiest of the evergreen ceanothus, it is suitable in an open border as well as for informal training on walls; best against a warm wall in cold areas.

CULTIVATION *Grow in well-drained, moderately fertile soil, in full sun with shelter from cold winds. Tolerates lime. Tip-prune young plants in spring, and trim established plants after flowering.*

☼ ♁ ♀H3 ↕↔ 3m (10ft)

CEANOTHUS 'BLUE MOUND'

This mound-forming, late spring-flowering ceanothus is an evergreen shrub carrying masses of rich dark blue flowers. The leaves are finely toothed and glossy dark green. Ideal for ground cover, for cascading over banks or low walls, or in a large, sunny rock garden. *Ceanothus* 'Burkwoodii' and 'Italian Skies' can be used similarly. May not survive in areas with cold winters.

CULTIVATION *Grow in well-drained, fertile soil, in full sun. Tip-prune young plants and trim established ones after flowering, in mid-summer.*

☼ ♁ ♀H3 ↕1.5m (5ft) ↔2m (6ft)

CEANOTHUS X *DELILEANUS* 'GLOIRE DE VERSAILLES'

This deciduous ceanothus is a fast-growing shrub. From mid-summer to early autumn, large spikes of tiny, pale blue flowers are borne amid broadly oval, finely toothed, mid-green leaves. 'Topaze' is very similar, with dark blue flowers. They benefit from harder annual pruning than evergreen ceanothus.

CULTIVATION *Grow in well-drained, fairly fertile, light soil, in sun. Tolerates lime. In spring, shorten the previous year's stems by half or more, or cut right back to a low framework.*

☼ ◊ 🏆H4 ↕↔ 1.5m (5ft)

CEANOTHUS THYRSIFLORUS VAR. *REPENS*

This low and spreading ceanothus is a mound-forming, evergreen shrub, bearing rounded clusters of tiny blue flowers in late spring and early summer. The leaves are dark green and glossy. A good shrub to clothe a sunny or slightly shaded bank, but needs the protection of a warm, sunny site in cold areas.

CULTIVATION *Best in light, well-drained, fertile soil, in sun or light shade. Trim back after flowering to keep compact.*

☼ ◐ ◊ 🏆H3 ↕1m (3ft) ↔2.5m (8ft)

CERATOSTIGMA PLUMBAGINOIDES

A spreading, woody-based, sub-shrubby perennial bearing clusters of brilliant blue flowers in late summer. The oval, bright green leaves, carried on upright, slender red stems, become red-tinted in autumn. Good for a rock garden, and also suitable for ground cover.

CULTIVATION *Grow in light, moist but well-drained, moderately fertile soil. Choose a sheltered site in full sun. Cut back stems to about 2.5–5cm (1–2in) from the ground in mid-spring.*

☼ ♢ ♈H3–4 ↕to 45cm (18in) ↔to 30cm (12in) or more

CERATOSTIGMA WILLMOTTIANUM

The Chinese plumbago is an open and spreading, deciduous shrub carrying pale to mid-blue flowers during late summer and autumn. The roughly diamond-shaped, mid-green leaves turn red in autumn. Dies right back in cold winters, but usually regenerates in spring. Suits a sheltered mixed border.

CULTIVATION *Grow in any fertile soil, including dry soil, in full sun. In mid-spring, cut out all dead wood and shorten the remaining stems to a low woody framework.*

☼ ♢ ♈H3–4 ↕1m (3ft) ↔1.5m (5ft)

CERCIS SILIQUASTRUM

The Judas tree is a handsome, broadly spreading, deciduous tree which gradually develops a rounded crown. Clusters of pea-like, bright pink flowers appear on the previous year's wood either before or with the heart-shaped leaves in mid-spring. The foliage is bronze when young, maturing to dark blue-green, then to yellow in autumn. Flowering is best after a long, hot summer.

CULTIVATION *Grow in deep, reliably well-drained, fertile soil, in full sun or light dappled shade. Prune young trees to shape in early summer, removing any frost-damaged growth.*

☼ ◐ ◊ ♀H4 ↕↔ 10m (30ft)

CHAENOMELES SPECIOSA 'MOERLOOSEI'

A fast-growing and wide-spreading, deciduous shrub bearing large white flowers, flushed dark pink, in early spring. Tangled, spiny branches carry oval, glossy dark green leaves. The flowers are followed in autumn by apple-shaped, aromatic, yellow-green fruits. Use as a free-standing shrub or train against a wall.

CULTIVATION *Grow in well-drained, moderately fertile soil, in full sun for best flowering, or light shade. If wall-trained, shorten sideshoots to 2 or 3 leaves in late spring. Free-standing shrubs require little pruning.*

☼ ◐ ◊ ♀H4 ↕2.5m (8ft) ↔5m (15ft)

CHAENOMELES X *SUPERBA* 'CRIMSON AND GOLD'

This spreading, deciduous shrub bears masses of dark red flowers with conspicuous golden-yellow anthers from spring until summer. The dark green leaves appear on the spiny branches just after the first bloom of flowers; these are followed by yellow-green fruits. Useful as ground cover or low hedging, but may not survive in cold areas.

CULTIVATION *Grow in well-drained, fertile soil, in sun. Trim lightly after flowering; shorten sideshoots to 2 or 3 leaves if grown against a wall.*

☼ ◊ ♀H3 ↕1m (3ft) ↔2m (6ft)

CHAENOMELES X *SUPERBA* 'PINK LADY'

A rounded, deciduous shrub with spiny, spreading branches that bear cup-shaped, dark pink flowers from early spring. The glossy, dark green leaves appear after the first bloom of flowers; aromatic, yellow-green fruits follow in autumn. Grow against a warm wall in cold areas.

CULTIVATION *Grow in any but water-logged soil, in full sun or partial shade. Trim back after flowering, as necessary. If grown against a wall, shorten side-shoots to 2 or 3 leaves, in summer.*

☼ ◊♦ ♀H3 ↕1.5m (5ft) ↔2m (6ft)

LAWSON CYPRESSES *(CHAMAECYPARIS LAWSONIANA)*

Cultivars of *C. lawsoniana* are popular evergreen conifers, available in many different shapes, sizes and foliage colours. All have red-brown bark and dense crowns of branches that droop at the tips. The flattened sprays of dense, aromatic foliage, occasionally bearing small, rounded cones, make the larger types of Lawson cypress very suitable for thick hedging, such as bright blue-grey 'Pembury Blue', or golden-yellow 'Lane'. Use compact cultivars in smaller gardens such as 'Ellwoodii' (to 3m/10ft) and 'Ellwood's Gold'; dwarf upright types such as 'Chilworth Silver' make eye-catching feature plants for containers, rock gardens or borders.

CULTIVATION *Grow in moist but well-drained soil, in sun. They tolerate chalky soil, but not exposed sites. Trim regularly from spring to autumn; do not cut into older wood. To train as formal hedges, pruning must begin on young plants.*

☼ ◊ 🏆H4

1 ↕1.5m (5ft) or more ↔ 60cm (24in)

2 ↕to 40m (130ft) ↔ to 5m (15ft)

3 ↕to 15m (50ft) ↔ 2–5m (6–15ft)

1 *C. lawsoniana* 'Ellwood's Gold' **2** *C. lawsoniana* 'Lane' **3** *C. lawsoniana* 'Pembury Blue'

CHAMAECYPARIS NOOTKATENSIS 'PENDULA'

This large and drooping conifer develops a gaunt, open crown as it matures. Hanging from the arching branches are evergreen sprays of dark green foliage with small, round cones which ripen in spring. Its unusual habit makes an interesting feature for a large garden.

CULTIVATION *Best in full sun, in moist but well-drained, neutral to slightly acid soil; will also tolerate dry, chalky soil. Regular pruning is not required.*

☼ ◊♦ ♈H4 ↕to 30m (100ft) ↔to 8m (25ft)

CHAMAECYPARIS OBTUSA 'NANA GRACILIS'

This dwarf form of Hinoki cypress is an evergreen, coniferous tree with a dense pyramidal habit. The aromatic, rich green foliage is carried in rounded, flattened sprays, bearing small cones which ripen to yellow-brown. Useful in a large rock garden, particularly to give Oriental style. 'Nana Aurea' looks very similar but grows to only half the size.

CULTIVATION *Grow in moist but well-drained, neutral to slightly acid soil, in full sun. Also tolerates dry, chalky soil. Regular pruning is not necessary.*

☼ ◊ ♈H4 ↕3m (10ft) ↔2m (6ft)

CHAMAECYPARIS OBTUSA 'TETRAGONA AUREA'

This cultivar of Hinoki cypress, with golden to bronze-yellow, evergreen foliage, is a conical, coniferous tree. The aromatic foliage, carried in flattened sprays on upward-sweeping stems, is greener in the shade, and bears small green cones which ripen to brown. Grow as a specimen tree.

CULTIVATION *Best in moist but well-drained, neutral to slightly acid soil. Tolerates chalky conditions. For the best foliage colour, position in full sun. No regular pruning is required.*

☼ ◊ H4 ↕to 10m (30ft) ↔to 3m (10ft)

CHAMAECYPARIS PISIFERA 'BOULEVARD'

A broad, evergreen, conifer that develops into a conical tree with an open crown. The soft, blue-green foliage is borne in flattened sprays with angular green cones, maturing to brown. Very neat and compact in habit; an excellent specimen tree for poorly drained, damp soil.

CULTIVATION *Grow in reliably moist, preferably neutral to acid soil, in full sun. No regular pruning is required.*

☼ H4 ↕10m (30ft) ↔to 5m (15ft)

CHIMONANTHUS PRAECOX 'GRANDIFLORUS'

Wintersweet is a vigorous, upright, deciduous shrub grown for the fragrant flowers borne on its bare branches in winter; on this cultivar they are large, cup-shaped and deep yellow with maroon stripes inside. The leaves are mid-green. Suitable for a shrub border or for training against a warm, sunny wall.

CULTIVATION *Grow in well-drained, fertile soil, in a sunny, sheltered site. Best left unpruned when young so that mature flowering wood can develop. Cut back flowered stems of wall-trained plants in spring.*

☼ ◊ 🏆H4 ↕4m (12ft) ↔3m (10ft)

CHIONODOXA LUCILIAE

Glory of the snow is a small, bulbous perennial bearing star-shaped, clear blue flowers with white eyes, in early spring. The glossy green leaves are usually curved backwards. Grow in a sunny rock garden, or naturalize under deciduous trees. Sometimes referred to as *C. gigantea* of gardens. *C. forbesii* is very similar, with more erect leaves.

CULTIVATION *Grow in any well-drained soil, with a position in full sun. Plant bulbs 8cm (3in) deep in autumn.*

☼ ◊ 🏆H4 ↕15cm (6in) ↔3cm (1¼in)

CHOISYA TERNATA

Mexican orange blossom is a fast-growing, rounded, evergreen shrub valued for its attractive foliage and fragrant flowers. The aromatic, dark green leaves are divided into three. Clusters of star-shaped white flowers appear in spring. A fine, pollution-tolerant shrub for town gardens, but prone to frost damage in exposed sites. 'Aztec Pearl' is similar, its flowers perhaps not quite as fragrant.

CULTIVATION *Grow in well-drained, fairly fertile soil, in full sun. Naturally forms a well-shaped bush without pruning. Cutting back flowered shoots encourages a second flush of flowers.*

☼ ◊ 🏆H4 ↕↔ 2.5m (8ft)

CHOISYA TERNATA 'SUNDANCE'

This slower-growing, bright yellow-leaved variety of Mexican orange blossom is a compact, evergreen shrub. The aromatic leaves, divided into three leaflets, are a duller yellow-green if positioned in shade. Flowers are rare. Grow against a warm wall, with extra protection in cold climates. Correctly named as *C. ternata* SUNDANCE 'Lich'.

CULTIVATION *Best in well-drained, fertile soil, in full sun for the best leaf colour. Provide shelter from cold winds. Trim wayward shoots in summer, removing any frost-damaged shoots in spring.*

☼ ◊ 🏆H3 ↕↔ 2.5m (8ft)

GARDEN CHRYSANTHEMUMS

These upright, bushy perennials are a mainstay of the late border, with bright, showy flowerheads traditionally used for display and cutting. The lobed or feathery leaves are aromatic and bright green. They flower from late summer to mid-autumn, depending on the cultivar; very late-flowering cultivars are best raised under glass in frost-prone climates (see p.125). The flower form, although always many-petalled, varies from the daisy-like 'Pennine Alfie' to the blowsy, reflexed blooms of 'George Griffith'. Lift in autumn and store over winter in frost-free conditions. Plant out after any risk of frost has passed.

CULTIVATION *Grow in moist but well-drained, neutral to slightly acid soil enriched with well-rotted manure, in a sunny, sheltered site. Stake tall flower stems. Apply a balanced fertilizer when in growth, until flower buds begin to show.*

☼ ◊◆ ♀H3

1 ↕ 1.2m (4ft) ↔ 75cm (30in) **2** ↕ 1.2m (4ft) ↔ 60–75cm (24–30in) **3** ↕ 1m (3ft) ↔ 60–75cm (24–30in)

4 ↕ 50cm (20in) ↔ 25cm (10in) **5** ↕ 30–60cm (12–24in) ↔ 60cm (24in) **6** ↕ 1.1m (3½ft) ↔ 60–75cm (24–30in)

1 *Chrysanthemum* 'Amber Enbee Wedding' **2** *C.* 'Amber Yvonne Arnaud' **3** *C.* 'Angora' **4** *C.* 'Bravo' **5** *C.* 'Bronze Fairie' **6** *C.* 'Cherry Nathalie'

7 ↕1.1m (3½ft) ↔ 60–75cm (24–30in)

8 ↕90cm (30in) ↔ 30cm (12in)

9 ↕1.3–1.5m (4½–5ft) ↔ 75cm (30in)

10 ↕1.2m (4ft) ↔ 75cm (30in)

11 ↕90cm (30in) ↔ 30cm (12in)

12 ↕60cm (24in) ↔ 30cm (12in)

13 ↕90cm (30in) ↔ 30cm (12in)

14 ↕1.2m (4ft) ↔ 75cm (30in)

15 ↕1.2m (4ft) ↔ 75cm (30in)

7 *C.* 'Eastleigh' **8** *C.* 'Flo Cooper' **9** *C.* 'George Griffiths' **10** *C.* 'Madeleine' **11** *C.* 'Mancetta Bride' **12** *C.* 'Mavis' **13** *C.* 'Myss Madi' **14** *C.* 'Pennine Alfie' **15** *C.* 'Pennine Flute'

16 ↕70cm (28in) ↔ 30cm (10in)

17 ↕65cm (25in) ↔ 30cm (12in)

18 ↕1.2m (4ft) ↔ 60–75cm (24–30in)

19 ↕1m (3ft) ↔ 45cm (18in)

20 ↕1.2m (4ft) ↔ 75cm (30in)

21 ↕30–60cm (12–24in) ↔ 60cm (24in)

22 ↕1.2m (4ft) ↔ 60–75cm (24–30in)

23 ↕85cm (34in) ↔ 45cm (18in)

24 ↕1.2m (4ft) ↔ 60–75cm (24–30in)

25 ↕1.2m (4ft) ↔ 60–75cm (24–30in)

MORE CHOICES

'Margaret' Pink, with reflexed petals.

'Max Riley' Yellow.

'Pennine Signal' Scarlet.

'Yellow Pennine Oriel' Sprays of small yellow flowers.

16 *Chrysanthemum* 'Pennine Lace' **17** *C.* 'Pennine Marie' **18** *C.* 'Pennine Oriel' **19** *C.* 'Primrose Allouise' **20** *C.* 'Purple Pennine Wine' **21** *C.* 'Salmon Fairie' **22** *C.* 'Salmon Margaret' **23** *C.* 'Southway Swan' **24** *C.* 'Wendy' **25** *C.* 'Yvonne Arnaud'

LATE-FLOWERING CHRYSANTHEMUMS

This group of herbaceous perennials comes into flower from autumn into winter, which means that the protection of a warm greenhouse during this time is essential in cold areas. None can tolerate frost, but they can be moved outdoors during summer. The showy flowerheads come in a wide range of shapes and sizes and are available in a blaze of golds, bronzes, yellows, oranges, pinks and reds. Protection from the elements enables perfect blooms for exhibition to be nurtured.

Grow late-flowering chrysanthemums in containers or in a greenhouse border.

CULTIVATION *Best in a good loam-based potting compost, kept slightly moist at all times, in bright, filtered light. Stake plants as they grow, and give liquid fertilizer until flower buds begin to form. Ensure adequate ventilation and a min. temp. of 10°C (50°F). In frost-free areas, grow outdoors in fertile, moist but well-drained soil in full sun.*

☼ ◊♦ ♀H2

1 ↕1.2m (4ft) ↔ 60cm (24in)

2 ↕1.2–1.5m (4–5ft) ↔ 60cm (24in)

3 ↕1.4m (4½ft) ↔ 60cm (24in)

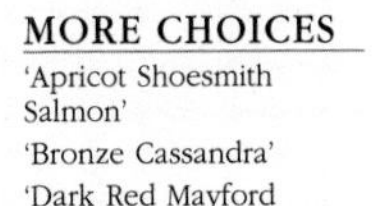

MORE CHOICES

'Apricot Shoesmith Salmon'

'Bronze Cassandra'

'Dark Red Mayford Perfection'

'Pink Gin' Light purple.

'Rose Mayford Perfection'

'Rynoon' Light pink.

4 ↕1.2m (4ft) ↔ 75–100cm (30–39in)

5 ↕1.2m (4ft) ↔ 60–75cm (24–30in)

1 *Chrysanthemum* 'Beacon' **2** *C.* 'Golden Cassandra' **3** *C.* 'Roy Coopland' **4** *C.* 'Satin Pink Gin' **5** *C.* 'Yellow John Hughes'

CIMICIFUGA RACEMOSA

Black snake root is a clump-forming perennial that produces long spikes of unpleasantly scented, tiny white flowers in mid-summer. The dark green leaves are broadly oval to lance-shaped. Plant in a bog garden, moist border or light woodland, away from paths and seating areas. *C. simplex* 'Elstead' does not have the scent problem but is much less imposing, to only 1m (3ft) tall.

CULTIVATION *Grow in reliably moist, fertile, preferably humus-rich soil, in partial shade. Provide clumps with support, using ring stakes or similar.*

☀ 💧 ♈H4 ↕1.2–2.2m (4–7ft) ↔60cm (24in)

CISTUS x *AGUILARII* 'MACULATUS'

This fast-growing, evergreen shrub bears large, solitary white flowers for a few weeks in early and mid-summer. At the centre of each flower is a mass of bright golden yellow stamens surrounded by five crimson blotches. The lance-shaped leaves are sticky, aromatic and bright green. Excellent on a sunny bank or in containers. May perish in cold areas.

CULTIVATION *Grow in well-drained, poor to moderately fertile soil, in a sunny, sheltered site. Lime-tolerant. If necessary, trim lightly in early spring or after flowering, but do not prune hard.*

☼ ○ ♈H3 ↕↔ 1.2m (4ft)

CISTUS X *HYBRIDUS*

A dense, spreading shrub producing white flowers with yellow blotches at the centres. These are borne singly or in clusters of two or three during late spring and early summer. The aromatic leaves are wrinkled, oval, and dark green. Suitable for a shrub border or rock garden; it may need the protection of a warm wall in cold, damp climates. Sometimes known as *C.* x *corbariensis.*

CULTIVATION *Grow in well-drained, poor to moderately fertile soil, in a sheltered site in full sun. Tolerates lime. If necessary, trim lightly after flowering, but do not prune hard.*

☼ ◊ H4 ↕1m (3ft) ↔1.5m (5ft)

CISTUS X *PURPUREUS*

This summer-flowering, rounded, evergreen shrub bears few-flowered clusters of dark pink flowers with maroon blotches at the base of each petal. The dark green leaves are borne on upright, sticky, red-flushed shoots. Good in a large rock garden, on a sunny bank or in a container. May not survive cold winters.

CULTIVATION *Grow in well-drained, poor to moderately fertile soil. Choose a sheltered site in full sun. Tolerates lime. Can be trimmed lightly after flowering, but avoid hard pruning.*

↕↔ 1m (3ft)

EARLY-FLOWERING CLEMATIS

The early-flowering "species" clematis are valued for their showy displays during spring and early summer. They are generally deciduous climbers with mid- to dark green, divided leaves. The flowers of the earliest species to bloom are usually bell-shaped; those of the later *C. montana* types are either flat or saucer-shaped. Flowers are often followed by decorative seedheads. Many clematis, especially *C. montana* types, are vigorous and will grow very quickly, making them ideal for covering featureless or unattractive walls. Allowed to grow through deciduous shrubs, they may flower before their host comes into leaf.

CULTIVATION *Grow in well-drained, fertile, humus-rich soil, in full sun or semi-shade. The roots and base of the plant should be shaded. Tie in young growth carefully and prune out shoots that exceed the allotted space, immediately after flowering.*

☼ ◑ ◊ ♆H4

1 ↕2–3m (6–10ft) ↔ 1.5m (5ft)

2 ↕2–3m (6–10ft) ↔ 1.5m (5ft)

3 ↕10m (30ft) ↔ 4m (12ft)

MORE CHOICES

C. alpina 'White Columbine' White flowers.

C. alpina 'Helsingborg' Deep blue and brown.

C. cirrhosa var. *balearica* and *C. cirrhosa* 'Freckles' Cream flowers speckled with pinky-brown.

4 ↕10m (30ft) ↔ 2–3m (6–10ft)

5 ↕5m (15ft) ↔ 2–3m (6–10ft)

1 *Clematis alpina* 'Frances Rivis' **2** *C. macropetala* 'Markham's Pink' **3** *C. montana* f. *grandiflora* **4** *C. montana* var. *rubens* **5** *C. montana* var. *rubens* 'Tetrarose'

MID-SEASON CLEMATIS

The mid-season, mainly hybrid clematis are twining and deciduous climbers, bearing an abundance of stunning flowers throughout the summer months. Their flowers are large, saucer-shaped and outward-facing with a plentiful choice of shapes and colours. Towards the end of the summer, blooms may darken. The leaves are pale to mid-green and divided into several leaflets. Mid-season clematis look very attractive scrambling through other shrubs, especially if they bloom before or after their host. Top growth may be damaged in severe winters, but plants are usually quick to recover.

CULTIVATION *Grow in well-drained, fertile, humus-rich soil, with the roots in shade and the heads in sun. Pastel flowers may fade in sun; better in semi-shade. Mulch in late winter, avoiding the immediate crown. Tie in young growth carefully, cutting back older stems to strong buds in late winter.*

☼ ◑ ◊ 🏆H4

1 ↕2.5m (8ft) ↔ 1m (3ft)

2 ↕2.5m (8ft) ↔ 1m (3ft)

3 ↕2–3m (6–10ft) ↔ 1m (3ft)

1 *Clematis* 'Bees' Jubilee' **2** *C.* 'Doctor Ruppel' **3** *C.* 'Elsa Späth'

5 ↕ to 3m (10ft) ↔ 1m (3ft)

6 ↕ 2.5m (8ft) ↔ 1m (3ft)

7 ↕ 2.5m (8ft) ↔ 1m (3ft)

8 ↕ 3m (10ft) ↔ 1m (3ft)

9 ↕ 2.5m (8ft) ↔ 1m (3ft)

10 ↕ 3m (10ft) ↔ 1m (3ft)

11 ↕ 2.5m (8ft) ↔ 1m (3ft)

12 ↕ 2–3m (6–10ft) ↔ 1m (3ft)

5 *C.* 'Fireworks' **6** *C.* 'Gillian Blades' **7** *C.* 'H. F. Young' **8** *C.* 'Henryi'
9 *C.* 'Lasurstern' **10** *C.* 'Marie Boisselot' **11** *C.* 'Miss Bateman' **12** *C.* 'Nelly Moser'

13 ↕2–3m (6–10ft) ↔ 1m (3ft)

14 ↕2–3m (6–10ft) ↔ 1m (3ft)

15 ↕2m (6ft) ↔ 1m (3ft)

16 ↕2m (6ft) ↔ 1m (3ft)

MORE CHOICES

'Lord Nevill' Deep blue flowers with purple-red anthers.

'Mrs Cholmondely' Lavender, brown anthers.

'Will Goodwin' Pale blue with yellow anthers.

17 ↕2–3m (6–10ft) ↔ 1m (3ft)

18 ↕2–3m (6–10ft) ↔ 1m (3ft)

13 *C.* 'Niobe' **14** *C.* 'Richard Pennell' **15** *C.* 'Royalty' **16** *C.* 'Silver Moon'
17 *C.* 'The President' **18** *C.* 'Vyvyan Pennell'

LATE-FLOWERING CLEMATIS

Many large-flowered hybrid clematis flower from mid- to late summer, when the season for the *C. viticella* types, characterised usually by smaller but more profuse flowers, also begins. These are followed by other late-flowering species. They may be deciduous or evergreen, with an enormous variety of flower and leaf shapes and colours. Many, like 'Perle d'Azur', are vigorous and will cover large areas of wall or disguise unsightly buildings. May develop decorative, silvery-grey seedheads which last well into winter. With the exception of 'Bill Mackenzie', most look good when trained up into small trees.

CULTIVATION *Grow in humus-rich, fertile soil with good drainage, with the base in shade and the upper part in sun or partial shade. Mulch in late winter, avoiding the immediate crown. Cut back hard each year before growth begins, in early spring.*

☼ ◑ ◊ ♀H4

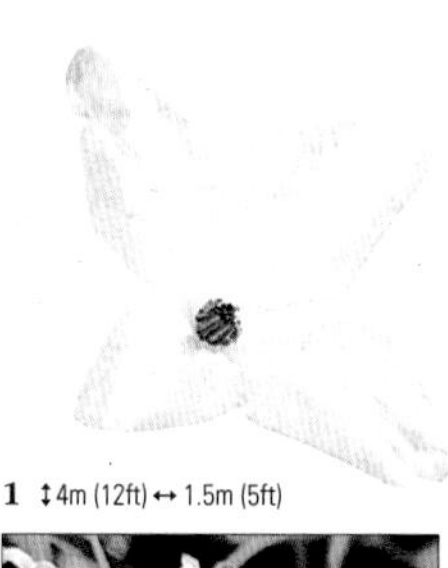

1 ↕4m (12ft) ↔ 1.5m (5ft)

2 ↕7m (22ft) ↔ 2–3m (6–10ft)

3 ↕2.5m (8ft) ↔ 1.5m (5ft)

1 *C.* 'Alba Luxurians' **2** *C.* 'Bill Mackenzie' **3** *C.* 'Duchess of Albany'

4 ↕2–3m (6–10ft) ↔ 1m (3ft)

5 ↕3–5m (10–15ft) ↔ 1.5m (5ft)

6 ↕3m (10ft) ↔ 1m (3ft)

7 ↕3m (10ft) ↔ 1.5m (5ft)

8 ↕3m (10ft) ↔ 1m (3ft)

9 ↕3m (10ft) ↔ 1m (3ft)

10 ↕3m (10ft) ↔ 1m (3ft)

11 ↕3m (10ft) ↔ 1m (3ft)

12 ↕6–7m (20–22ft) ↔ 2–3m (6–10ft)

4 *C.* 'Comtesse de Bouchaud' **5** *C.* 'Etoile Violette' **6** *C.* 'Jackmanii' **7** *C.* 'Madame Julia Correvon' **8** *C.* 'Minuet' **9** *C.* 'Perle d'Azur' **10** *C.* 'Venosa Violacea' **11** *C. viticella* 'Purpurea Plena Elegans' **12** *C. rehderiana*

CLIANTHUS PUNICEUS

Lobster claw is an evergreen, woody-stemmed, climbing shrub with scrambling shoots. Drooping clusters of claw-like, brilliant red flowers are seen in spring and early summer. (The cultivar 'Albus' has pure white flowers.) The mid-green leaves are divided into many oblong leaflets. Suitable for wall-training; it needs the protection of a cool conservatory in frost-prone climates.

CULTIVATION *Grow in well-drained, fairly fertile soil, in sun with shelter from wind. Pinch-prune young plants to promote bushiness; otherwise, keep pruning to a minimum.*

☼ ◊ ♈H2 ↕4m (12ft) ↔3m (10ft)

CODONOPSIS CONVOLVULACEA

A slender, herbaceous, summer-flowering climber with twining stems that bears delicate, bell- to saucer-shaped, soft blue-violet flowers. The leaves are lance-shaped to oval and bright green. Grow in a herbaceous border or woodland garden, scrambling through other plants.

CULTIVATION *Grow in moist but well-drained, light, fertile soil, ideally in dappled shade. Provide support or grow through neighbouring shrubs. Cut to the base in spring.*

◐ ◊◆ ♈H4 ↕to 2m (6ft)

COLCHICUM SPECIOSUM 'ALBUM'

While *Colchicum speciosum*, the autumn crocus, has pink flowers, this cultivar is white. A cormous perennial, it produces thick, weather-resistant, goblet-shaped, snow-white blooms. Its narrow, mid-green leaves appear in winter or spring after the flowers have died down. Grow at the front of a border, at the foot of a bank, or in a rock garden. All parts are highly toxic if ingested.

CULTIVATION *Grow in moist but well-drained soil, in full sun. Plant bulbs in early autumn, 10cm (4in) below the surface of soil that is deep and fertile.*

☼ ◊ 🏆H4 ↕18cm (7in) ↔10cm (4in)

CONVALLARIA MAJALIS

Lily-of-the-valley is a creeping perennial bearing small, very fragrant white flowers which hang from arching stems in late spring or early summer. The narrowly oval leaves are mid- to dark green. An excellent ground cover plant for woodland gardens and other shady areas; it will spread rapidly under suitable conditions.

CULTIVATION *Grow in reliably moist, fertile, humus-rich, leafy soil in deep or partial shade. Top-dress with leaf mould in autumn.*

◐ ● ♦ 🏆H4 ↕23cm (9in) ↔30cm (12in)

CONVOLVULUS CNEORUM

This compact, rounded, evergreen shrub bears masses of funnel-shaped, shining white flowers with yellow centres, which open from late spring to summer. The narrowly lance-shaped leaves are silvery-green. Excellent as a largish plant in a rock garden, or on a sunny bank. In areas with cold, wet winters, grow in a container and move into a cool greenhouse in winter.

CULTIVATION *Grow in gritty, very well-drained, poor to moderately fertile soil, in a sunny, sheltered site. Trim back after flowering, if necessary.*

☼ ◊ ♀H3 ↕60cm (24in) ↔90cm (36in)

CONVOLVULUS SABATIUS

A small, trailing perennial bearing trumpet-shaped, vibrant blue-purple flowers from summer into early autumn. The slender stems are clothed with small, oval, mid-green leaves. Excellent in crevices between rocks; in cold areas, best grown in containers with shelter under glass during the winter. Sometimes seen as *C. mauritanicus.*

CULTIVATION *Grow in well-drained, gritty, poor to moderately fertile soil. Provide a sheltered site in full sun.*

☼ ◊ ♀H3 ↕15cm (6in) ↔50cm (20in)

CORDYLINE AUSTRALIS 'ALBERTII'

This New Zealand cabbage palm is a palm-like, evergreen tree carrying lance-shaped, matt green leaves with red midribs, cream stripes and pink margins. Clusters of creamy-white flowers appear on mature specimens in summer, followed by white or blue-tinted berries. Unlikely to grow very tall in cool climates, where it is best in a container for standing out during the warmer months.

CULTIVATION *Best in well-drained, fertile soil or compost, in sun or partial shade. Remove dead leaves and cut out faded flower stems as necessary.*

☼◑ ◊ 🏆H3 ↕to 10m (30ft) ↔to 4m (12ft)

CORNUS ALBA 'SIBIRICA'

A deciduous shrub that is usually grown for the winter effect of its bright coral-red, bare young stems. Small clusters of creamy-white flowers appear in late spring and early summer amid oval, dark green leaves which turn red in autumn. Particularly effective in a waterside planting or any situation where the winter stems show up well.

CULTIVATION *Grow in any moderately fertile soil, in sun. For the best stem effect, cut back hard and feed every spring once established, although this will be at the expense of the flowers.*

 ↕↔ 3m (10ft)

CORNUS ALBA 'SPAETHII'

A vigorous and upright, deciduous shrub bearing bright green, elliptic leaves which are margined with yellow. It is usually grown for the effect of its bright red young shoots in winter. In late spring and early summer, small clusters of creamy-white flowers appear amid the foliage. Very effective wherever the stems show up well in winter.

CULTIVATION *Grow in any moderately fertile soil, preferably in full sun. For the best stem effect, but at the expense of any flowers, cut back hard and feed each year in spring, once established.*

☼ ◊◊ ♈H4 ↕↔ 3m (10ft)

CORNUS CANADENSIS

The creeping dogwood is a superb ground cover perennial for under-planting a shrub border or wood-land garden. Flower clusters with prominent white bracts appear above the oval, bright green leaves during late spring and early summer. These are followed by round, bright red berries.

CULTIVATION *Best in moist, acid, leafy soil, in partial shade. Divide plants in spring or autumn to restrict spread or increase plants.*

☀ ◊ ♈H4 ↕to 15cm (6in) ↔indefinite

CORNUS KOUSA VAR. *CHINENSIS*

This broadly conical, deciduous tree with flaky bark is valued for its dark green, oval leaves which turn an impressive, deep crimson-purple in autumn. The early summer flowers have long white bracts, fading to red-pink. An effective specimen tree, especially in a woodland setting. *C. kousa* 'Satomi' has even deeper autumn colour, and pink flower bracts.

CULTIVATION *Best in well-drained, fertile, neutral to acid soil that is rich in humus, in full sun or partial shade. Keep pruning to a minimum.*

☼◑ 💧 🏆H4 ↕7m (22ft) ↔5m (15ft)

CORNUS MAS

The cornelian cherry is a vigorous and spreading, deciduous shrub or small tree. Clusters of small yellow flowers provide attractive late winter colour on the bare branches. The oval, dark green leaves turn red-purple in autumn, giving a display at the same time as the fruits ripen to red. Particularly fine as a specimen tree for a woodland garden.

CULTIVATION *Tolerates any well-drained soil, in sun or partial shade. Pruning is best kept to a minimum.*

☼◑ 💧 🏆H4 ↕↔ 5m (15ft)

CORNUS STOLONIFERA 'FLAVIRAMEA'

This vigorous, deciduous shrub makes a bright display of its bare yellow-green young shoots in winter, before the oval, dark green leaves emerge in spring. Clusters of white flowers appear in late spring and early summer. The leaves redden in autumn. Excellent in a bog garden or in wet soil near water.

CULTIVATION *Grow in reliably moist soil, in full sun. Restrict spread by cutting out 1 in 4 old stems annually. Prune all stems hard and feed each year in early spring for the best display of winter stems.*

↕2m (6ft) ↔4m (12ft)

CORREA BACKHOUSEANA

The Australian fuchsia is a dense, spreading, evergreen shrub with small clusters of tubular, pale red-green or cream flowers during late autumn to late spring. The hairy, rust-red stems are clothed with oval, dark green leaves. In cold climates, grow against a warm wall or over-winter in frost-free conditions. *Correa* 'Mannii', with red flowers, is also recommended.

CULTIVATION *Grow in well drained, fertile, acid to neutral soil, in full sun. Trim back after flowering, if necessary.*

☼ ◊ ♀H2 ↕1–2m (3–6ft) ↔1.5–2.5m (5–8ft)

CORTADERIA SELLOANA 'AUREOLINEATA'

This pampas grass, with rich yellow-margined, arching leaves which age to dark golden-yellow, is a clump-forming, evergreen perennial, also known as 'Gold Band'. Feathery plumes of silvery flowers appear on tall stems in late summer. May not survive in cold areas. The flower-heads can be dried for decoration.

CULTIVATION *Grow in well-drained, fertile soil, in full sun. In late winter, cut out all dead foliage and remove the previous year's flower stems: wear gloves to protect hands from the sharp foliage.*

☼ ◊ ♀H3 ↕to 2.2m (7ft) ↔1.5m (5ft) or more

CORTADERIA SELLOANA 'SUNNINGDALE SILVER'

This sturdy pampas grass is a clump-forming, evergreen, weather-resistant perennial. In late summer, silky plumes of silvery-cream flowers are borne on strong, upright stems above the narrow, arching, sharp-edged leaves. In cold areas, protect the crown with a dry winter mulch. The flowerheads can be dried.

CULTIVATION *Grow in well-drained, fertile, not too heavy soil, in full sun. Remove old flower stems and any dead foliage in late winter: wear gloves to protect hands from the sharp foliage.*

☼ ◊ ♀H3 ↕3m (10ft) or more ↔to 2.5m (8ft)

CORYDALIS SOLIDA 'GEORGE BAKER'

A low, clump-forming, herbaceous perennial bearing upright spires of deep salmon-rose flowers. These appear in spring above the delicate, finely cut, greyish-green leaves. Excellent in a rock garden or in an alpine house.

CULTIVATION *Grow in sharply drained, moderately fertile soil or compost. Site in full sun, but tolerates some shade.*

☼ ◑ ◇ ♈H4 ↕to 25cm (10in) ↔to 20cm (8in)

CORYLOPSIS PAUCIFLORA

This deciduous shrub bears hanging, catkin-like clusters of small, fragrant, pale yellow flowers on its bare branches during early to mid-spring. The oval, bright green leaves are bronze when they first emerge. Often naturally well-shaped, it makes a beautiful shrub for sites in dappled shade. The flowers may be damaged by frost.

CULTIVATION *Grow in moist but well-drained, humus-rich, acid soil, in partial shade with shelter from wind. Allow room for the plant to spread. The natural shape is easily spoilt, so prune only to remove dead wood.*

◑ ◇◆ ♈H4 ↕1.5m (5ft) ↔2.5m (8ft)

CORYLUS AVELLANA 'CONTORTA'

The corkscrew hazel is a deciduous shrub bearing strongly twisted shoots that are particularly striking in winter; they can also be useful in flower arrangements. Winter interest is enhanced later in the season with the appearance of pale yellow catkins. The mid-green leaves are almost circular and toothed.

CULTIVATION *Grow in any well-drained, fertile soil, in sun or semi-shade. Once established, the twisted branches tend to become congested and may split, so thin out in late winter.*

☼ ◑ 💧 🏆H4 ↕↔ 15m (50ft)

CORYLUS MAXIMA 'PURPUREA'

The purple filbert is a vigorous, open, deciduous shrub that, left unpruned, will grow into a small tree. In late winter, purplish catkins appear before the rounded, purple leaves emerge. The edible nuts ripen in autumn. Effective as a specimen tree, in a shrub border or as part of a woodland planting.

CULTIVATION *Grow in any well-drained, fertile soil, in sun or partial shade. For the best leaf effect, but at the expense of the nuts, cut back hard in early spring.*

☼ ◑ 💧 🏆H4 ↕to 6m (20ft) ↔5m (15ft)

COSMOS BIPINNATUS 'SONATA WHITE'

A branching but compact annual bearing single, saucer-shaped white flowers with yellow centres from summer to autumn at the tips of upright stems. The leaves are bright green and feathery. Excellent for exposed gardens. The flowers are good for cutting.

CULTIVATION *Grow in moist but well-drained, fertile soil, in full sun. Deadhead to prolong flowering.*

☼ ◊♦ H4 ↕↔ 30cm (12in)

COTINUS COGGYGRIA 'ROYAL PURPLE'

This deciduous shrub is grown for its rounded, red-purple leaves which turn a brilliant scarlet in autumn. Smoke-like plumes of tiny, pink-purple flowers are produced on older wood, but only in areas with long, hot summers. Good in a shrub border or as a specimen tree; where space permits, plant in groups.

CULTIVATION *Grow in moist but well-drained, fairly fertile soil, in full sun or partial shade. For the best foliage effect, cut back hard to a framework of older wood each spring, before growth begins.*

☼ ☀ ◊♦ H4 ↕↔ 5m (15ft)

COTINUS 'GRACE'

A fast-growing, deciduous shrub or small tree carrying oval, purple leaves which turn a brilliant, translucent red in late autumn. The smoke-like clusters of tiny, pink-purple flowers only appear in abundance during hot summers. Effective on its own or in a border. For green leaves during the summer but equally brilliant autumn foliage colour, look for *Cotinus* 'Flame'.

CULTIVATION *Grow in moist but well-drained, reasonably rich soil, in sun or partial shade. The best foliage is seen after hard pruning each spring, just before new growth begins.*

☼◑ ◊♦ 🏆H4 ↕6m (20ft) ↔5m (15ft)

COTONEASTER ATROPURPUREUS 'VARIEGATUS'

This compact, low-growing shrub, sometimes seen as *C. horizontalis* 'Variegatus', has fairly inconspicuous red flowers in summer, followed in autumn by a bright display of orange-red fruits. The small, oval, white-margined, deciduous leaves also give autumn colour, turning pink and red before they drop. Effective as ground cover or in a rock garden.

CULTIVATION *Grow in well-drained, moderately fertile soil, in full sun. Tolerates dry soil and partial shade. Pruning is best kept to a minimum.*

☼◑ ◊ 🏆H4 ↕45cm (18in) ↔90cm (36in)

COTONEASTER CONSPICUUS 'DECORUS'

A dense, mound-forming, evergreen shrub that is grown for its shiny red berries. These ripen in autumn and will often persist until late winter. Small white flowers appear amid the dark green leaves in summer. Good in a shrub border or under a canopy of deciduous trees.

CULTIVATION *Grow in well-drained, moderately fertile soil, ideally in full sun, but tolerates shade. If necessary, trim lightly to shape after flowering.*

☼☀ ◊ 🏆H4 ↕1.5m (5ft) ↔2–2.5m (6–8ft)

COTONEASTER HORIZONTALIS

A deciduous shrub with spreading branches that form a herringbone pattern. The tiny, pinkish-white flowers, which appear in summer, are attractive to bees. Bright red berries ripen in autumn, and the glossy, dark green leaves redden before they fall. Good as ground cover, but most effective when grown flat up against a wall.

CULTIVATION *Grow in any but water-logged soil. Site in full sun for the best berries, or semi-shade. Keep pruning to a minimum; if wall-trained, shorten outward-facing shoots in late winter.*

☼☀ ◊♦ 🏆H4 ↕1m (3ft) ↔1.5m (5ft)

COTONEASTER LACTEUS

This dense, evergreen shrub has arching branches that bear clusters of small, cup-shaped, milky-white flowers from early to mid-summer. These are followed by brilliant red berries in autumn. The oval leaves are dark green and leathery, with grey-woolly undersides. Ideal for a wildlife garden, since it provides food for bees and birds; also makes a good windbreak or informal hedge.

CULTIVATION *Grow in well-drained, fairly fertile soil, in sun or semi-shade. Trim hedges lightly in summer, if necessary; keep pruning to a minimum.*

☼ ◑ ◊ 🏆H4 ↕↔ 4m (12ft)

COTONEASTER SIMONSII

An upright, deciduous or semi-evergreen shrub with small, cup-shaped white flowers in summer. The bright orange-red berries that follow ripen in autumn and persist well into winter. Autumn colour is also seen in the glossy leaves which redden from dark green. Good for hedging, and can also be clipped fairly hard to a semi-formal outline.

CULTIVATION *Grow in any well-drained soil, in full sun or partial shade. Clip hedges to shape in late winter or early spring, or allow to grow naturally.*

☼ ◑ ◊ 🏆H4 ↕3m (10ft) ↔2m (6ft)

COTONEASTER STERNIANUS

This graceful, evergreen or semi-evergreen shrub bears arching branches which produce clusters of pink-tinged white flowers in summer, followed by a profusion of large, orange-red berries in autumn. The grey-green leaves have white undersides. Good grown as a hedge.

CULTIVATION *Grow in any well-drained soil, in sun or semi-shade. Trim hedges lightly after flowering, if necessary; pruning is best kept to a minimum.*

☼☀ ◊ ♀H4 ↕↔ 3m (10ft)

COTONEASTER x *WATERERI* 'JOHN WATERER'

A fast-growing, evergreen or semi-evergreen shrub or small tree valued for the abundance of red berries that clothe its branches in autumn. In summer, clusters of white flowers are carried among the lance-shaped, dark green leaves. Good on its own or at the back of a shrub border.

CULTIVATION *Grow in any but water-logged soil, in sun or semi-shade. When young, cut out any badly placed shoots to develop a framework of well-spaced branches. Thereafter, keep pruning to an absolute minimum.*

☼☀ ◊♦ ♀H4 ↕↔ 5m (15ft)

CRAMBE CORDIFOLIA

A tall, clump-forming, vigorous perennial grown for its stature and its fragrant, airy, billowing sprays of small white flowers which are very attractive to bees. These are borne on strong stems in summer above the large, elegant, dark green leaves. Magnificent in a mixed border, but allow plenty of space.

CULTIVATION *Grow in any well-drained, preferably deep, fertile soil, in full sun or partial shade. Provide shelter from strong winds.*

☼☼◑ ◊ ♈H4 ↕to 2.5m (8ft) ↔1.5m (5ft)

CRATAEGUS LAEVIGATA 'PAUL'S SCARLET'

This rounded, thorny, deciduous tree is valued for its long season of interest. Abundant clusters of double, dark pink flowers appear from late spring to summer, followed in autumn by small red fruits. The leaves, divided into three or five lobes, are a glossy mid-green. A particularly useful specimen tree for a town, coastal or exposed garden. The very similar 'Rosea Flore Pleno' makes a good substitute.

CULTIVATION *Grow in any but waterlogged soil, in full sun or partial shade. Pruning is best kept to a minimum.*

☼☼◑ ◊◊ ♈H4 ↕↔ 8m (25ft)

CRATAEGUS x *LAVALLEI* 'CARRIEREI'

This vigorous hawthorn is a broadly spreading, semi-evergreen tree with thorny shoots and leathery green leaves which turn red in late autumn and winter. Flattened clusters of white flowers appear in early summer, followed in autumn by round red fruits that persist into winter. Tolerates pollution, so is good for a town garden. Often listed just as *C.* x *lavallei*.

CULTIVATION *Grow in any but water-logged soil, in full sun or partial shade. Pruning is best kept to a minimum.*

☼☀ ◊◊ ♈H4 ↕7m (22ft) ↔10m (30ft)

CRINODENDRON HOOKERIANUM

The lantern tree is an upright, evergreen shrub, so-called because of its large, scarlet to carmine-red flowers, which hang from the upright shoots during late spring and early summer. The leaves are narrow and glossy dark green. It dislikes alkaline soils, and may not survive in areas with cold winters.

CULTIVATION *Grow in moist but well-drained, fertile, humus-rich, acid soil, in partial shade with protection from cold winds. Tolerates a sunny site if the roots are kept cool and shaded. Trim lightly after flowering, if necessary.*

☀ ◊ ♈H3 ↕6m (20ft) ↔5m (15ft)

CRINUM x *POWELLII* 'ALBUM'

A sturdy, bulbous perennial bearing clusters of up to ten large, fragrant, widely flared, pure white flowers on upright stems in late summer and autumn. *C.* x *powellii* is equally striking, with pink flowers. The strap-shaped leaves, to 1.5m (5ft) long, are mid-green and arch over. In cold areas, choose a sheltered site and protect the dormant bulb over winter with a deep, dry mulch.

CULTIVATION *Grow in deep, moist but well-drained, fertile soil that is rich in humus, in full sun with shelter from frost and cold, drying winds.*

☼ ◊ 🏆H3 ↕1.5m (5ft) ↔30cm (12in)

CROCOSMIA x *CROCOSMIIFLORA* 'SOLFATERRE'

This clump-forming perennial produces spikes of funnel-shaped, apricot-yellow flowers on arching stems in mid-summer. The bronze-green, deciduous leaves are strap-shaped, emerging from swollen corms at the base of the stems. Excellent in a border; the flowers are good for cutting. May be vulnerable to frost in cold areas.

CULTIVATION *Grow in moist but well-drained, fertile, humus-rich soil, in full sun. Provide a dry mulch over winter.*

☼ ◊ 🏆H3 ↕60–70cm (24–28in) ↔8cm (3in)

CROCOSMIA 'LUCIFER'

A robust, clump-forming perennial with swollen corms at the base of the stems. These give rise to pleated, bright green leaves and, in summer, arching spikes of upward-facing red flowers. Particularly effective at the edge of a shrub border or by a pool.

CULTIVATION *Grow in moist but well-drained, moderately fertile, humus-rich soil. Site in full sun or dappled shade.*

☼☀ 💧💧 🏆H4 ↕1–1.2m (3–4ft) ↔8cm (3in)

CROCOSMIA MASONIORUM

A robust, late-summer-flowering perennial bearing bright vermilion, upward-facing flowers. These are carried above the dark green foliage on arching stems. The flowers and foliage emerge from a swollen, bulb-like corm. Thrives in coastal gardens. Where frosts are likely, grow in the shelter of a warm wall.

CULTIVATION *Best in moist but well-drained, fairly fertile, humus-rich soil, in full sun or partial shade. In cold areas, provide a dry winter mulch.*

☼☀ 💧 🏆H3 ↕1.2m (4ft) ↔8cm (3in)

SPRING-FLOWERING CROCUS

Spring-flowering crocus are indispensable dwarf perennials as they bring a welcome splash of early spring colour into the garden. Some cultivars of *C. sieberi*, such as 'Tricolor' or 'Hubert Edelstein', bloom even earlier, in late winter. The goblet-shaped flowers emerge from swollen, underground corms at the same time as or just before the narrow, almost upright foliage. The leaves are mid-green with silver-green central stripes, and grow markedly as the blooms fade. Very effective in drifts at the front of a mixed or herbaceous border, or in massed plantings in rock gardens or raised beds.

CULTIVATION *Grow in gritty, well-drained, poor to moderately fertile soil, in full sun. Water freely during the growing season, and apply a low-nitrogen fertiliser each month.* C. corsicus *must be kept completely dry over summer. Can be naturalized under the right growing conditions.*

☼ 💧 🏆H4

1 ↕8–10cm (3–4in) ↔ 4cm (1½in)

2 ↕7cm (3in) ↔ 5cm (2in)

3 ↕5–8cm (2–3in) ↔ 2.5cm (1in)

4 ↕5–8cm (2–3in) ↔ 2.5cm (1in)

5 ↕5–8cm (2–3in) ↔ 2.5cm (1in)

1 *C. corsicus* **2** *C. chrysanthus* 'E.A. Bowles' **3** *C. sieberi* 'Albus' **4** *C. sieberi* 'Tricolor' **5** *C. sieberi* 'Hubert Edelstein'

AUTUMN-FLOWERING CROCUS

These crocus are invaluable for their late-flowering, goblet-shaped flowers with showy insides. They are dwarf perennials with underground corms which give rise to the foliage and autumn flowers. The leaves are narrow and mid-green with silver-green central stripes, appearing at the same time or just after the flowers. All types are easy to grow in the right conditions, and look excellent when planted in groups in a rock garden. Rapid-spreading crocus, like *C. ochroleucus*, are useful for naturalizing in grass or under deciduous shrubs. *C. banaticus* is effective planted in drifts at the front of a border, but do not allow it to become swamped by larger plants.

CULTIVATION *Grow in gritty, well-drained, poor to moderately fertile soil, in full sun. Reduce watering during the summer for all types except* C. banaticus, *which prefers damper soil and will tolerate partial shade.*

☼ ◊ 🏆H4

1 ↕10cm (4in) ↔ 5cm (2in)

2 ↕10cm (4in) ↔ 5cm (2in)

3 ↕6–8cm (2½–3in) ↔ 5cm (2in)

4 ↕8cm (3in) ↔ 2.5cm (1in)

5 ↕5cm (2in) ↔ 2.5cm (1in)

6 ↕10–12cm (4–5in) ↔ 4cm (1½in)

1 *C. banaticus* **2** *C. goulimyi* **3** *C. kotschyanus* **4** *C. medius* **5** *C. ochroleucus* **6** *C. pulchellus*

CRYPTOMERIA JAPONICA 'ELEGANS COMPACTA'

This small, slow-growing conifer looks good in a heather bed or rock garden with other dwarf conifers. It has feathery sprays of slender, soft green leaves, which turn a rich bronze-purple in winter. Mature trees usually have a neat cone shape. *C. elegans* 'Bandai-Sugu' is very similar, but grows to only 2m (6ft).

CULTIVATION *Grow in any well-drained soil, in full sun or partial shade. Needs no formal pruning, but to renovate untidy trees, cut back to within about 70cm (28in) of ground level, in spring.*

☼ ◐ ◊ ♀H4 ↕2–4m (6–12ft) ↔2m (6ft)

CUPHEA IGNEA

The cigar flower is a spreading, evergreen shrub or subshrub grown as an annual in frost-prone climates. It bears small, slender, dark orange-red flowers over a long period from spring to autumn, amid the lance-shaped, dark green leaves. In frost-free areas it suits a shrub border; otherwise, treat as a bedding plant or grow in a cool conservatory.

CULTIVATION *Grow in well-drained, moderately fertile soil or compost, in full sun or partial shade. Pinch-prune young plants to encourage bushiness. Minimum temperature 2°C (35°F).*

☼ ◐ ◊ ♀H1 ↕30–75cm (12–30in) ↔30–90cm (12–36in)

x *CUPRESSOCYPARIS LEYLANDII* 'HAGGERSTON GREY'

This popular cultivar of the Leyland cypress is a fast-growing coniferous tree with a tapering, columnar shape. It has dense, grey-green foliage, and will establish quickly planted as a screen or windbreak. There are no flowers.

CULTIVATION *Grow in deep, well-drained soil in full sun or partial shade. Needs no formal pruning, unless grown as a hedge, when it should be trimmed 2 or 3 times during each growing season.*

☼ ◐ ◊ ♀H4 ↕to 35m (120ft) ↔to 5m (15ft)

CYANANTHUS LOBATUS

A spreading, mat-forming perennial grown for its late summer display of bright blue-purple, broadly funnel-shaped flowers, on single stems with hairy brown calyces. The leaves are fleshy and dull green, with deeply cut lobes. Perfect in a rock garden or trough.

CULTIVATION *Grow in poor to moderately fertile, moist but well-drained soil, preferably neutral to slightly acid, and rich in organic matter. Choose a site in partial shade.*

◐ ◊◆ ♀H4 ↕5cm (2in) ↔ to 30cm (12in)

CYCAS REVOLUTA

The Japanese sago palm is a robust cycad – a slow-growing tree with a palm-like appearance. Large leaves, up to 1.5m (5ft) long, are divided into glossy leaflets. Flowers and fruit are rare in container-grown plants, but it makes an excellent foliage, house or conservatory plant that can be brought outside in summer.

CULTIVATION *Grow in a mix of equal parts loam, garden compost, and coarse bark, with added grit, charcoal, and slow-release fertilizer. Provide full light, with shade from hot sun, and moderate humidity. Water sparingly in winter. Minimum temperature 7°C (45°F).*

☼ 💧 🏆H1 ↕↔1–2m (3–6ft)

CYCLAMEN CILICIUM

A tuberous perennial valued for its slender, nodding, white or pink flowers, borne in autumn and often into winter among patterned, rounded or heart-shaped, mid-green leaves. This elegant plant needs a warm and dry summer, when it is dormant, and in cool-temperate climates it may be better grown in a cool greenhouse, or under trees or shrubs, to avoid excessive summer moisture.

CULTIVATION *Grow in moderately fertile, well-drained soil enriched with organic matter, in partial shade. Mulch annually with leafmould after flowering when the leaves wither.*

☀ 💧 🏆H2–4 ↕5cm (2in) ↔8cm (3in)

CYCLAMEN COUM PEWTER GROUP

This winter- to spring-flowering, tuberous perennial is excellent for naturalizing beneath trees or shrubs. The compact flowers have upswept petals that vary from white to shades of pink and carmine-red. These emerge from swollen, underground tubers at the same time as the rounded, silver-green leaves. Provide a deep, dry mulch in cold areas.

CULTIVATION *Grow in gritty, well-drained, fertile soil that dries out in summer, in sun or light shade. Mulch annually when the leaves wither.*

H2–4 ↕5–8cm (2–3in) ↔10cm (4in)

CYCLAMEN HEDERIFOLIUM

An autumn-flowering, tuberous perennial bearing shuttlecock-like flowers which are pale to deep pink and flushed deep maroon at the mouths. The ivy-like leaves, mottled with green and silver, appear after the flowers from a swollen, underground tuber. It self-seeds freely, forming extensive colonies under trees and shrubs, especially where protected from summer rainfall.

CULTIVATION *Grow in well-drained, fertile soil, in sun or partial shade. Mulch each year after the leaves wither.*

H4 ↕10–13cm (4–5in) ↔15cm (6in)

CYNARA CARDUNCULUS

The cardoon is a clump-forming, statuesque, late-summer-flowering perennial which looks very striking in a border. The large, thistle-like purple flowerheads, which are very attractive to bees, are carried above the deeply divided, silvery leaves on stout, grey-woolly stems. The flowerheads dry well for indoor display and are very attractive to bees; when blanched, the leaf stalks can be eaten.

CULTIVATION *Grow in any well-drained, fertile soil, in full sun with shelter from cold winds. For the best foliage effect, remove the flower stems as they emerge.*

☼ ◊ 🏆H4 ↕1.5m (5ft) ↔1.2m (4ft)

CYTISUS BATTANDIERI

Pineapple broom develops a loose and open-branched habit. It is a semi-evergreen shrub bearing dense clusters of pineapple-scented, bright yellow flowers from early to mid-summer. The silvery-grey leaves, divided into three, make an attractive backdrop to herbaceous and mixed plantings. Best by a sunny wall in cold areas. The cultivar 'Yellow Tail' is recommended.

CULTIVATION *Grow in any well-drained, not too rich soil, in full sun. Very little pruning is necessary, but old wood can be cut out after flowering, and will be replaced with strong, young growth. Resents transplanting.*

☼ ◊ 🏆H4 ↕↔5m (15ft)

CYTISUS x *BEANII*

This low-growing and semi-trailing, deciduous, spring-flowering shrub carries an abundance of pea-like, rich yellow flowers on arching stems. The leaves are small and dark green. It is a colourful bush for a rock garden or raised bed, and is also effective if allowed to cascade over a wall. *Cytisus* x *ardoinei* is similar but slightly less spreading where space is limited.

CULTIVATION *Grow in well-drained, poor to moderately fertile soil, in full sun. Trim lightly after flowering, but avoid cutting into old wood.*

☼ ♢ ♀H4 ↕to 60cm (24in) ↔to 1m (3ft)

CYTISUS x *PRAECOX* 'ALLGOLD'

This compact, deciduous shrub is smothered by a mass of pea-like, dark yellow flowers from mid- to late spring. The tiny, grey-green leaves are carried on arching stems. Suitable for a sunny shrub border or large rock garden. 'Warminster' is very similar, with paler, cream-yellow flowers.

CULTIVATION *Grow in well-drained, acid to neutral soil, in sun. Pinch out the growing tips to encourage bushiness, then cut back new growth by up to two-thirds after flowering; avoid cutting into old wood. Replace old, leggy specimens.*

☼ ♢ ♀H4 ↕1.2m (4ft) ↔1.5m (5ft)

DABOECIA CANTABRICA 'BICOLOR'

This straggling, heather-like shrub bears slender spikes of urn-shaped flowers from spring to autumn. They are white, pink or beetroot-red, sometimes striped with two colours. The leaves are small and dark green. Good in a heather bed or among other acid-soil-loving plants. 'Waley's Red', with glowing deep magenta flowers, is also recommended.

CULTIVATION *Best in sandy, well-drained, acid soil, in full sun; tolerates neutral soil and some shade. Clip over lightly in early spring to remove spent flowers, but do not cut into old wood.*

☼ ◐ ◊ ♀H4 ↕45cm (18in) ↔60cm (24in)

DABOECIA CANTABRICA 'WILLIAM BUCHANAN'

This vigorous, compact, heather-like shrub bears slender spikes of bell-shaped, purple-crimson flowers from late spring to mid-autumn. The narrow leaves are dark green above with silver-grey undersides. Good with other acid-soil-loving plants, or among conifers in a rock garden. Try planting with 'Silverwells', more ground-hugging, with white flowers.

CULTIVATION *Grow in well-drained, sandy, acid to neutral soil, preferably in sun, but tolerates light shade. Trim in early to mid-spring to remove old flowers, but do not cut into old wood.*

☼ ◐ ◊ ♀H4 ↕45cm (18in) ↔60cm (24in)

DAHLIAS

Dahlias are showy, deciduous perennials, grown as annuals in cold climates, with swollen underground tubers which should be stored in frost-free conditions in climates with cold winters. They are valued for their massive variety of brightly coloured flowers, which bloom from mid-summer to autumn when many other plants are past their best. The leaves are mid- to dark green and divided. Very effective in massed plantings, wherever space allows: in small gardens, choose a selection of dwarf types to fill gaps in border displays, or grow in containers. The flowers are ideal for cutting.

CULTIVATION *Best in well-drained soil, in sun. In frost-prone climates, lift tubers and store in cool peat over winter. Plant out tubers once the danger of frost has passed. Feed with high-nitrogen fertiliser every week in early summer. Taller varieties need staking. Deadhead to prolong flowering.*

☼ ◊ 🏆H3

1 ↕1.1m (3½ft) ↔ 45cm (18in)

2 ↕1.1m (3½ft) ↔ 60cm (24in)

3 ↕60cm (24in) ↔ 45cm (18in)

4 ↕1.2m (4ft) ↔ 60cm (24in)

5 ↕1.1m (3½ft) ↔ 60cm (24in)

1 *D.* 'Bishop of Llandaff' **2** *D.* 'Clair de Lune' **3** *D.* 'Fascination' (dwarf)
4 *D.* 'Hamari Accord' **5** *D.* 'Conway'

6 ↕1.2m (4ft) ↔ 60cm (24in)
7 ↕1.1m (3½ft) ↔ 60cm (24in)
8 ↕1.2m (4ft) ↔ 60cm (24in)
9 ↕1m (3ft) ↔ 45cm (18in)
10 ↕↔ 45–50cm (18–20in)
11 ↕1.1m (3½ft) ↔ 60cm (24in)
12 ↕1.1m (3½ft) ↔ 60cm (24in)
13 ↕60cm (24in) ↔ 45cm (18in)
14 ↕1.2m (4ft) ↔ 60cm (24in)
15 ↕1.2m (4ft) ↔ 60cm (24in)

6 *D.* 'Hamari Gold' **7** *D.* 'Hillcrest Royal' **8** *D.* 'Kathryn's Cupid' **9** *D.* 'Rokesley Mini' **10** *D.* 'Sunny Yellow' (dwarf) **11** *D.* 'So Dainty' **12** *D.* 'Wootton Cupid' **13** *D.* 'Yellow Hammer' (dwarf) **14** *D.* 'Zorro' **15** *D.* 'Wootton Impact'

DAPHNE BHOLUA 'GURKHA'

An upright, deciduous shrub bearing clusters of strongly fragrant, tubular, white and purplish-pink flowers on its bare stems in late winter. They open from deep pink-purple buds, and are followed by round, black-purple fruits. The lance-shaped leaves are leathery and dark green. A fine plant for a winter garden, but needs protection in cold areas. All parts are highly toxic if ingested.

CULTIVATION *Grow in well-drained but moist soil, in sun or semi-shade. Mulch to keep the roots cool. Best left unpruned.*

☼ ◑ ♢ ♈H3 ↕2–4m (6–12ft) ↔1.5m (5ft)

DAPHNE PETRAEA 'GRANDIFLORA'

A very compact, slow-growing, evergreen shrub bearing clusters of fragrant, deep rose-pink flowers in late spring. The spoon-shaped leaves are leathery and glossy dark green. Ideal for a sheltered rock garden; may not survive in cold climates. All parts of the plant are highly toxic.

CULTIVATION *Grow in reasonably moist but well-drained, fairly fertile soil that is rich in humus, in sun or semi-shade. Regular pruning is not necessary.*

☼ ◑ ♢ ♈H2–4 ↕10cm (4in) ↔25cm (10in)

DAPHNE TANGUTICA RETUSA GROUP

These dwarf forms of *D. tangutica*, sometimes listed simply as *D. retusa*, are evergreen shrubs valued for their clusters of very fragrant, white to purple-red flowers which are borne during late spring and early summer. The lance-shaped leaves are glossy and dark green. Useful in a variety of sites, such as a large rock garden, shrub border or mixed planting. All parts are toxic.

CULTIVATION *Grow in well-drained, moderately fertile, humus-rich soil that does not dry out, in full sun or dappled shade. Pruning is not necessary.*

☼ ◑ ◊ ♀H4 ↕↔ 75cm (30in)

DARMERA PELTATA

A handsome, spreading perennial, sometimes included in the genus *Peltiphyllum*, that forms an imposing, umbrella-like clump with large, round, mid-green leaves, to 60cm (24in) across, which turn red in autumn. The foliage appears after the compact clusters of star-shaped, white to bright pink, spring flowers. Ideal for a bog garden or by the edge of a pond or stream.

CULTIVATION *Grow in reliably moist, moderately fertile soil, in sun or partial shade; tolerates drier soil in shade.*

☼ ◑ ♦ ♀H4 ↕to 2m (6ft) ↔1m (3ft)

DELPHINIUMS

Delphiniums are clump-forming perennials cultivated for their towering spikes of exquisite, shallowly cup-shaped, spurred, single or double flowers. These appear in early to mid-summer, and are available in a range of colours from creamy-whites through lilac-pinks and clear sky-blues to deep indigo-blue. The toothed and lobed, mid-green leaves are arranged around the base of the stems. Grow tall delphiniums in a mixed border or island bed with shelter to prevent them being blown over in strong winds; shorter ones suit a rock garden. The flowers are good for cutting.

CULTIVATION *Grow in well-drained, fertile soil, in full sun. For quality blooms, feed weekly with a balanced fertiliser in spring, and thin out the young shoots when they reach 7cm (3in) tall. Most cultivars need staking. Remove spent flower spikes, and cut back all growth in autumn.*

☼ ♢ ♈H4

1 ↕1.7m (5½ft) ↔ 60–90cm (24–36in)

2 ↕2m (6ft) ↔ 60–90cm (24–36in)

3 ↕1.7m (5½ft) ↔ 60–90cm (24–36in)

1 *Delphinium* 'Blue Nile' **2** *D.* 'Bruce' **3** *D.* 'Cassius'

4 ↕ to 2m (6ft) ↔ 60–90cm (24–36in)

5 ↕ to 1.5m (5ft) ↔ 60–90cm (24–36in)

6 ↕ 1.7m (5½ft) ↔ 60–90cm (24–36in)

7 ↕ 2m (6ft) ↔ 60–90cm (24–36in)

8 ↕ 1.7m (5½ft) ↔ 60–90cm (24–36in)

4 *D.* 'Claire' **5** *D.* 'Conspicuous' **6** *D.* 'Emily Hawkins' **7** *D.* 'Fanfare'
8 *D.* 'Giotto'

MORE CHOICES

'Blue Dawn' Pinky-blue flowers with brown eyes.

'Constance Rivett' White.

'Faust' Cornflower blue with indigo eyes.

'Fenella' Deep blue with black eyes.

'Gillian Dallas' White.

'Loch Leven' Mid-blue with white eyes.

'Michael Ayres' Violet.

'Min' Mauve veined with deep purple, brown eyes.

'Oliver' Light blue/mauve with dark eyes.

'Spindrift' Cobalt blue with cream eyes.

'Tiddles' Slate-blue.

9 ↕ to 2m (6ft) ↔ 60–90cm (24–36in)

10 ↕ to 1.5m (5ft) ↔ 60–90cm (24–36in) **11** ↕ to 1.5m (5ft) ↔ 60–90cm (24–36in)

9 *D.* 'Kathleen Cooke' **10** *D.* 'Langdon's Royal Flush' **11** *D.* 'Lord Butler'

12 ↕2m (6ft) ↔ 75cm (30in)

13 ↕1.7m (5½ft) ↔ 60–90cm (24–36in)

14 ↕to 1.5m (5ft) ↔ 60–90cm (24–36in)

15 ↕to 1.5m (5ft) ↔ 75cm (30in)

16 ↕to 1.5m (5ft) ↔ 60–90cm (24–36in)

17 ↕1.2m (4ft) ↔ 60–90cm (24–36in)

18 ↕1.7m (5½ft) ↔ 60–90cm (24–36in)

12 *D.* 'Mighty Atom' **13** *D.* 'Our Deb' **14** *D.* 'Rosemary Brock' **15** *D.* 'Sandpiper' **16** *D.* 'Sungleam' **17** *D.* 'Thamesmead' **18** *D.* 'Walton Gemstone'

DEUTZIA x *ELEGANTISSIMA* 'ROSEALIND'

This compact, rounded, deciduous shrub bears profuse clusters of small, deep carmine-pink flowers. These are carried from late spring to early summer amid the oval, dull green leaves. Very suitable for a mixed border.

CULTIVATION *Grow in any well-drained, fertile soil that does not dry out, in full sun or partial shade. Tip-prune when young to encourage bushiness; after flowering, thin out by cutting some older stems right back to the ground.*

☼☀ 💧 ♆H4 ↕1.2m (4ft) ↔1.5m (5ft)

DEUTZIA x *HYBRIDA* 'MONT ROSE'

A dense, upright shrub, very similar to *D.* x *elegantissima* 'Rosealind' (above), but with smaller flower clusters. These consist of small, star-shaped, light pink or pinkish-purple flowers, borne in early summer amid the narrow, dark green, deciduous leaves. Good as a specimen shrub or in a mixed border.

CULTIVATION *Grow in any fertile, well-drained but not too dry soil. Best in sun, but tolerates light shade. Tip-prune on planting; in subsequent years, prune young shoots below the flowered wood.*

☼☀ 💧 ♆H4 ↕↔ 1.2m (4ft)

BORDER CARNATIONS (*DIANTHUS*)

This group of *Dianthus* are annuals or evergreen perennials of medium height, suitable for mixed or herbaceous borders. They are grown for their mid-summer flowers, which are good for cutting. Each flower stem bears five or more double flowers to 8cm (3in) across, with no fewer than 25 petals. These may be of one colour only, as in 'Golden Cross'; gently flecked, like 'Grey Dove'; or often with white petals margined and striped in strong colours. Some have clove-scented flowers. The linear leaves of all carnations are blue-grey or grey-green with a waxy bloom.

CULTIVATION *Ideal in well-drained, neutral to alkaline soil enriched with well-rotted manure or garden compost; apply a balanced fertilizer in spring. Plant in full sun. Provide support in late spring using thin canes or twigs, or wire rings. Deadhead to prolong flowering and maintain a compact habit.*

☼ ◊ ♀H4

1 ↕45–60cm (18–24in) ↔ 45cm (18in)

2 ↕45–60cm (18–24in) ↔ 45cm (18in)

3 ↕45–60cm (18–24in) ↔ 45cm (18in)

4 ↕45–60cm (18–24in) ↔ 45cm (18in)

5 ↕45–60cm (18–24in) ↔ 45cm (18in)

6 ↕45–60cm (18–24in) ↔ 45cm (18in)

1 *Dianthus* 'David Russell' **2** *D.* 'Devon Carla' **3** *D.* 'Golden Cross' **4** *D.* 'Grey Dove' **5** *D.* 'Ruth White' **6** *D.* 'Spinfield Wizard'

PINKS (*DIANTHUS*)

Summer-flowering pinks belong, like carnations, to the genus *Dianthus* and are widely grown for their charming, often clove-scented flowers and narrow, blue-grey leaves. They flower profusely over long periods; when cut, the stiff-stemmed blooms last exceptionally well. Thousands are available, usually in shades of pink, white, carmine, salmon, or mauve with double or single flowers; they may be plain (self), marked with a contrasting colour or laced around the margins. Most are excellent border plants; tiny alpine pinks like 'Pike's Pink' and magenta-flowered 'Joan's Blood' are well suited to a rock garden or raised bed.

CULTIVATION *Best in well-drained, neutral to alkaline soil, in an open, sunny site. Alpine pinks need very sharp drainage. Feed with a balanced fertiliser in spring. Deadhead all types to prolong flowering and to maintain a compact habit.*

☼ ◊ ♀H4

1 ↕25–45cm (10–18in) ↔ 40cm (16in)

2 ↕25–45cm (10–18in) ↔ 40cm (16in)

3 ↕25–45cm (10–18in) ↔ 40cm (16in)

4 ↕25–45cm (10–18in) ↔ 40cm (16in)

5 ↕25–45cm (10–18in) ↔ 40cm (16in)

6 ↕25–45cm (10–18in) ↔ 40cm (16in)

1 *Dianthus* 'Becky Robinson' **2** *D.* 'Devon Pride' **3** *D.* 'Doris' **4** *D.* 'Gran's Favourite' **5** *D.* 'Haytor White' **6** *D.* 'Houndspool Ruby'

7 ↕25–45cm (10–18in) ↔ 40cm (16in)

8 ↕8–10cm (3–4in) ↔ 20cm (8in)

9 ↕25–45cm (10–18in) ↔ 40cm (16in)

10 ↕25–45cm (10–18in) ↔ 40cm (16in)

11 ↕25–45cm (10–18in) ↔ 40cm (16in)

12 ↕25–45cm (10–18in) ↔ 40cm (16in)

MORE CHOICES

D. alpinus The original alpine pink, with pale-flecked pink flowers.

D. alpinus 'Joan's Blood' Dark-centred deep pink flowers.

'Bovey Belle' Clove-scented, fuchsia-pink.

'Coronation Ruby' Clove-scented, warm pink flowers with ruby markings.

'Inschriach Dazzler' Alpine pink with fringed carmine flowers.

'Valda Wyatt' Clove-scented, double lavender flowers.

13 ↕8–10cm (3–4in) ↔ 20cm (8in)

14 ↕25–45cm (10–18in) ↔ 40cm (16in)

15 ↕25–45cm (10–18in) ↔ 40cm (16in)

16 ↕25–45cm (10–18in) ↔ 40cm (16in)

7 *D.* 'Kathleen Hitchcock' **8** *D.* 'La Bourboule' **9** *D.* 'Monica Wyatt' **10** *D.* 'Natalie Saunders' **11** *D.* 'Oakwood Romance' **12** *D.* 'Oakwood Splendour' **13** *D.* 'Pike's Pink' **14** *D.* 'Suffolk Pride' **15** *D.* 'Trisha's Choice' **16** *D.* 'White Joy'

DIASCIA BARBERAE 'BLACKTHORN APRICOT'

A mat-forming perennial bearing loose spikes of apricot flowers. These are produced over a long period from summer to autumn, above the narrowly heart-shaped, mid-green leaves. Good in a rock garden, at the front of a mixed border or on a sunny bank. *D. barbarae* 'Fisher's Flora' and 'Ruby Field' are also recommended.

CULTIVATION *Grow in moist but well-drained, fertile soil, in full sun. In cold climates, shelter from heavy frost and overwinter young plants under glass.*

☼ ◊ ♈H3–4 ↕25cm (10in) ↔to 50cm (20in)

DIASCIA RIGESCENS

This trailing perennial is valued for its tall spires of salmon-pink flowers. These appear above the mid-green, heart-shaped leaves during summer. Excellent in a rock garden or raised bed, or at the front of a border. In cold areas, grow at the base of a warm, sunny wall.

CULTIVATION *Grow in moist but well-drained, fertile soil that is rich in humus. Site in full sun. Overwinter young plants under glass.*

☼ ◊ ♈H3 ↕30cm (12in) ↔to 50cm (20in)

DICENTRA SPECTABILIS

Bleeding heart is an elegant, hummock-forming perennial. In late spring and early summer, it bears rows of hanging, red-pink, heart-shaped flowers on arching stems. The leaves are deeply cut and mid-green. A beautiful plant for a shady border or woodland garden.

CULTIVATION *Grow in reliably moist, fertile, humus-rich, neutral or slightly alkaline soil. Prefers a site in partial shade, but tolerates full sun.*

☼ ◐ 💧 ♀H4 ↕to 1.2m (4ft) ↔45cm (18in)

DICENTRA SPECTABILIS 'ALBA'

This white-flowered form of bleeding heart is otherwise very similar to the species (above), so to avoid possible confusion, purchase plants during the flowering period. The leaves are deeply cut and light green, and the heart-shaped flowers are borne along arching stems in late spring and early summer.

CULTIVATION *Grow in reliably moist but well-drained soil that is enriched with well-rotted organic matter. Prefers partial shade, but tolerates full sun.*

☼ ◐ 💧 ♀H4 ↕to 1.2m (4ft) ↔45cm (18in)

DICENTRA 'STUART BOOTHMAN'

A spreading perennial that bears heart-shaped pink flowers along the tips of arching stems during late spring and summer. The very finely cut foliage is fern-like and grey-green. Very attractive for a shady border or woodland planting.

CULTIVATION *Best in moist but well drained, humus-rich, neutral to slightly alkaline soil, in partial shade. Divide and replant clumps in early spring, after the leaves have died down.*

☀ ◊♦ ♀H4 ↕30cm (12in) ↔40cm (16in)

DICTAMNUS ALBUS

Burning bush is a tall, woody-based perennial that suits both mixed and herbaceous borders. Dense spikes of star-shaped, fragrant white flowers appear above the lemon-scented, light green foliage in early summer. The aromatic oils produced by the flowers and ripening seedpods can be ignited in hot weather, hence the common name.

CULTIVATION *Grow in well-drained, fertile soil, in sun or partial shade. Does not respond well to root disturbance, so avoid transplanting.*

☀☼ ◊ ♀H4 ↕40–90cm (16–36in) ↔60cm (24in)

DICTAMNUS ALBUS VAR. *PURPUREUS*

This purple-flowered perennial is otherwise identical to the species (see facing page, below). Upright spikes of flowers are borne in early summer, above the highly aromatic, light green leaves. Striking in a mixed or herbaceous border.

CULTIVATION *Grow in any dry, well-drained, moderately fertile soil, in full sun or partial shade.*

☼◐ ◊ 🏆H4 ↕40–90cm (16–36in) ↔60cm (24in)

DIGITALIS GRANDIFLORA

The yellow foxglove is a short-lived, evergreen perennial producing upright spikes of tubular, pale yellow flowers from early to mid-summer. The leaves are oval and mid-green. Smaller and more subtle than many cultivated foxgloves, it looks attractive in woodland gardens or "natural" planting schemes.

CULTIVATION *Grow in moist but well-drained soil that is not allowed to dry out. Site in partial shade. Self-seeds freely, so deadhead after flowering to prevent unwanted seedlings.*

◐ ◊💧🏆H4 ↕to 1m (3ft) ↔45cm (18in)

DIGITALIS x *MERTONENSIS*

An evergreen, clump-forming, short-lived perennial grown for its spires of flared, strawberry-pink flowers which are borne from late spring to early summer. The leaves are dark green and lance-shaped. A beautiful plant for mixed or herbaceous borders or woodland plantings. The dusky pink cultivar 'Raspberry' is also recommended.

CULTIVATION *Best in moist but well-drained soil, in partial shade, but tolerates full sun and dry soil.*

☼☀ 💧💧 ♈H4 ↕to 90cm (36in) ↔30cm (12in)

DIGITALIS PURPUREA F. *ALBIFLORA*

This ghostly form of the common foxglove is a tall biennial producing robust, stately spires of tubular, pure white flowers during summer. The coarse, lance-shaped leaves are bright green. A lovely addition to a woodland garden or mixed border. May be sold as *D. purpurea* 'Alba'.

CULTIVATION *Grow in moist but well-drained soil, in partial shade, but also tolerates dry soil in full sun.*

☼☀ 💧💧 ♈H4 ↕1–2m (3–6ft) ↔to 60cm (24in)

DODECATHEON MEADIA F. *ALBUM*

This herbaceous perennial is valued for its open clusters of creamy-white flowers with strongly reflexed petals. These are borne on arching stems during mid- to late spring, above the rosettes of oval, pale green, toothed leaves. Suits a woodland or shady rock garden; flowering is followed by a period of dormancy. *D. dentatum* looks very similar, but is only half the height, ideal for a small rock garden.

CULTIVATION *Grow in moist, humus-rich soil, in partial shade. May be prone to slug and snail damage in spring.*

☀ 💧 🏆H4 ↕40cm (16in) ↔25cm (10in)

DRYOPTERIS FILIX-MAS

The male fern is a deciduous foliage perennial forming large clumps of lance-shaped, mid-green fronds which emerge from a stout, scaly crown in spring. No flowers are produced. Ideal for a shady border or corner, by the side of a stream or pool, or in woodland. 'Cristata', with crested fronds, is a handsome cultivar of this fern.

CULTIVATION *Grow in reliably moist, humus-rich soil. Site in partial shade with shelter from cold, drying winds.*

☀ 💧 🏆H4 ↕↔ 1m (3ft)

DRYOPTERIS WALLICHIANA

Wallich's wood fern is a deciduous foliage perennial with a strongly upright, shuttlecock-like habit. The fronds are yellow-green when they emerge in spring, becoming dark green in summer. A fine architectural specimen for a moist, shady site.

CULTIVATION *Best in damp soil that is rich in humus. Choose a sheltered position in partial shade.*

☀ 💧 ♈H4 ↕1–2m (3–6ft) ↔75cm (30in)

ECCREMOCARPUS SCABER

The Chilean glory flower is a fast-growing, scrambling climber with clusters of brilliant orange-red, tubular flowers in summer. The leaves are divided and mid-green. Grow as a short-lived perennial to clothe an arch or pergola, or up into a large shrub. In frost-prone areas, use as a trailing annual.

CULTIVATION *Grow in free-draining, fertile soil, in a sheltered, sunny site. Cut back to within 30–60cm (12–24in) of the base in spring.*

☼ ◊ ♈H3 ↕15cm (6in) ↔20cm (8in)

ECHEVERIA AGAVOIDES

A lovely, star-shaped, succulent perennial forming rosettes of chunky, sharply pointed, pale green leaves. Clusters of yellow-tipped red flowers appear above the foliage in spring and early summer. Usually grown in a conservatory or as a houseplant in cold areas, but where the climate is reliably warm, it will form clumps in a succulent border.

CULTIVATION *Grow in well-drained, fairly fertile soil, or standard cactus compost, in sun. Keep barely moist in winter; increase watering in spring. Minimum temperature 2°C (35°F).*

☼ 💧 🏆H1 ↕15cm (6in) ↔30cm (12in) or more

ECHINOPS RITRO

This globe thistle is a compact perennial forming clumps of eye-catching flower-heads, metallic blue at first, turning a brighter blue as the flowers open. The leathery green leaves have white-downy undersides. Excellent for a wild garden. The flowers dry well if cut before fully open. *E. bannaticus* 'Taplow Blue' makes a good substitute if this plant cannot be found; it may grow a little taller.

CULTIVATION *Grow in well-drained, poor to moderately fertile soil. Site in full sun, but tolerates partial shade. Deadhead to prevent self-seeding.*

☼ ◐ 💧 🏆H4 ↕60cm (24in) ↔45cm (18in)

ELAEAGNUS x *EBBINGEI* 'GILT EDGE'

A large, dense, evergreen shrub grown for its oval, dark green leaves which are edged with rich golden-yellow. Inconspicuous yet highly scented, creamy-white flowers are produced in autumn. Makes an excellent, fast-growing, well-shaped specimen shrub; can also be planted as an informal hedge.

CULTIVATION *Grow in well-drained, fertile soil, in full sun. Dislikes very chalky soil. Trim to shape in late spring; completely remove any shoots with plain green leaves as soon as seen.*

☼ ◊ ♈H4 ↕↔ 4m (12ft)

ELAEAGNUS PUNGENS 'MACULATA'

This large, evergreen foliage shrub bears oval, dark green leaves that are generously splashed in the centre with dark yellow. Small but very fragrant flowers are produced from mid-autumn, followed by red fruits. Tolerates coastal conditions.

CULTIVATION *Best in well-drained, fairly fertile soil, in sun. Dislikes very chalky soil. Trim lightly in spring, as needed. Remove completely any shoots with plain green leaves as soon as seen.*

☼ ◊ ♈H4 ↕4m (12ft) ↔5m (15ft)

ELAEAGNUS 'QUICKSILVER'

A fast-growing shrub with an open habit bearing small yellow flowers in summer which are followed by yellow fruits. The silver, deciduous leaves are lance-shaped, and are carried on silvery shoots. Makes an excellent specimen shrub, or can be planted to great effect with other silver-leaved plants. May be known as *E. angustifolia* var. *caspica*.

CULTIVATION *Grow in any but chalky soil that is fertile and well-drained, in full sun. Tolerates dry soil and coastal winds. Keep pruning to a minimum.*

☼ ◊ H4 ↕↔ 4m (12ft)

ENKIANTHUS CAMPANULATUS

A spreading, deciduous shrub grown for its dense, hanging clusters of bell-shaped, creamy-yellow flowers in late spring. The dull green leaves give a fine autumn display when they change to orange-red. Suits an open site in a woodland garden; can become tree-like with age. *E. cernuus* var. *rubens*, with deep pink flowers, is more suited to a small garden, to only 2.5m (8ft) tall.

CULTIVATION *Grow in reliably moist but well-drained, peaty or humus-rich, acid soil. Best in full sun, but tolerates some shade. Keep pruning to a minimum.*

↕↔ 4–5m (12–15ft)

EPIMEDIUM X *PERRALCHICUM*

A robust, clump-forming, evergreen perennial that produces spikes of delicate, bright yellow flowers above the glossy dark green foliage in mid- and late spring. The leaves are tinged bronze when young. Very useful as ground cover under trees or shrubs.

CULTIVATION *Grow in moist but well-drained, moderately fertile soil that is enriched with humus, in partial shade. Provide shelter from cold, drying winds.*

☀ ◊♦ ♀H4 ↕40cm (16in) ↔60cm (24in)

EPIMEDIUM X *RUBRUM*

A compact perennial bearing loose clusters of pretty crimson flowers with yellow spurs, in spring. The divided leaves are tinted bronze-red when young, ageing to mid-green, then reddening in autumn. Clump together to form drifts in a damp, shady border or woodland garden. *E. grandiflorum* 'Rose Queen' is very similar.

CULTIVATION *Grow in moist but well-drained, moderately fertile, humus-rich soil, in partial shade. Cut back old, tatty foliage in late winter so that the flowers can be seen in spring.*

☀ ◊♦ ♀H4 ↕↔ 30cm (12in)

EPIMEDIUM X *YOUNGIANUM* 'NIVEUM'

A clump-forming perennial that bears dainty clusters of small white flowers in late spring. The bright green foliage is bronze-tinted when young and, despite being deciduous, persists well into winter. Makes good ground cover in damp, shady borders and woodland areas.

CULTIVATION *Grow in moist but well-drained, fertile, humus-rich soil, in partial shade. In late winter, cut back old, tatty foliage so that the new flowers can be seen in spring.*

H4 ↕20–30cm (8–12in) ↔30cm (12in)

ERANTHIS HYEMALIS

Winter aconite is one of the earliest spring-flowering bulbs, bearing buttercup-like, bright yellow flowers. These sit on a ruff of light green leaves, covering the ground from late winter until early spring. Ideal for naturalizing beneath deciduous trees and large shrubs, as is the cultivated variety, 'Guinea Gold', with similar flowers and bronze-green leaves.

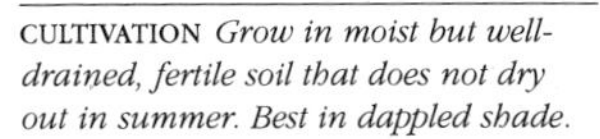

CULTIVATION *Grow in moist but well-drained, fertile soil that does not dry out in summer. Best in dappled shade.*

H4 ↕5–8cm (2–3in) ↔5cm (2in)

ERICA ARBOREA VAR. *ALPINA*

This tree heath is an upright shrub, much larger than other heathers, densely clothed with clusters of small, honey-scented white flowers from late winter to late spring. The evergreen leaves are needle-like and dark green. A fine centrepiece for a heather garden.

CULTIVATION *Grow in well-drained, ideally sandy, acid soil, in an open, sunny site. Tolerates alkaline conditions. Cut back young plants in early spring by about two-thirds to promote bushy growth; in later years, pruning is unnecessary. Tolerates hard pruning to renovate.*

☼ ◊ ♀H4 ↕2m (6ft) ↔85cm (34in)

ERICA X *VEITCHII* 'EXETER'

An upright, open, evergreen shrub that bears masses of highly scented, tubular to bell-shaped white flowers from mid-winter to spring. The needle-like leaves are bright green. Good in a large rock garden with conifers or as a focal point among low-growing heathers, but may not survive winter in cold climates.

CULTIVATION *Grow in sandy soil that is well-drained, in sun. Best in acid soil, but tolerates slightly alkaline conditions. When young, cut back by two-thirds in spring to encourage a good shape; reduce pruning as the plant gets older.*

☼ ◊ ♀H3 ↕2m (6ft) ↔65cm (26in)

EARLY-FLOWERING ERICAS

The low-growing, early flowering heaths are evergreen shrubs valued for their urn-shaped flowers in winter and spring. The flowers come in white and a wide range of pinks, bringing invaluable early interest during the winter months. This effect can be underlined by choosing cultivars with colourful foliage; *E. erigena* 'Golden Lady', for example, has bright golden-yellow leaves, and *E. carnea* 'Foxhollow' carries yellow, bronze-tipped foliage which turns a deep orange in cold weather. Excellent as ground cover, either in groups of the same cultivar, or with other heathers and dwarf conifers.

CULTIVATION *Grow in open, well-drained, preferably acid soil, but tolerate alkaline conditions. Choose a site in full sun. Cut back flowered stems in spring to remove most of the previous year's growth; cultivars of* E. erigena *may be scorched by frosts; remove any affected growth in spring.*

☼ ◊ ♈H4

1 ↕15cm (6in) ↔ 25cm (10in)
2 ↕15cm (6in) ↔ 40cm (16in)
3 ↕15cm (6in) ↔ 45cm (18in)
4 ↕15cm (6in) ↔ 35cm (14in)
5 ↕30cm (12in) ↔ 60cm (24in)
6 ↕30cm (12in) ↔ 40cm (16in)

1 *E. carnea* 'Ann Sparkes' **2** *E. carnea* 'Foxhollow' **3** *E. carnea* 'Springwood White' **4** *E. carnea* 'Vivellii' **5** *E.* x *darleyensis* 'Jenny Porter' **6** *E. erigena* 'Golden Lady'

LATE-FLOWERING ERICAS

Mostly low and spreading in form, the summer-flowering heaths are fully hardy, evergreen shrubs. They look well on their own or mixed with other heathers and dwarf conifers. Flowers are borne over a very long period; *E.* x *stuartii* 'Irish Lemon' starts in late spring, and cultivars of *E. ciliaris* and *E. vagans* bloom well into autumn. Their season of interest can be further extended by choosing types with colourful foliage; the young shoots of *E. williamsii* 'P.D. Williams' are tipped with yellow in spring, and the golden foliage of *E. cinerea* 'Windlebrooke' turns a deep red in winter.

CULTIVATION *Grow in well-drained, acid soil, although* E. vagans *and* E. williamsii *tolerate alkaline conditions. Choose an open site in full sun. Prune or trim lightly in early spring, cutting back to strong shoots below the flower clusters, or trim over with shears en masse.*

☼ ◊ ♕H4

1 ↕22cm (9in) ↔ 35cm (14in)

2 ↕to 40cm (16in) ↔ 45cm (18in)

3 ↕20cm (8in) ↔ 50cm (20in)

4 ↕25cm (10in) ↔ 50cm (20in)

5 ↕15cm (6in) ↔ 45cm (18in)

1 *E. ciliaris* 'Corfe Castle' **2** *E. ciliaris* 'David McClintock' **3** *E. cinerea* 'Eden Valley' **4** *E. cinerea* 'C.D. Eason' **5** *E. cinerea* 'Windlebrooke'

6 ↕25cm (10in) ↔ 45cm (18in)

7 ↕25cm (10in) ↔ 50cm (20in)

8 ↕20cm (8in) ↔ 30cm (12in)

9 ↕30cm (12in) ↔ 50cm (20in)

10 ↕25cm (10in) ↔ 50cm (20in)

11 ↕30cm (12in) ↔ 45cm (18in)

12 ↕15cm (6in) ↔ 30cm (12in)

13 ↕20cm (8in) ↔ to 85cm (34in)

14 ↕30cm (12in) ↔ 45cm (18in)

6 *E. cinerea* 'Fiddler's Gold' **7** *E.* x *stuartii* 'Irish Lemon' **8** *E. tetralix* 'Alba Mollis'
9 *E. vagans* 'Birch Glow' **10** *E. vagans* 'Lyonesse' **11** *E. vagans* 'Mrs. D.F. Maxwell'
12 *E. vagans* 'Valerie Proudley' **13** *E. watsonii* 'Dawn' **14** *E.* x *williamsii* 'P.D. Williams'

ERIGERON KARVINSKIANUS

This carpeting, evergreen perennial with grey-green foliage produces an abundance of yellow-centred, daisy-like flowerheads in summer. The outer petals are initially white, maturing to pink and purple. Ideal for wall crevices, or cracks in paving, but may not survive through winter in frost-prone climates. Sometimes sold as *E. mucronatus.*

CULTIVATION *Grow in well-drained, fertile soil. Choose a site in full sun, ideally with some shade at midday.*

☼ ◊ ♈H3 ↕15–30cm (6–12in) ↔1m (3ft) or more

ERINUS ALPINUS

The fairy foxglove is a tiny, short-lived, evergreen perennial producing short spikes of pink, purple or white, daisy-like flowers in late spring and summer. The lance- to wedge-shaped leaves are soft, sticky and mid-green. Ideal for a rock garden or in crevices in old walls. 'Mrs Charles Boyle' is a recommended cultivar.

CULTIVATION *Grow in light, moderately fertile soil that is well-drained. Tolerates semi-shade, but best in full sun.*

☼ ◐ ◊ ♈H4 ↕8cm (3in) ↔10cm (4in)

ERYNGIUM ALPINUM

This spiky, upright, thistle-like perennial bears cone-shaped, purple-blue flowerheads in summer; these are surrounded by prominent, feathery bracts. The deeply toothed, mid-green leaves are arranged around the base of the stems. An excellent textural plant for a sunny garden. The flower heads can be cut, and dry well for arrangements.

CULTIVATION *Grow in free-draining but not too dry, poor to moderately fertile soil, in full sun. Choose a site not prone to excessive winter wet.*

☼ ◊ 🏆H4 ↕70cm (28in) ↔45cm (18in)

ERYNGIUM x *OLIVERIANUM*

An upright, herbaceous perennial bearing cone-shaped, bright silver-blue flowerheads with a flat ring of silvery, dagger-like bracts around the base, produced from mid-summer to early autumn, above spiny-toothed, dark green leaves. Longer-lived than its parent *E. giganteum*, and an essential architectural addition to a dry, sunny border with a theme of silver- or grey-leaved plants.

CULTIVATION *Grow in free-draining, fairly fertile soil, in full sun. Choose a site not prone to excessive winter wet.*

☼ ◊ 🏆H4 ↕90cm (36in) ↔45cm (18in)

ERYNGIUM X *TRIPARTITUM*

This delicate but spiky, upright perennial bears cone-like heads of tiny, metallic blue flowers in late summer and autumn. The flowerheads sit on a ring of pointed bracts, and are carried above rosettes of grey-green foliage on the tips of wiry, blue-tinted stems. The flowers dry well if cut before fully open.

CULTIVATION *Grow in free-draining, moderately fertile soil. Choose a position not prone to excessive winter wet, in full sun. Trim lightly after flowering to prevent legginess.*

☼ ◊ ♀H4 ↕60–90cm (24–36in) ↔50cm (20in)

ERYSIMUM 'BOWLES' MAUVE'

This vigorous, shrubby wallflower is one of the longest-flowering of all perennials. It forms a rounded, evergreen bush of narrow, grey-green leaves and produces dense spikes of small, four-petalled, rich mauve flowers all year, most freely in spring and summer. An excellent border plant that benefits from the shelter of a warm wall in frost-prone climates.

CULTIVATION *Grow in any well-drained, preferably alkaline soil, in full sun. Trim lightly after flowering to keep compact. Often short-lived, but easily propagated by cuttings in summer.*

☼ ◊ ♀H3 ↕to 75cm (30in) ↔60cm (24in)

ERYSIMUM CHEIRI 'HARPUR CREWE'

A short-lived, upright and bushy, evergreen perennial that brings long-lasting spring colour to the front of a warm, sunny border or raised bed. Spikes of sweetly scented, bright yellow flowers are borne from late winter to early summer above the narrow, grey-green leaves. May be sold as *Cheiranthus* 'Harpur Crewe'.

CULTIVATION *Grow in well-drained, poor to moderately fertile, preferably alkaline soil, in full sun. Trim lightly after flowering to prevent legginess. Usually dies after just a few years, but easily propagated by cuttings in summer.*

☼ ♢ 🏆H3 ↕30cm (12in) ↔60cm (24in)

ERYSIMUM 'WENLOCK BEAUTY'

A bushy, evergreen perennial that produces clusters of bluish-pink and salmon, bronze-shaded flowers from early to late spring. The lance-shaped leaves are softly hairy and mid-green. Good for early-season colour in a sunny rock garden or dry wall; in areas with cold winters, choose a warm, sheltered spot.

CULTIVATION *Grow in poor or fairly fertile, ideally alkaline soil that has good drainage, in full sun. Trim lightly after flowering to keep compact.*

☼ ♢ 🏆H3 ↕↔ 45cm (18in)

ERYTHRONIUM 'PAGODA'

This very vigorous, clump-forming, bulbous perennial is related to the dog's tooth violet, *E. dens-canis* and, like it, looks good planted in groups under deciduous trees and shrubs. Clusters of pale sulphur-yellow flowers droop from slender stems in spring, above the large, oval, bronze-mottled, glossy dark green leaves.

CULTIVATION *Grow in moist but well-drained soil that is rich in humus. Choose a position in partial shade.*

◑ ◊♦ ♆H4 ↕15–35cm (6–14in) ↔10cm (4in)

ESCALLONIA 'APPLE BLOSSOM'

A compact, evergreen shrub carrying dense, glossy dark foliage and, from early to mid-summer, a profusion of small, pink-flushed white flowers. Valuable in a shrub border, it can also be grown as a hedge, barrier or windbreak. Very useful in coastal areas, if prevailing winds are not harshly cold. *E.* 'Donard Seedling', also usually readily available, looks similar and is a little hardier.

CULTIVATION *Grow in any well-drained, fertile soil, in full sun. In especially cold areas, shelter from wind. Cut out old or damaged growth after flowering.*

☼ ◊ ♆H4 ↕↔2.5m (8ft)

ESCALLONIA 'IVEYI'

A vigorous, upright, evergreen shrub bearing large clusters of fragrant, pure white flowers from mid- to late summer. The rounded leaves are glossy dark green. Grow in a shrub border, or use as a hedge where winters are reliably mild; the foliage often takes on bronze tints in cold weather, but choose a sheltered position in frost-prone areas.

CULTIVATION *Grow in any fertile soil with good drainage, in full sun with shelter from cold, drying winds. Remove damaged growth in autumn, or in spring if flowering finishes late.*

☼ ◇ H3 ↕↔ 3m (10ft)

ESCALLONIA 'LANGLEYENSIS'

This graceful, semi-evergreen shrub produces abundant clusters of small, rose-pink flowers. These are borne from early to mid-summer above the oval, glossy bright green leaves. Thrives in relatively mild coastal gardens as an informal hedge or in a shrub border.

CULTIVATION *Best in well-drained, fertile soil, in a sunny site. Protect from cold, drying winds in frost-prone areas. Cut out dead or damaged growth after flowering; old plants can be renovated by hard pruning in spring.*

☼ ◇ H4 ↕2m (6ft) ↔3m (10ft)

ESCHSCHOLZIA CAESPITOSA

A tufted annual bearing a profusion of scented, bright yellow flowers in summer. The blue-green leaves are finely divided and almost thread-like. Suitable for a sunny border, rock garden or gravel patch. The flowers close up in dull weather.

CULTIVATION *Grow in well-drained, poor soil. Choose a site in full sun. For early flowers the following year, sow seed directly outdoors in autumn.*

☼ ♢ ♈H4 ↕↔ to 15cm (6in)

ESCHSCHOLZIA CALIFORNICA

The California poppy is a mat-forming annual with cup-shaped flowers borne on slender stems throughout summer. Colours are mixed, including white, red or yellow, but most are usually orange. The cultivar 'Dali' flowers in scarlet only. The leaves are finely cut and greyish-green. Grow in a sunny border or rock garden; the flowers last well when cut.

CULTIVATION *Best in light, poor soil with good drainage, in full sun. For early flowers the following year, sow seed directly outdoors in autumn.*

☼ ♢ ♈H4 ↕to 3cm (12in) ↔to 15cm (6in)

EUCALYPTUS GUNNII

The cider gum is a vigorous, evergreen tree useful as a fast-growing feature in a new garden. The new yellow- to greyish-green bark is revealed in late summer as the old, whitish-green layer is shed. Young plants have rounded, grey-blue leaves; on adult growth they are lance-shaped. Protect in cold areas with a thick, dry, winter mulch, especially when young.

CULTIVATION *Grow in well-drained, fertile soil, in sun. To keep compact, and for the best display of young foliage, cut back hard each spring.*

☼ ◊ 🏆H3 ↕10–25m (30–80ft) ↔6–15m (20–50ft)

EUCALYPTUS PAUCIFLORA SUBSP. *NIPHOPHILA*

The snow gum is a handsome, silvery, evergreen tree with open, spreading branches and attractively peeling, white and grey bark. Young leaves are oval and dull blue-green; on mature stems they are lance-shaped and deep blue-green. The flowers are less significant. Popularly grown as a shrub, pruned hard back at regular intervals.

CULTIVATION *Grow in well-drained, fertile soil, in full sun. For the best display of young foliage, cut back hard each year in spring.*

☼ ◊ 🏆H4 ↕↔ to 6m (20ft)

EUCRYPHIA × *NYMANSENSIS* 'NYMANSAY'

A column-shaped, evergreen tree that bears clusters of large, fragrant, glistening white flowers with yellow stamens. These are borne in late summer to early autumn amid the oval, glossy dark green leaves. Makes a magnificent flowering specimen tree; best in mild and damp, frost-free climates.

CULTIVATION *Grow in well-drained, reliably moist soil, preferably in full sun with shade at the roots, but will tolerate semi-shade. Shelter from cold winds. Remove damaged growth in spring.*

☼ ◊◊ ♀H3 ↕15m (50ft) ↔5m (15ft)

EUONYMUS ALATUS

The winged spindle is a dense, deciduous shrub with winged stems, much-valued for its spectacular autumn display; small, purple and red fruits split to reveal orange seeds as the oval, deep green foliage turns to scarlet. The flowers are much less significant. Excellent in a shrub border or light woodland. The fruits are poisonous. 'Compactus' is a dwarf version of this shrub, only half its height.

CULTIVATION *Grow in any well-drained, fertile soil. Tolerates light shade, but fruiting and autumn colour are best in full sun. Keep pruning to a minimum.*

☼☀ ◊ ♀H4 ↕2m (6ft) ↔3m (10ft)

EUONYMUS EUROPAEUS 'RED CASCADE'

A tree-like, deciduous shrub that produces colourful autumn foliage. Inconspicuous flowers in early summer are followed by rosy-red fruits that split to reveal orange seeds. The oval, mid-green leaves turn scarlet-red at the end of the growing season. The fruits are toxic.

CULTIVATION *Grow in any fertile soil with good drainage, but thrives on chalky soil. Tolerates light shade, but fruiting and autumn colour are best in full sun. Two or more specimens are required to guarantee a good crop of fruits. Very little pruning is necessary.*

☼ ◐ ◊ 🏆H4 ↕3m (10ft) ↔2.5m (8ft)

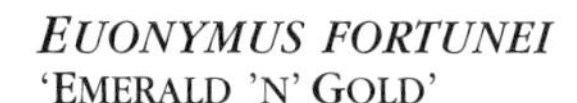

EUONYMUS FORTUNEI 'EMERALD 'N' GOLD'

A small and scrambling, evergreen shrub that will climb if supported. The bright green, oval leaves have broad, bright golden-yellow margins, and take on a pink tinge in cold weather. The spring flowers are insignificant. Use to fill gaps in a shrub border, or wall-train.

CULTIVATION *Grow in any but water-logged soil. The leaves colour best in full sun, but tolerates light shade. Trim over in mid-spring. Trained up a wall, it may reach a height of up to 5m (15ft).*

☼ ◐ ◊◆ 🏆H4 ↕60cm (24in) or more ↔90cm (36in)

EUONYMUS FORTUNEI 'SILVER QUEEN'

A compact, upright or scrambling, evergreen shrub that looks most effective when grown as a climber against a wall or up into a tree. The dark green leaves have broad white edges that become pink-tinged in prolonged frost. The greenish-white flowers in spring are insignificant.

CULTIVATION *Grow in any but water-logged soil, in sun or light shade. Leaf colour is best in full sun. Trim shrubs in mid-spring. Allowed to climb, it can grow up to 6m (20ft) tall.*

☼ ◑ ◊ ♦ ♀H4 ↕2.5m (8ft) ↔1.5m (5ft)

EUPHORBIA AMYGDALOIDES VAR. *ROBBIAE*

Mrs. Robb's bonnet, also known simply as *E. robbiae*, is a spreading, evergreen perennial bearing open heads of yellowish-green flowers in spring. The long, dark green leaves are arranged in rosettes at the base of the stems. Particularly useful in shady areas. Can be invasive.

CULTIVATION *Grow in well-drained but moist soil, in full sun or partial shade. Tolerates poor, dry soil. Dig up invasive roots to contain spread. The milky sap can irritate skin.*

☼ ◑ ◊ ♦ ♀H4 ↕75–80cm (30–32in) ↔30cm (12in)

EUPHORBIA CHARACIAS

A shrubby, evergreen perennial that forms clumps of narrow, dark blue-green leaves. Large, rounded heads of dark-eyed, pale yellowish-green flowers are carried at the tips of the upright stems in spring and early summer. A dramatic structural plant, bringing long-lasting colour to a spacious Mediterranean-style garden. May be damaged by heavy frost.

CULTIVATION *Grow in well-drained, light soil, in full sun with shelter from cold winds. Cut the flowered stems back to the base in autumn; wear gloves, since the milky sap can irritate skin.*

☼ ◊ ♀H3 ↕↔ 1.2m (4ft)

EUPHORBIA CHARACIAS SUBSP. *WULFENII* 'JOHN TOMLINSON'

This billowing, shrubby perennial bears larger flowerheads than the species (above). The flowers are bright yellow-green, without dark eyes, and are borne in rounded heads above the narrow, grey-green leaves. In frost-prone climates, best grown at the base of a warm, sunny wall.

CULTIVATION *Grow in light soil that has good drainage, in full sun. Shelter from cold winds. Cut the flowered stems back to the base in autumn; wear gloves, since the milky sap can irritate skin.*

 ↕↔ 1.2m (4ft)

EUPHORBIA X *MARTINII*

This upright, clump-forming, evergreen subshrub bears spikes of yellow-green flowers with very distinctive, dark red nectar glands. These are carried on red-tinged shoots from spring to mid-summer, above lance-shaped, mid-green leaves which are often tinged purple when young. A choice plant with an architectural look for a hot, dry site. May suffer in cold winters.

CULTIVATION *Grow in well-drained soil, in a sheltered, sunny site. Deadhead after flowering, wearing gloves to protect hands from the milky sap which may irritate skin.*

☼ ◊ ♈H3 ↕↔ 1m (3ft)

EUPHORBIA MYRSINITES

A small, evergreen perennial bearing sprawling stems which are densely clothed with a spiral arrangement of fleshy, elliptic, blue-green leaves. Clusters of yellow-green flowers brighten the tips of the stems in spring. Excellent in a dry, sunny rock garden, or trailing over the edge of a raised bed.

CULTIVATION *Grow in well-drained, light soil, in full sun. Deadhead after flowering as it self-seeds freely. Wear gloves to avoid contact with the milky sap, which is a potential skin irritant.*

☼ ◊ ♈H3–4 ↕10cm (4in) ↔to 30cm (12in)

EUPHORBIA POLYCHROMA

This evergreen perennial forms a neat, rounded clump of softly hairy, mid-green foliage. It is covered with clusters of brilliant greenish-yellow flowers over long periods in spring. Excellent with spring bulbs, in a border or light woodland. Tolerates a wide range of soil types. Its cultivar 'Major' grows a little taller.

CULTIVATION *Grow in either well-drained, light soil, in full sun, or moist, humus-rich soil, in light dappled shade. Deadhead after flowering. The milky sap can irritate skin.*

☼ ◑ ◊ ♦ 🏆H4 ↕40cm (16in) ↔60cm (24in)

EUPHORBIA SCHILLINGII

This vigorous, clump-forming perennial which, unlike many euphorbias, dies back in winter, bears clusters of long-lasting, yellowish-green flowers from mid-summer to mid-autumn. The lance-shaped leaves are dark green and have pale green or white central veins. Ideal for lighting up a woodland planting or shady wild garden.

CULTIVATION *Grow in reliably moist, humus-rich soil, in light dappled shade. Deadhead after flowering. The milky sap can irritate skin.*

◑ ♦ 🏆H4 ↕1m (3ft) ↔30cm (12in)

EXOCHORDA X *MACRANTHA* 'THE BRIDE'

A dense, spreading, deciduous shrub grown for its gently arching habit and abundant clusters of fragrant, pure white flowers. These are borne in late spring and early summer amid the oval, fresh green leaves. An elegant foil to other plants in a mixed border.

CULTIVATION *Grow in any well-drained soil, in full sun or light dappled shade. Does not like shallow, chalky soil. Very little pruning is necessary.*

☼☀ ◊ ♀H4 ↕2m (6ft) ↔3m (10ft)

FALLOPIA BALDSCHUANICA

Russian vine, sometimes still included in *Polygonum*, is an extremely vigorous, woody-stemmed, deciduous climber that bears hanging clusters of tiny, pink-tinted white flowers during summer and autumn. The leaves are oval and mid-green. Use to cover unsightly buildings or in a wild garden where there is plenty of space. In other situations, it can be difficult to control.

CULTIVATION *Grow in any well-drained, poor to moderately fertile soil, in sun or partial shade. Flowering is best in sun. Cut back as necessary in early spring.*

☼☀ ◊ ♀H4 ↕12m (40ft)

FARGESIA NITIDA

Fountain bamboo is a slow-growing perennial which forms a dense clump of upright, dark purple-green canes. In the second year after planting, cascades of narrow, dark green leaves are produced from the top of the clump. Handsome in a wild garden; to restrict spread, grow in a large container. May be sold as *Sinarundinaria nitida. F. murielae* is equally attractive, similar but with yellow stems and brighter leaves.

CULTIVATION *Grow in reliably moist, fertile soil, in light dappled shade. Shelter from cold, drying winds.*

◐ 💧 🏆H4 ↕to 5m (15ft) ↔1.5m (15ft) or more

X *FATSHEDERA LIZEI*

The tree ivy is a mound-forming, evergreen shrub cultivated for its large, handsome, glossy dark green, ivy-like leaves. Small white flowers appear in autumn. Excellent in shady areas; in cold climates, grow in a sunny, sheltered site. Given support, it can be trained against a wall; it also makes a fine house- or conservatory plant. For leaves with cream edges, look for 'Variegata'.

CULTIVATION *Best in moist but well-drained, fertile soil, in sun or partial shade. No regular pruning is required. Tie in to grow against a support.*

☼ ◐ 💧💧 🏆H3 ↕1.2–2m (4–6ft) or more ↔3m (10ft)

FATSIA JAPONICA

The Japanese aralia is a spreading, evergreen shrub grown for its large, palm-shaped, glossy green leaves. Broad, upright clusters of rounded, creamy-white flowerheads appear in autumn. An excellent architectural plant for a shady border. Tolerates atmospheric pollution and thrives in sheltered city gardens. 'Variegata' has cream-edged leaves.

CULTIVATION *Grow in any well-drained soil, in sun or shade. Provide shelter from cold winds and hard frosts, especially when young. Little pruning is necessary, except to cut out wayward shoots and damaged growth in spring.*

☼ ● ◊ ♀H4 ↕↔ 1.5–4m (5–12ft)

FELICIA AMELLOIDES 'SANTA ANITA'

This blue daisy, with white-marked, bright green foliage, is a rounded, evergreen subshrub. Large, daisy-like flowers, with blue petals and bright yellow centres, open from late spring to autumn. Good in a sunny rock garden, or in hanging baskets or other containers. Best treated as an annual in climates with frosty winters. There is a version with variegated leaves, also recommended.

CULTIVATION *Grow in well-drained, fairly fertile soil, in sun. Dislikes damp. Pinch-prune to encourage bushiness, and deadhead to prolong flowering.*

☼ ◊ ♀H2–3 ↕↔ 30–60cm (12–24in)

FESTUCA GLAUCA 'BLAUFUCHS'

This bright blue fescue is a densely tufted, evergreen, perennial grass. It is excellent in a border or rock garden, as a foil to other plants. Spikes of not particularly striking, violet-flushed, blue-green flowers are borne in early summer, above the foliage. The narrow, bright blue leaves are its chief attraction.

CULTIVATION *Grow in dry, well-drained, poor to moderately fertile soil, in sun. For the best foliage colour, divide and replant clumps every 2 or 3 years.*

☼ ◊ ♀H4 ↕to 30cm (12in) ↔25cm (10in)

FILIPENDULA PURPUREA

An upright, clump-forming perennial that looks well planted in groups to form drifts of elegant, dark green foliage. Feathery clusters of red-purple flowers are carried above the leaves on purple-tinged stems in summer. Suitable for a waterside planting or bog garden; can also be naturalized in damp woodland.

CULTIVATION *Grow in reliably moist, moderately fertile, humus-rich soil, in partial shade. Can be planted in full sun where the soil does not dry out.*

☼☀ ♦♦ ♀H4 ↕1.2m (4ft) ↔60cm (24in)

FILIPENDULA RUBRA 'VENUSTA'

A vigorous, upright perennial that produces feathery plumes of tiny, soft pink flowers on tall branching stems in mid-summer. The large, dark green leaves are jaggedly cut into several lobes. Excellent in a bog garden, in moist soil by the side of water or in a damp wild garden.

CULTIVATION *Grow in reliably moist, moderately fertile soil, in partial shade. Thrives in wet or boggy soil, where it will tolerate full sun.*

◑ 💧💧 ♈H4 ↕2–2.5m (6–8ft) ↔1.2m (4ft)

FORSYTHIA x *INTERMEDIA* 'LYNWOOD'

A vigorous, deciduous shrub with upright stems that arch slightly at the tips. Its golden-yellow flowers appear in profusion on bare branches in early spring. Mid-green, oval leaves emerge after flowering. A reliable shrub for a mixed border or as an informal hedge: particularly effective with spring bulbs. The flowering stems are good for cutting.

CULTIVATION *Grow in well-drained, fertile soil, in full sun. Tolerates partial shade, although flowering will be less profuse. On established plants, cut out some old stems after flowering.*

☼◑ 💧 ♈H4 ↕↔3m (10ft)

FORSYTHIA SUSPENSA

This upright, deciduous shrub, sometimes called golden bell, is less bushy than *F.* x *intermedia* (see facing page, below), with more strongly arched stems. It is valued for the nodding, bright yellow flowers which open on its bare branches from early to mid-spring. The leaves are oval and mid-green. Good planted on its own, as an informal hedge or in a shrub border.

CULTIVATION *Best in well-drained, fertile soil, ideally in full sun; flowering is less spectacular in shade. Cut out some of the older stems on established plants after flowering.*

☼ ◑ ◊ 🏆H4 ↕↔ 3m (10ft)

FOTHERGILLA MAJOR

This slow-growing, upright shrub bears spikes of bottlebrush-like, fragrant white flowers in late spring. It is also valued for its blaze of autumn foliage. The deciduous leaves are glossy dark green in summer, then turn orange, yellow and red before they fall. An attractive addition to a shrub border or in light woodland.

CULTIVATION *Best in moist but well-drained, acid soil that is rich in humus. Choose a position in full sun for the best flowers and autumn colour. Requires very little pruning.*

☼ ◊◑ 🏆H4 ↕2.5m (8ft) ↔2m (6ft)

FREMONTODENDRON 'CALIFORNIA GLORY'

A vigorous, upright, semi-evergreen shrub producing large, cup-shaped, golden-yellow flowers from late spring to mid-autumn. The rounded leaves are lobed and dark green. Excellent for wall-training; in frost-prone areas, grow against walls that receive plenty of sun.

CULTIVATION *Best in well-drained, poor to moderately fertile, neutral to alkaline soil, in sun. Shelter from cold winds. Best with no pruning, but wall-trained plants can be trimmed in spring.*

☼ ◊ ♀H3 ↕6m (20ft) ↔4m (12ft)

FRITILLARIA ACMOPETALA

This bulbous perennial is grown for its drooping, bell-shaped, pale green flowers which are stained red-brown on the insides. They appear singly or in clusters of two or three in late spring, at the same time as the narrow, blue-green leaves. Suitable for a rock garden or sunny border.

CULTIVATION *Grow in fertile soil that has good drainage, in full sun. Divide and replant bulbs in late summer.*

☼ ◊ ♀H4 ↕to 40cm (16in) ↔5–8cm (2–3in)

FRITILLARIA MELEAGRIS

The snake's head fritillary is a bulbous perennial, which naturalizes well in grass where summers are cool and damp. The drooping, bell-shaped flowers are pink, pinkish-purple or white, and are strongly chequered. They are carried singly or in pairs during spring. The narrow leaves are grey-green. There is a white-flowered form, f. *alba*, which is also recommended.

CULTIVATION *Grow in any moist but well-drained soil that is rich in humus, in full sun or light shade. Divide and replant bulbs in late summer.*

☼◑ ◊♦ 🏆H4 — ↕to 30cm (12in) ↔5–8cm (2–3in)

FRITILLARIA PALLIDIFLORA

In late spring, this robust, bulbous perennial bears bell-shaped, foul-smelling flowers above the grey-green, lance-shaped leaves. They are creamy-yellow with green bases, chequered brown-red inside. Suits a rock garden or border in areas with cool, damp summers. Naturalizes easily in damp meadows.

CULTIVATION *Grow in moist but well-drained, moderately fertile soil, in sun or partial shade. Divide and replant bulbs in late summer.*

☼◑ ◊♦ 🏆H4 — ↕to 40cm (16in) ↔5–8cm (2–3in)

HARDY FUCHSIAS

The hardy fuchsias are wonderfully versatile shrubs: as well as being useful in mixed borders and as flowering hedges, they can be trained as espaliers or fans against warm walls, or grown as free-standing standards or pillars. Hanging flowers, varying in form from single to double, appear throughout summer and into autumn. During cold winters, even hardy fuchsias may lose some of their upper growth after severe frosts. They are fast to recover, however, and most will retain their leaves if over-wintered in a cool greenhouse, or temperatures stay above 4°C (39°F). The hardiest cultivar is 'Genii' (H4).

CULTIVATION *Grow in well-drained but moist, fertile soil. Choose a position in sun or semi-shade with shelter from cold winds. Provide a deep winter mulch. Pinch-prune young plants to encourage a bushy habit. In early spring, remove frost damaged stems; cut back healthy growth to the lowest buds.*

☼ ◐ 💧 🏆 H3, H4

1 ↕↔ 75–90cm (30–36in)

2 ↕ 15–30cm (6–12in) ↔ 45cm (18in)

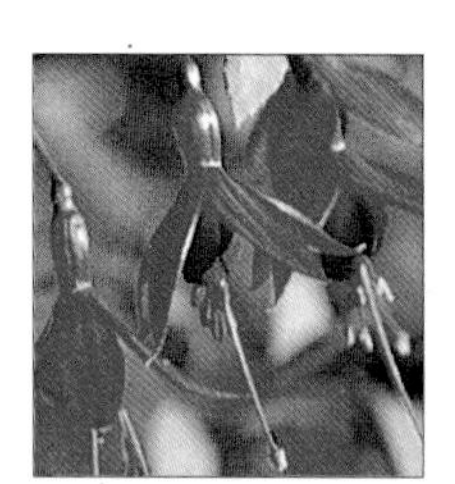

3 ↕↔ 1–1.1m (3–3½ft)

4 ↕ 2–3m (6–10ft) ↔ 1–2m (3–6ft)

5 ↕↔ 15–30cm (6–12in)

6 ↕ to 3m (10ft) ↔ 2–3m (6–10ft)

1 *F.* 'Genii' (H4) **2** *F.* 'Lady Thumb' (H3) **3** *F.* 'Mrs. Popple' (H3) **4** *F.* 'Riccartonii' (H3) **5** *F.* 'Tom Thumb' (H3) **6** *F. magellanica* 'Versicolor' (H3)

HALF-HARDY AND TENDER FUCHSIAS

The half-hardy and tender fuchsias are flowering shrubs that require at least some winter protection in cold climates. The compensation for this extra care is an increased range of beautiful summer flowers for the garden, greenhouse and conservatory. All can be grown out of doors in the summer months and make superb patio plants when grown in containers, whether pinch-pruned into dense bushes or trained as columns or standards. In frost-prone areas, shelter half-hardy fuchsias in a greenhouse or cool conservatory over the winter months; the tender species fuchsias, including 'Thalia', need a minimum temperature of 10°C (50°C) at all times.

CULTIVATION *Grow in moist but well-drained, fertile soil or compost, in sun or partial shade. Pinch-prune young plants to promote bushiness, and trim after flowering to remove spent blooms. Prune back to an established framework in early spring.*

☼ ◐ 💧💧 ♀H1–3

1 ↕↔ 30–60cm (12–24in)

2 ↕ to 60cm (24in) ↔ to 45cm (18in)

3 ↕ to 4m (12ft) ↔ 1–1.2m (3–4ft)

1 *Fuchsia* 'Annabel' (H3) **2** *F.* 'Billy Green' (H1+3) **3** *F. boliviana* var. *alba* (H1+3)

4 ↕↔ 45–75cm (18–30in)

5 ↕ to 90cm (36in) ↔ to 75cm (30in)

6 ↕ to 90cm (36in) ↔ to 60cm (24in)

7 ↕ 1.5m (5ft) ↔ to 80cm (32in)

8 ↕ to 45cm (18in) ↔ to 60cm (24in)

9 ↕ to 75cm (18in) ↔ to 60cm (to 24in)

4 *Fuchsia* 'Celia Smedley' (H3) **5** *F.* 'Checkerboard' (H3) **6** *F.* 'Coralle' (H2)
7 *F. fulgens* (H1+3) **8** *F.* 'Joy Patmore' (H3) **9** *F.* 'Leonora' (H3–4)

10 ↕↔ 30–60cm (12–24in)

11 ↕to 45cm (18in) ↔ 45cm (18in)

12 ↕to 75cm (30in) ↔ to 60cm (24in)

14 ↕ to 60cm (24in) ↔ 75cm (30in)

15 ↕ ↔ 45–90cm (18–36in)

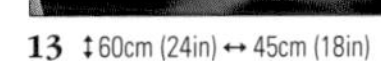

13 ↕60cm (24in) ↔ 45cm (18in)

MORE CHOICES

'Brookwood Belle' (H3) Cerise and white flowers.

'Pacquesa' (H3) Red and red-veined white.

'Winston Churchill' (H3) Lavender and pink.

10 *F.* 'Mary' (H2) **11** *F.* 'Nellie Nuttall' (H3) **12** *F.* 'Royal Velvet' (H3)
13 *F.* 'Snowcap' (H3–4) **14** *F.* 'Swingtime' (H3) **15** *F.* 'Thalia' (H1+3)

TRAILING FUCHSIAS

Fuchsia cultivars with a trailing or spreading habit are the perfect plants to have trailing over the edge of a tall container, window box or hanging basket, where their pendulous flowers will be shown to great effect. They bloom continuously throughout summer and into early autumn, and can be left in place undisturbed right up until the end of the season. After this, they are best discarded and new plants bought the following year. An alternative, longer-lived planting is to train trailing fuchsias into attractive weeping standards for container plantings, but they must be kept frost-free in winter.

CULTIVATION *Grow in fertile, moist but well- drained soil or compost, in full sun or partial shade. Shelter from cold, drying winds. Little pruning is necessary, except to remove wayward growth. Regularly pinch out the tips of young plants to encourage bushiness and a well-balanced shape.*

☼ ◐ ◊ ♦ ♈H3

1 ↕15–30cm (6–12in) ↔ 45cm (18in)

2 ↕15–30cm (6–12in) ↔ 45cm (18in)

3 ↕45cm (18in) ↔ 60cm (24in)

4 ↕to 60cm (24in) ↔ 75cm (30in)

1 *F.* 'La Campanella' **2** *F.* 'Golden Marinka' **3** *F.* 'Jack Shahan' **4** *F.* 'Lena'

GAILLARDIA 'DAZZLER'

A bushy, short-lived perennial that bears large, daisy-like flowers over a long period in summer. These have yellow-tipped, bright orange petals surrounding an orange-red centre. The leaves are soft, lance-shaped and mid-green. Effective in a sunny, mixed or herbaceous border; the flowers are good for cutting. May not survive winter in cold climates.

CULTIVATION *Grow in well-drained, not too fertile soil, in full sun. May need staking. Often short-lived, but can be reinvigorated by division in winter.*

☼ ◊ H3 ↕60–85cm (24–34in) ↔ 45cm (18in)

GALANTHUS ELWESII

This robust snowdrop is a bulbous perennial that produces slender, honey-scented, pure white flowers in late winter, above the bluish-green foliage. The inner petals have green markings. Good for borders and rock gardens; naturalizes easily in light woodland.

CULTIVATION *Grow in moist but well-drained, humus-rich soil that does not dry out in summer. Choose a position in partial shade.*

☀ ◊♦ H4 ↕10–15cm (4–6in) ↔8cm (3in)

GALANTHUS 'MAGNET'

This tall snowdrop is a vigorous bulbous perennial bearing drooping, pear-shaped, pure white flowers during late winter and early spring; the inner petals have a deep green, V-shaped mark at the tips. The strap-shaped, grey-green leaves are arranged around the base of the plant. Good for naturalizing in grass or in a woodland garden.

CULTIVATION *Grow in moist but well-drained, fertile soil that does not dry out in summer. Choose a position in partial shade.*

☀ 💧 🏆H4 ↕20cm (8in) ↔8cm (3in)

GALANTHUS NIVALIS 'FLORE PLENO'

This double-flowered form of the common snowdrop, *G. nivalis*, is a robust, bulbous perennial. Drooping, pear-shaped, pure white flowers appear from late winter to early spring, with green markings on the tips of the inner petals. The narrow leaves are grey-green. Good for naturalizing under deciduous trees or shrubs.

CULTIVATION *Grow in reliably moist but well-drained, fertile soil, in light shade. Divide and replant every few years after flowering to maintain vigour.*

☀ 💧 🏆H4 ↕↔ 10cm (4in)

GALANTHUS 'S. ARNOTT'

This honey-scented snowdrop, which has even larger flowers than 'Magnet' (facing page, above), is a fast-growing, bulbous perennial. Nodding, pear-shaped, pure white flowers appear in late winter to early spring, and have a green, V-shaped mark at the tip of each inner petal. The narrow leaves are grey-green. Suits a rock garden or raised bed. 'Atkinsii' is another recommended large snowdrop, similar to this one.

CULTIVATION *Grow in moist but well-drained, fertile soil, in dappled shade. Keep reliably moist in summer. Divide and replant clumps after flowering.*

H4 ↕20cm (8in) ↔8cm (3in)

GARRYA ELLIPTICA 'JAMES ROOF'

This silk-tassel bush is an upright, evergreen shrub, becoming tree-like with age. It is grown for its long, silver-grey catkins, which dangle from the branches in winter and early spring. The leaves are dark sea-green and have wavy margins. Excellent in a shrub border, against a shady wall or as hedging; tolerates coastal conditions.

CULTIVATION *Grow in well-drained, moderately fertile soil, in full sun or partial shade. Tolerates poor, dry soil. Trim after flowering, as necessary.*

 H4 ↕↔ 4m (12ft)

GAULTHERIA MUCRONATA 'MULBERRY WINE' (FEMALE)

This evergreen, spreading shrub, sometimes included in *Pernettya*, is much-valued for its autumn display of large, rounded, magenta to purple berries which show off well against the toothed, glossy dark green leaves. Small white flowers are borne throughout summer.

CULTIVATION *Grow in reliably moist, peaty, acid to neutral soil, in partial shade or full sun. Plant close to male varieties to ensure a reliable crop of berries. Chop away spreading roots with a spade to restrict the overall size.*

☼ ◑ 💧 ♆H4 ↕↔ 1.2m (4ft)

GAULTHERIA MUCRONATA 'WINTERTIME' (FEMALE)

An evergreen, spreading shrub, sometimes included in the genus *Pernettya*, bearing large, showy white berries which persist well into winter. Small white flowers are borne from late spring into early summer. The glossy dark green leaves are elliptic to oblong and toothed.

CULTIVATION *Grow in reliably moist, acid to neutral, peaty soil. Best in light shade, but tolerates sun. Plant close to male varieties to ensure a good crop of berries. Dig out spreading roots as necessary to restrict the overall size.*

☼ ◑ 💧 ♆H4 ↕↔ 1.2m (4ft)

GAULTHERIA PROCUMBENS

Checkerberry is a creeping shrub which bears drooping, urn-shaped, white or pink flowers in summer, followed by aromatic scarlet fruits. As these usually persist until spring, they give winter colour. The glossy, dark green leaves have a strong fragrance when crushed, giving the plant its other common name, wintergreen. Good ground cover in shade.

CULTIVATION *Grow in acid to neutral, peaty, moist soil in partial shade; full sun is tolerated only where the soil is always moist. Trim after flowering.*

◐ ♦ ♀H4 ↕15cm (6in) ↔to 1m (3ft) or more

GAURA LINDHEIMERI

A tall, clump-forming perennial with basal leaves and slender stems. From late spring to autumn, it bears loose spires of pinkish-white buds, which open in the morning to display white flowers. A graceful plant for a mixed flower border, tolerating both heat and drought.

CULTIVATION *Grow in fertile, moist but well-drained soil, ideally in full sun, but some part-day shade is tolerated. If necessary, divide clumps in spring.*

☼ ◊♦ ♀H3 ↕to 1.5m (5ft) ↔90cm (36in)

GAZANIAS

These useful summer bedding plants are vigorous, spreading perennials, usually grown as annuals. They are cultivated for their long display of large and very colourful, daisy-like flowers, being quite similar to sunflowers, although they close in dull or cool weather. Flowers may be orange, white, golden yellow, beige, bronze, or bright pink, often with striking contrasting central zones. Hybrid selections with variously coloured flowerheads are also popular. The lance-shaped leaves are dark green with white-silky undersides. Gazanias grow well in containers and tolerate coastal conditions.

CULTIVATION *Grow in light, sandy, well-drained soil in full sun. Remove old or faded flowerheads to prolong flowering. Water freely during the growing season. The plants will die back on arrival of the first frosts.*

☼ ◊ 🏆H2

1 ↕20cm (8in) ↔ 25cm (10in)

2 ↕20cm (8in) ↔ 25cm (10in)

MORE CHOICES

G. Chansonette Series Mixed colours, zoned in a contrasting shade.

G. Mini-star Series Mixed or single zoned colours.

G. Talent Series Mixed or single colours.

3 ↕20cm (8in) ↔ 25cm (10in)

4 ↕20cm (8in) ↔ 25cm (10in)

5 ↕20cm (8in) ↔ 25cm (10in)

1 *Gazania* 'Aztec' **2** *G.* 'Cookei' **3** *G.* 'Daybreak Garden Sun' **4** *G.* 'Michael' **5** *G. rigens* var. *uniflora*

GENISTA AETNENSIS

The Mount Etna broom is an upright, almost leafless, deciduous shrub which is excellent on its own or at the back of a border in hot, dry situations. Masses of fragrant, pea-like, golden-yellow flowers cover the weeping, mid-green stems during mid-summer.

CULTIVATION *Grow in well-drained, light, poor to moderately fertile soil, in full sun. Keep pruning to a minimum; old, straggly plants are best replaced.*

☼ ◊ 🏆H4 ↕↔ 8m (25ft)

GENISTA LYDIA

This low, deciduous shrub forms a mound of arching, prickle-tipped branches which are covered with yellow, pea-like flowers in early summer. The small, narrow leaves are blue-green. Ideal for hot, dry sites such as a rock garden or raised bed.

CULTIVATION *Grow in well-drained, light, poor to moderately fertile soil, in full sun. Keep pruning to a minimum; old, straggly plants are best replaced.*

☼ ◊ 🏆H4 ↕to 60cm (24in) ↔to 1m (3ft)

GENISTA TENERA 'GOLDEN SHOWER'

This graceful cultivar of broom is a deciduous shrub with slender shoots and narrow grey-green leaves. It is valued for its mass of fragrant, bright yellow flowers, which appear in profusion during the first half of summer. It is usually grown in a shrub or mixed border. It may not survive a very cold winter.

CULTIVATION *Grow in light, poor to moderately fertile, well-drained soil, in full sun. Trim in late winter or early spring, but do not cut back to old wood.*

☼ ♢ ♀H4 — ↕3m (10ft) ↔5m (15ft)

GENTIANA ACAULIS

The trumpet gentian is a mat-forming, evergreen perennial which bears large, trumpet-shaped, vivid deep blue flowers in spring. The leaves are oval and glossy dark green. Good in a rock garden, raised bed or trough; thrives in areas with cool, damp summers.

CULTIVATION *Grow in reliably moist but well-drained, humus-rich soil, in full sun or partial shade. Protect from hot sun in areas with warm, dry summers.*

☼ ◑ ♢♦ ♀H4 — ↕8cm (3in) ↔to 30cm (12in)

GENTIANA ASCLEPIADEA

The willow gentian is an arching, clump-forming perennial producing trumpet-shaped, dark blue flowers in late summer and autumn; these are often spotted or striped with purple on the insides. The leaves are lance-shaped and fresh green. Suitable for a border or large rock garden.

CULTIVATION *Grow in moist, fertile, humus-rich soil. Best in light shade, but tolerates sun if the soil is reliably moist.*

◐ 💧 🏆H4 ↕60–90cm (24–36in) ↔45cm (18in)

GENTIANA SEPTEMFIDA

This late-summer-flowering gentian is a spreading to upright, clump-forming herbaceous perennial. Clusters of narrowly bell-shaped, bright blue flowers with white throats are borne amid the oval, mid-green leaves. Good for a rock garden or raised bed; thrives in cool, damp summers. The variety var. *lagodechiana*, also recommended, has just one flower per stem.

CULTIVATION *Grow in moist but well-drained, humus-rich soil, in full sun or partial shade. Protect from hot sun in areas with warm, dry summers.*

☼ ◐ 💧💧 🏆H4 ↕to 15–20cm (6–8in) ↔to 30cm (12in)

SMALL HARDY GERANIUMS

The low-growing members of the genus *Geranium* (often incorrectly referred to as pelargoniums, see pages 319–321) are versatile plants, useful not only at the front of borders but also in rock gardens or as ground cover. These fully hardy, evergreen perennials are long-lived and undemanding, tolerating a wide range of sites and soil types. The lobed and toothed leaves are often variegated or aromatic. In summer, they bear typically saucer-shaped flowers ranging in colour from white through soft blues such as 'Johnson's Blue' to the intense pink of *G. cinereum* var. *caulescens*, often with contrasting veins, eyes, or other markings.

CULTIVATION *Grow in sharply drained, humus-rich soil, in full sun. Feed with a balanced fertilizer every month during the growing season, but water sparingly in winter. Remove withered flower stems and old leaves to encourage fresh growth.*

☼ ◊ ♈H4

1 ↕to 45cm (18in) ↔ 1m (3ft) or more

2 ↕to 15cm (6in) ↔ to 30cm (12in)

1 *G.* 'Ann Folkard' **2** *G. cinereum* 'Ballerina'

3 ↕ to 15cm (6in) ↔ to 30cm (12in)

4 ↕ to 45cm (18in) ↔ indefinite

5 ↕ to 15cm (6in) ↔ 50cm (20in)

6 ↕ 30cm (12in) ↔ 60cm (24in)

7 ↕ to 45cm (18in) ↔ 75cm (30in)

8 ↕ to 30cm (12in) ↔ to 1m (3ft)

9 ↕ 30cm (12in) ↔ 1.2m (4ft)

3 *G. cinereum* var. *subcaulescens* **4** *G. clarkei* 'Kashmir White' **5** *G. dalmaticum*
6 *G. himalayense* 'Gravetye' **7** *G.* 'Johnson's Blue' **8** *G.* x *riversleaianum* 'Russell Prichard'
9 *G. wallichianum* 'Buxton's Variety'

LARGE HARDY GERANIUMS

The taller, clump-forming types of hardy perennial geranium – not to be confused with pelargoniums (see pages 319–321) – make effective, long-lived border plants or infill among shrubs, requiring a minimum of attention. They are especially suited to cottage garden plantings and between roses. Their lobed, evergreen leaves may be coloured or aromatic, and give a long season of interest. Throughout summer, this is heightened by an abundance of saucer-shaped flowers, in white and shades of blue, pink, and purple. Markings on flowers vary, from the dramatic, contrasting dark eyes of *G. psilostemon* to the delicate venation of *G. sanguineum* var. *striatum*.

CULTIVATION *Best in well-drained, fairly fertile soil, in full sun or partial shade, but tolerant of any soil that is not waterlogged. Remove old leaves and withered flower stems to encourage new growth.*

☼☀ ◊◊ 🏆H4

1 ↕45cm (18in) ↔60cm (24in)

2 ↕↔60cm (24in)

3 ↕50cm (20in) ↔60cm (24in)

1 *G. endressii* **2** *G.* × *magnificum* **3** *G. macrorrhizum* 'Ingwersen's Variety'

4 ↕60–90cm (24–36in) ↔ 60cm (24in)

5 ↕60–120cm (24–48in) ↔ 60cm (24in)

6 ↕↔ 30cm (12in)

7 ↕10cm (4in) ↔ 30cm (12in)

8 ↕60cm (24in) ↔ 90cm (36in)

9 ↕60cm (24in) ↔ 90cm (36in)

4 *G. pratense* 'Mrs. Kendall Clark' **5** *G. psilostemon* (syn. *G. armenum*) **6** *G. renardii*
7 *G. sanguineum* var. *striatum* **8** *G. sylvaticum* 'Mayflower' **9** *G.* x *oxonianum* 'Wargrave Pink'

GEUM 'LADY STRATHEDEN'

A clump-forming perennial bearing double, bright yellow flowers on arching stems over a long period in summer. The mid-green leaves are large and lobed. An easy, long-flowering plant for brightening up a mixed or herbaceous border. The taller 'Fire Opal' is closely related, with reddish-orange flowers on purple stems.

CULTIVATION *Grow in moist but well-drained, fertile soil, in full sun. Avoid sites that become waterlogged in winter.*

☼ ◊◆ ♈H4 ↕40–60cm (16–24in) ↔60cm (24in)

GEUM MONTANUM

A small, clump-forming perennial grown for its solitary, cup-shaped, deep golden-yellow flowers. These are produced in spring and early summer above large, lobed, dark green leaves. Excellent for a rock garden, raised bed or trough.

CULTIVATION *Grow in well-drained, preferably gritty, fertile soil, in full sun. Will not tolerate waterlogging in winter.*

☼ ◊ ♈H4 ↕15cm (6in) ↔to 30cm (12in)

GILLENIA TRIFOLIATA

An upright, graceful perennial that forms clumps of olive green leaves which turn red in autumn. Delicate white flowers with slender petals are borne on wiry red stems in summer. Effective in a shady border or light woodland; the cut flowers last well.

CULTIVATION *Grow in moist but well drained, fertile soil. Best in partial shade, but tolerates some sun if shaded during the hottest part of the day.*

☀ ◊◊ ♀H4 ↕to 1m (3ft) ↔60cm (24in)

GLADIOLUS CALLIANTHUS

An upright cormous perennial that bears strongly scented white flowers with purple-red throats. These hang from elegant stems during summer, above fans of narrow, mid-green leaves. Ideal for mixed borders; the flowers are suitable for cutting. Will not survive cold winters.

CULTIVATION *Grow in fertile soil, in full sun. Plant on a layer of sand to improve drainage. In frost-prone climates lift the corms when the leaves turn yellow-brown, snap off the leaves, and store the corms in frost-free conditions.*

☼ ◊ ♀H2 ↕70–100cm (28–39in) ↔5cm (2in)

GLADIOLUS COMMUNIS SUBSP. *BYZANTINUS*

In late spring, this upright, cormous perennial produces a blaze of magenta flowers with purple-marked lips. These are arranged in spikes above fans of narrow, mid-green leaves. An elegant subject for a mixed or herbaceous border; the flowers are good for cutting.

CULTIVATION *Grow in fertile soil, in full sun. Plant corms on a bed of sharp sand to improve drainage. No need to lift corms in winter, but benefits from a winter mulch in frost-prone climates.*

☼ ◊ ♈H4 ↕to 1m (3ft) ↔8cm (3in)

GLEDITSIA TRIACANTHOS 'SUNBURST'

This fast-growing honey locust is a broadly conical, deciduous tree valued for its beautiful foliage and light canopy. The finely divided leaves are bright gold-yellow when they emerge in spring, maturing to dark green, then yellowing again before they fall. A useful, pollution-tolerant tree for a small garden.

CULTIVATION *Grow in any well-drained, fertile soil, in full sun. Prune only to remove dead, damaged or diseased wood, from late summer to mid-winter.*

☼ ◊ ♈H4 ↕12m (40ft) ↔10m (30ft)

GOMPHRENA HAAGEANA 'STRAWBERRY FAYRE'

An upright, bushy annual bearing brilliant red flowerheads throughout summer to early autumn. These are carried on upright stems above narrow, mid-green leaves which are covered with white hairs when young. Good for summer bedding; the flowers can also be cut and dry well for winter decoration. For bronze flowers on a similar plant, look for 'Amber Glow'.

CULTIVATION *Grow in any moderately fertile soil with good drainage. Choose a position in full sun.*

☼ ◊ 🏆H3 ↕75–80cm (30–32in) ↔to 30cm (12in)

GUNNERA MANICATA

A massive, clump-forming perennial that produces the largest leaves of any hardy garden plant, to 2m (6ft) long. These are rounded, lobed, sharply toothed and dull green, with stout, prickly stalks. Tall spikes of tiny, greenish-red flowers appear in summer. An imposing plant by water or in a bog garden. Needs protection to survive winter in cold climates.

CULTIVATION *Best in permanently moist, fertile soil, in full sun or partial shade. Shelter from cold winds. Protect from frost by folding the dead leaves over the dormant crown before winter.*

☼ ◐ ◊◆ 🏆H3 ↕2.5m (8ft) ↔3–4m (10–12ft) or more

GYMNOCALYCIUM ANDREAE

This prickly cactus forms clusters of spherical, dark blue-green or black-green stems with warty ribs and pale yellow-white spines. Bright yellow flowers are borne in early summer. Ideal for a cactus garden; must be grown in a warm conservatory or heated greenhouse in cold climates.

CULTIVATION *Best in sharply drained, poor soil, or standard cactus compost, in sun. Keep dry in winter. Minimum temperature 2–10°C (35–50°F).*

☼ ◊ ♈H1 ↕to 6cm (2½in) ↔to 15cm (6in)

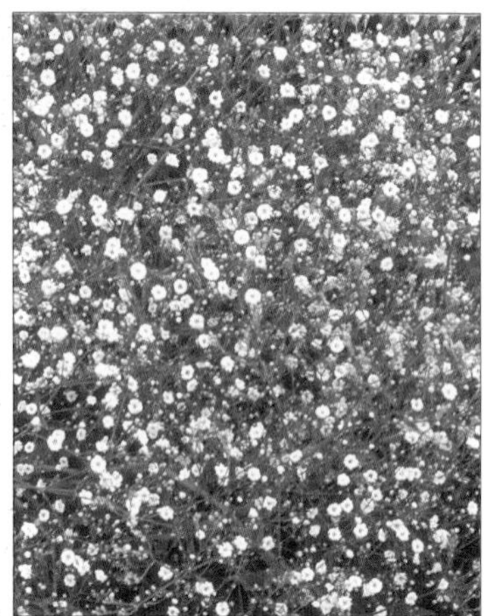

GYPSOPHILA PANICULATA 'BRISTOL FAIRY'

This herbaceous perennial forms a mound of slightly fleshy, lance-shaped, blue-green leaves on very slender, wiry stems. The profusion of tiny, double white flowers in summer, forms a cloud-like display. Very effective cascading over a low wall or as a foil to more upright, sharply defined flowers.

CULTIVATION *Grow in well-drained, deep, moderately fertile, preferably alkaline soil, in full sun. Resents being disturbed after planting.*

☼ ◊ ♈H4 ↕↔ to 1.2m (4ft)

GYPSOPHILA 'ROSENSCHLEIER'

A mound-forming perennial, also sold as 'Rosy Veil' or 'Veil of Roses', carrying trailing stems which look well cascading over a low wall. In summer, tiny, double white flowers which age to pale pink are carried in airy sprays, forming a dense cloud of blooms. The slightly fleshy leaves are lance-shaped and blue-green. The flowers dry well for decoration.

CULTIVATION *Grow in well-drained, deep, moderately fertile, preferably alkaline soil. Choose a position in full sun. Resents root disturbance.*

H4 ↕40–50cm (16–20in) ↔1m (3ft)

HAKONECHLOA MACRA 'AUREOLA'

This colourful grass is a deciduous perennial forming a clump of narrow, arching, bright yellow leaves with cream and green stripes. They flush red in autumn, and persist well into winter. Reddish-brown flower spikes appear in late summer. A versatile plant that can be used in a border, rock garden or containers.

CULTIVATION *Grow in moist but well-drained, fertile, humus-rich soil. Leaf colour is best in partial shade, but tolerates full sun.*

H4 ↕35cm (14in) ↔40cm (16in)

x *HALIMIOCISTUS SAHUCII*

A compact shrub that forms mounds of linear, dark green leaves with downy undersides. Masses of saucer-shaped white flowers are produced throughout summer. Good in a border, at the base of a warm wall or in a rock garden.

CULTIVATION *Best in freely draining, poor to moderately fertile, light, gritty soil, in a sunny site. Shelter from excessive winter wet.*

☼ ◊ ♈H4 ↕45cm (18in) ↔90cm (36in)

x *HALIMIOCISTUS WINTONENSIS* 'MERRIST WOOD CREAM'

A spreading, evergreen shrub bearing creamy-yellow flowers with red bands and yellow centres in late spring and early summer. The lance-shaped leaves are grey-green. Good at the front of a mixed border or at the foot of a warm wall. Also suits a raised bed or rock garden. May not survive cold winters.

CULTIVATION *Grow in freely draining, poor to moderately fertile soil, in full sun. Choose a position protected from excessive winter wet.*

☼ ◊ ♈H3 ↕60cm (24in) ↔90cm (36in)

HALIMIUM LASIANTHUM

A spreading bush with clusters of saucer-shaped, golden-yellow flowers in late spring and early summer. Each petal normally has a brownish-red mark at the base. The foliage is grey-green. Halimiums flower best in regions with long, hot summers, and this one is suited to a coastal garden.

CULTIVATION *Best in well-drained, moderately fertile sandy soil in full sun, with shelter from cold, drying winds. Established plants dislike being moved. After flowering, trim to maintain symmetry.*

☼ ◊ 🏆H3 ↕1m (3ft) ↔1.5m (5ft)

HALIMIUM 'SUSAN'

A small, spreading, evergreen shrub valued for its single or semi-double summer flowers, bright yellow with deep purple markings. The leaves are oval and grey-green. Good for rock gardens in warm, coastal areas. Provide shelter at the foot of a warm wall where winters are cold. Flowers are best during long, hot summers. *H. ocymoides*, similar but more upright in habit, is also recommended.

CULTIVATION *Grow in freely draining, fairly fertile, light, sandy soil, in full sun. Provide shelter from cold winds. Trim lightly in spring, as necessary.*

☼ ◊ 🏆H3 ↕45cm (18in) ↔60cm (24in)

WITCH HAZELS (*HAMAMELIS*)

These spreading, deciduous shrubs are grown for their large clusters of usually yellow spidery flowers, which appear on the branches when the shrubs are bare in winter; those of *H. vernalis* may coincide with the unfurling of the leaves. Each flower has four narrow petals and an enchanting fragrance. Most garden species also display attractive autumn foliage; the broad, bright green leaves turn red and yellow before they fall. Witch hazels bring colour and scent to the garden in winter; they are good as specimen plants or grouped in a shrub border or woodland garden.

CULTIVATION *Grow in moderately fertile, moist but well-drained, acid to neutral soil in full sun or partial shade, in an open but not exposed site. Witch hazels also tolerate deep, humus-rich soil over chalk. Remove wayward or crossing shoots when dormant in late winter or early spring to maintain a healthy, permanent framework.*

☼ ◐ ♁ ♀H4

1 ↕↔ 4m (12ft)

2 ↕↔ 4m (12ft)

1 *Hamamelis* x *intermedia* 'Arnold Promise' **2** *H.* x *intermedia* 'Barmstedt Gold'

3 *H.* x *intermedia* 'Diane' **4** *H.* x *intermedia* 'Jelena' **5** *H.* x *intermedia* 'Pallida' **6** *H. mollis*
7 *H. vernalis* 'Sandra'

HEBE ALBICANS

A neat, mound-forming shrub with tightly packed, glaucous grey-green foliage. It bears short, tight clusters of white flowers at the ends of the branches in the first half of summer. A useful evergreen hedging plant in mild coastal areas, or grow in a shrub border.

CULTIVATION *Grow in poor to moderately fertile, moist but well-drained, neutral to slightly alkaline soil in full sun or partial shade. Shelter from cold, drying winds. It will need little or no pruning.*

H4 ↕60cm (24in) ↔90cm (36in)

HEBE CUPRESSOIDES 'BOUGHTON DOME'

This dwarf, evergreen shrub is grown for its neat shape and dense foliage which forms a pale green dome. Flowers are infrequent. The congested, slender, greyish-green branches carry scale-like, pale green leaves. Excellent in a rock garden; gives a topiary effect without any clipping. Thrives in coastal gardens.

CULTIVATION *Grow in moist but well-drained, poor to moderately fertile soil, in full sun or partial shade. No regular pruning is necessary.*

H3–4 ↕30cm (12in) ↔60cm (24in)

HEBE X *FRANCISCANA* 'VARIEGATA'

A dense, rounded, evergreen shrub bearing colourful, oval leaves; these are mid-green with creamy-white margins. Purple flowers that contrast well with the foliage are carried in dense spikes during summer and autumn. A fine, pollution-tolerant plant for a mixed border or rock garden. In cold areas, shelter at the foot of a warm wall. 'Blue Gem' is a similar plant with plain leaves.

CULTIVATION *Grow in moist but well-drained, poor to moderately fertile soil, in sun or light shade. Shelter from cold, drying winds. No pruning is necessary.*

H2 ↕↔ 60–120cm (24–48in)

HEBE 'GREAT ORME'

An open, rounded, evergreen shrub that carries slender spikes of small, deep pink flowers which fade to white. These are borne from mid-summer to mid-autumn amid the lance-shaped, glossy dark green leaves. Good in a mixed or shrub border; shelter at the base of a warm wall in cold climates.

CULTIVATION *Grow in moist but well-drained, poor to moderately fertile soil, in sun or light shade. Shelter from cold winds. Pruning is unnecessary, but leggy plants can be cut back in spring.*

 H3 ↕↔ 1.2m (4ft)

HEBE MACRANTHA

This upright, spreading shrub has leathery green leaves. It bears relatively large flowers for a hebe, produced in clusters of three in early summer. A useful evergreen edging plant for seaside gardens in mild areas.

CULTIVATION *Grow in moist but well-drained, reasonably fertile, neutral to slightly alkaline soil in full sun or partial shade. Shelter from cold, drying winds. It needs little or no pruning.*

☼☀ ◊◊ ♀H3 ↕60cm (24in) ↔90cm (36in)

HEBE OCHRACEA 'JAMES STIRLING'

A compact shrub that bears its medium-sized white flowers in clusters from late spring to early summer. Like other whipcord hebes, it has small, scale-like leaves that lie flat against the stems to give the appearance of a dwarf conifer. An excellent evergreen for a rock garden, the rich ochre-yellow foliage looking very attractive in winter.

CULTIVATION *Best in moist but well-drained, neutral to slightly alkaline soil. Site in full sun or partial shade, and protect from cold, drying winds. Do not prune unless absolutely necessary.*

☼☀ ◊◊ ♀H4 ↕45cm (18in)↔60cm (24in)

HEBE PINGUIFOLIA 'PAGEI'

A low-growing, evergreen shrub bearing purple stems with four ranks of leathery, oval, blue-green leaves. Abundant clusters of white flowers appear at the tips of the shoots in late spring and early summer. Plant in groups as ground cover, or in a rock garden.

CULTIVATION *Grow in moist but well-drained, poor to moderately fertile soil, in sun or partial shade. Best with some shelter from cold, drying winds. Trim to neaten in early spring, if necessary.*

☼ ◑ ◊ ♦ ♀H4 ↕30cm (12in) ↔90cm (36in)

HEBE RAKAIENSIS

A rounded, evergreen shrub bearing spikes of white flowers from early to mid-summer. The leaves are elliptic and glossy bright green. Ideal either as a small, spreading specimen shrub or as a focal point in a large rock garden.

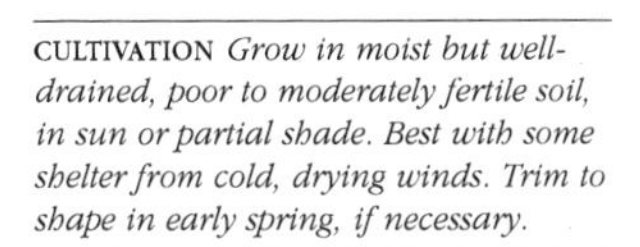

CULTIVATION *Grow in moist but well-drained, poor to moderately fertile soil, in sun or partial shade. Best with some shelter from cold, drying winds. Trim to shape in early spring, if necessary.*

☼ ◑ ◊ ♦ ♀H4 ↕1m (3ft) ↔1.2m (4ft)

HEDERA CANARIENSIS 'RAVENSHOLST'

This vigorous cultivar of Canary Island ivy is a self-clinging, evergreen climber useful as ground cover or to mask a bare wall. The leaves are shallowly lobed, glossy, and dark green. There are no flowers. It may be damaged in areas with severe winters, but it usually grows back quickly.

CULTIVATION *Best in fertile, moist but well-drained soil that is rich in organic matter. It tolerates shade, and can be pruned or trimmed at any time of year.*

☼☀ 💧💧 ♈H3 ↕5m (15ft)

HEDERA COLCHICA

Persian ivy is a vigorous, self-clinging, evergreen climber with large, heart-shaped leaves, which give it its other common name of bullock's heart ivy. They are leathery in texture and dark green. There are no flowers. A particularly useful climber for quickly covering an unsightly wall, as ground cover, or for training up into a large, deciduous tree.

CULTIVATION *Grow in fertile, moist but well-drained, preferably alkaline soil that is rich in organic matter. It tolerates shade and can be pruned at any time of year.*

☼☀ 💧💧 ♈H4 ↕10m (30ft)

HEDERA COLCHICA 'DENTATA'

This Persian ivy is a very vigorous, evergreen, self-clinging climber carrying large, heart-shaped, drooping, glossy green leaves. The stems and leaf-stalks are flushed purple. A handsome plant for covering an unattractive wall in shade; also effective as ground cover. 'Dentata Variegata' has mottled grey-green leaves edged with cream.

CULTIVATION *Best in moist but well-drained, fertile, ideally alkaline soil, in partial to deep shade. Prune at any time of the year to restrict size.*

◑● 💧💧 🏆H4 ↕10m (30ft)

HEDERA COLCHICA 'SULPHUR HEART'

This coloured-leaf Persian ivy is a very vigorous, self-clinging evergreen climber which can also be grown as ground cover. The large, heart-shaped leaves are dark green suffused with creamy-yellow; as they mature, the colour becomes more even. Will quickly cover a wall in shade.

CULTIVATION *Grow in moist but well-drained, fertile, preferably alkaline soil. Tolerates partial shade, but leaf colour is more intense in sun. Prune at any time of the year to restrict size.*

☼◑ 💧💧 🏆H4 ↕5m (15ft)

ENGLISH IVIES (*HEDERA HELIX*)

Hedera helix, the common or English ivy, is an evergreen, woody-stemmed, self-clinging climber and the parent of an enormous selection of cultivars. Leaf forms vary from heart-shaped to deeply lobed, ranging in colour from the bright gold 'Buttercup' to the deep purple 'Atropurpurea'. They make excellent ground cover, tolerating even dry shade, and will quickly cover featureless walls; they can damage paintwork or invade gutters if not kept in check. Variegated cultivars are especially useful for enlivening dark corners and shaded walls. Small ivies make good houseplants and can be trained over topiary frames.

CULTIVATION *Best in moist but well-drained, humus-rich, alkaline soil. Choose a position in full sun or shade; ivies with variegated leaves may lose their colour in shade. Trim as necessary to keep under control. 'Goldchild' and 'Little Diamond' may suffer in areas with heavy frosts.*

☼☀ 💧 ♀H4, some H3–4

1 ↕8m (25ft) 2 ↕2m (6ft) 3 ↕2m (6ft)
4 ↕1m (3ft) 5 ↕1m (3ft) 6 ↕30cm (12in)

1 *H. helix* 'Atropurpurea' (H4) **2** *H. helix* 'Buttercup' (H4) **3** *H. helix* 'Glacier' (H4)
4 *H. helix* 'Goldchild' (H3–4) **5** *H. helix* 'Ivalace' (H4) **6** *H. helix* 'Little Diamond' (H3–4)

HEDERA HIBERNICA

Irish ivy is a vigorous, evergreen, self-clinging climber valued for its broadly oval, dark green leaves which have grey-green veins and five triangular lobes. Useful for a wall or against a robust tree, or as fast-growing ground cover under trees or shrubs.

CULTIVATION *Best in moist but well-drained, fertile, ideally alkaline soil, in partial to full shade. Prune at any time of the year to restrict spread.*

◐● ◊◊ ♈H4 ↕to 10m (30ft)

HELIANTHEMUM 'FIRE DRAGON'

This rock rose, also called 'Mrs Clay' is a small, spreading, evergreen shrub bearing a profusion of saucer-shaped, bright orange-red flowers. These open in succession during late spring and summer amid the oblong, grey-green leaves. Ideal for a rock garden or raised bed, or as ground cover in groups on a sunny bank.

CULTIVATION *Grow in well-drained, slightly alkaline soil, in full sun. Trim after flowering to keep compact. Often short-lived, but easily propagated by softwood cuttings taken in late spring.*

☼ ◊ ♈H4 ↕20–30cm (8–12in) ↔30cm (12in) or more

HELIANTHEMUM 'HENFIELD BRILLIANT'

This rock rose is a small, spreading, evergreen shrub bearing saucer shaped, brick-red flowers in late spring and summer. The leaves are narrow and grey-green. Effective in groups on a sunny bank; also good in a rock garden or raised bed, or at the front of a border.

CULTIVATION *Grow in moderately fertile, well-drained, neutral to alkaline soil, in sun. Trim after flowering to keep bushy. Often short-lived, but easily propagated by softwood cuttings in late spring.*

☼ ◊ ♀H4 ↕20–30cm (8–12in) ↔30cm (12in) or more

HELIANTHEMUM 'RHODANTHE CARNEUM'

This long-flowering rock rose, also sold as 'Wisley Pink', is a low and spreading, evergreen shrub. Pale pink, saucer-shaped flowers with yellow-flushed centres appear from late spring to summer amid narrow, grey-green leaves. Good in a rock garden, raised bed or mixed border.

CULTIVATION *Best in well-drained, moderately fertile, neutral to alkaline soil, in full sun. Trim over after flowering to encourage further blooms.*

☼ ◊ ♀H4 ↕to 30cm (12in) ↔to 45cm (18in) or more

HELIANTHEMUM 'WISLEY PRIMROSE'

This primrose-yellow rock rose is a fast-growing, spreading, evergreen shrub. It bears a profusion of saucer-shaped flowers with golden centres over long periods in late spring and summer. The leaves are narrowly oblong and grey-green. Group in a rock garden, raised bed or sunny bank. For paler, creamy flowers, look for 'Wisley White'.

CULTIVATION *Grow in well-drained, moderately fertile, preferably neutral to alkaline soil, in full sun. Trim after flowering to encourage further blooms.*

☼ ◊ ♀H4 ↕to 30cm (12in) ↔to 45cm (18in) or more

HELIANTHUS 'LODDON GOLD'

This double-flowered sunflower is a tall, spreading perennial with coarse, oval, mid-green leaves which are arranged along the upright stems. Grown for its large, bright yellow flowers, which open during late summer and last into early autumn. Use to extend the season of interest in herbaceous and mixed borders.

CULTIVATION *Grow in moist to well-drained, moderately fertile soil that is rich in humus. Choose a sheltered site in full sun. Flowers are best during long, hot summers. Stake flower stems.*

☼ ◊♦ ♀H4 ↕to 1.5m (5ft) ↔90cm (36in)

HELIANTHUS 'MONARCH'

This semi-double sunflower is a tall, spreading perennial with sturdy, upright stems bearing oval and toothed, mid-green leaves. Large, star-like, bright golden-yellow flowers with yellow-brown centres are produced from late summer to autumn. A statuesque plant for late-summer interest in a herbaceous or mixed border. 'Capenoch Star' is another recommended perennial sunflower, lemon-yellow with golden centres, to 1.5m (5ft) tall.

CULTIVATION *Grow in any well-drained, moderately fertile soil. Choose a sunny, sheltered site. The stems need support.*

☼ ◊ ♀H4 ↕to 2m (6ft) ↔1.2m (4ft)

HELICHRYSUM PETIOLARE

The liquorice plant is a silvery, mound-forming, evergreen shrub with trailing stems. Its small leaves are densely felted and silver-grey. The inconspicuous summer flowers are often pinched off, as they spoil the foliage effect. In frost-prone areas, grow as an annual in a hanging basket or other container; in warm areas it is useful as ground cover.

CULTIVATION *Grow in any well-drained soil, in full sun. Pinch-prune young stems to promote bushiness. Minimum temperature 2°C (35°F).*

☼ ◊ ♀H1+3 ↕to 50cm (20in) ↔2m (6ft) or more

HELICHRYSUM PETIOLARE 'VARIEGATUM'

This variegated liquorice plant is a trailing, evergreen shrub grown for its densely felted, silver-grey and cream leaves. Small, creamy-yellow flowers appear in summer, but are of little ornamental interest and may be removed. Excellent in hanging baskets or other containers; in cold areas, grow as an annual or over-winter in frost-free conditions.

CULTIVATION *Grow in any well-drained soil, in full sun. Trim regularly and pinch off unwanted flowers as they form. Minimum temperature 2°C (35°F).*

☼ ◊ 🏆H1+3 ↕to 50cm (20in) ↔2m (6ft) or more

HELICHRYSUM SPLENDIDUM

A compact, white-woolly, evergreen perennial bearing linear, aromatic, silver-grey foliage. Small, bright yellow flowerheads open at the tips of the upright stems from mid-summer to autumn, and last into winter. Suitable for a mixed border or rock garden; the flowers can be dried for winter decoration. May not survive outside during cold winters.

CULTIVATION *Grow in well-drained, poor to moderately fertile, neutral to alkaline soil, in full sun. Remove dead or damaged growth in spring, cutting back leggy shoots into old wood.*

☼ ◊ 🏆H3 ↕↔ 1.2m (4ft)

HELIOTROPIUM 'PRINCESS MARINA'

A compact, evergreen shrub that is usually grown as an annual or as a conservatory plant in cold climates. Much-valued for its fragrant heads of deep violet-blue flowers which appear in summer above the oblong, wrinkled, mid- to dark green, often purple-tinged leaves. Good at the front of a border, in containers or as summer bedding; will not survive in areas with cold winters.

CULTIVATION *Grow in any moist but well-drained, fertile soil or compost, in sun. Minimum temperature 2°C (35°F).*

☼ ◊◊ H1 ↕↔ to 30cm (12in)

HELLEBORUS ARGUTIFOLIUS

The large Corsican hellebore, sometimes known as *H. corsicus*, is an early-flowering, clump-forming, evergreen perennial bearing large clusters of nodding, pale green flowers. These appear in winter and early spring above the handsome dark green leaves which are divided into three sharply toothed leaflets. Excellent for early interest in a woodland garden or mixed border.

CULTIVATION *Grow in moist, fertile, preferably neutral to alkaline soil, in full sun or partial shade. Often short-lived, but self-seeds readily.*

☼☼ ◊ H4 ↕to 1.2m (4ft) ↔90cm (36in)

HELLEBORUS FOETIDUS

The stinking hellebore is an upright, evergreen perennial forming clumps of dark green, divided leaves which smell unpleasant when crushed. In winter and early spring, clusters of small, nodding, cup-shaped flowers appear above the foliage; their green petals are edged with red. A striking specimen for a winter border.

CULTIVATION *Grow in moist, fertile, neutral to alkaline soil. Choose a position in sun or partial shade.*

☼ ◑ 💧 🏆H4 ↕to 80cm (32in) ↔45cm (18in)

HELLEBORUS NIGER

A clump-forming, usually evergreen perennial valued for its nodding clusters of cup-shaped white flowers in winter and early spring. The dark green leaves are divided into several leaflets. Effective with snowdrops beneath winter-flowering shrubs, but can be difficult to naturalize. 'Potter's Wheel' is a pretty cultivar of this hellebore, with large flowers; for heavier soil, look for *H.* x *nigercors*, of which *H. niger* is a parent.

CULTIVATION *Grow in deep, fertile, neutral to alkaline soil that is reliably moist. Site in dappled shade with shelter from cold, drying winds.*

◑ 💧 🏆H4 ↕to 30cm (12in) ↔45cm (18in)

DAYLILIES (*HEMEROCALLIS*)

Daylilies are clump-forming, herbaceous perennials, so-called because each of their showy flowers only lasts for about a day; in nocturnal daylilies, such as 'Green Flutter', the flowers open in late afternoon and last through the night. The blooms are abundant and rapidly replaced, some appearing in late spring while other cultivars flower into late summer. Flower shapes vary from circular to spider-shaped, in shades of yellow and orange. The leaves are strap-like and evergreen. Tall daylilies make a dramatic contribution to a mixed or herbaceous border; dwarf types, such as 'Stella de Oro', are useful for small gardens or in containers.

CULTIVATION *Grow in well-drained but moist, fertile soil, in sun; 'Corky' and 'Green Flutter' tolerate semi-shade. Mulch in spring, and feed with a balanced fertiliser every two weeks until buds form. Divide and replant every few years, in spring or autumn.*

☼ ◑ ♢♦ ♀H4

1 ↕70cm (28in) ↔ 40cm (16in)

2 ↕90cm (36in) ↔ 45cm (18in)

3 ↕50cm (20in) ↔ 1m (3ft)

4 ↕↔ 1m (3ft)

5 ↕60cm (24in) ↔ 1m (3ft)

6 ↕30cm (12in) ↔ 45cm (18in)

1 *H.* 'Corky' **2** *H.* 'Golden Chimes' **3** *H.* 'Green Flutter' **4** *H. lilioasphodelus* **5** *H.* 'Nova' **6** *H.* 'Stella de Oro'

HEPATICA NOBILIS

A small, slow-growing, anemone-like, semi-evergreen perennial bearing saucer-shaped, purple, white or pink flowers. These appear in early spring, usually before the foliage has emerged. The mid-green, sometimes mottled leaves are leathery and divided into three lobes. Suits a shady rock garden. *H.* x *media* 'Ballardii' is a very similar plant, with reliably deep blue flowers.

CULTIVATION *Grow in moist but well-drained, fertile, neutral to alkaline soil, in partial shade. Provide a mulch of leaf mould in autumn or spring.*

H4 ↕10cm (4in) ↔15cm (6in)

HEUCHERA MICRANTHA VAR. *DIVERSIFOLIA* 'PALACE PURPLE'

A clump-forming perennial valued for its glistening, dark purple-red, almost metallic foliage, which is topped by airy sprays of white flowers in summer. The leaves have five pointed lobes. Plant in groups as ground cover for a shady site, but can be slow to spread.

CULTIVATION *Grow in moist but well-drained, fertile soil, in sun or partial shade. Tolerates full shade where the ground is reliably moist. Lift and divide clumps every few years, after flowering.*

↕↔ 45–60cm (18–24in)

HEUCHERA 'RED SPANGLES'

This clump-forming, evergreen perennial is valued for its sprays of small, bell-shaped, crimson-scarlet flowers. These are borne in early summer, with a repeat bloom in late summer, on dark red stems above the lobed, heart-shaped, purplish-green leaves. Effective as ground cover when grouped together.

CULTIVATION *Grow in moist but well-drained, fertile soil, in sun or partial shade. Tolerates full shade where the ground is reliably moist. Lift and divide clumps every three years, after flowering.*

☼☀ 00 ♀H4 ↕50cm (20in) ↔25cm (10in)

HIBISCUS SYRIACUS 'OISEAU BLEU'

Also known as 'Blue Bird', this vigorous, upright, deciduous shrub bears large, mallow-like, lilac-blue flowers with red centres. These are borne from mid- to late summer amid deep green leaves. Ideal for a mixed or shrub border. Flowers are best in hot summers; plant against a warm wall in cold areas.

CULTIVATION *Grow in moist but well-drained, fertile, neutral to slightly alkaline soil, in full sun. Prune young plants hard in late spring to encourage branching at the base; keep pruning to a minimum once established.*

☼ 00 ♀H4 ↕3m (10ft) ↔2m (6ft)

HIBISCUS SYRIACUS 'WOODBRIDGE'

A fast-growing, upright, deciduous shrub producing large, deep rose-pink flowers with maroon blotches around the centres. These are borne from late summer to mid-autumn amid the lobed, dark green leaves. Valuable for its late season of interest. In cold areas, plant against a warm wall for the best flowers.

CULTIVATION *Grow in moist but well-drained, fertile, slightly alkaline soil, in full sun. Prune young plants hard to encourage branching; keep pruning to a minimum once established.*

☼ ◊♦ ♈H4 ↕3m (10ft) ↔2m (6ft)

HIPPOPHAE RHAMNOIDES

Sea buckthorn is a spiny, deciduous shrub with attractive fruits and foliage. Small yellow flowers in spring are followed by orange berries on female plants, which persist well into winter. The silvery-grey leaves are narrow and claw-like. Good for hedging, especially in coastal areas.

CULTIVATION *Best in sandy, moist but well-drained soil, in full sun. For good fruiting, plants of both sexes must grow together. Little pruning is required; trim hedges in late summer, as necessary.*

↕↔ 6m (20ft)

HOHERIA GLABRATA

This deciduous, spreading tree is grown for its graceful habit and clusters of cup-shaped, fragrant white flowers in mid-summer, which are attractive to butterflies. The broad, dark green leaves turn yellow in autumn before they fall. Best in a shrub border in a maritime climate, against a sunny wall in cold areas.

CULTIVATION *Grow in moderately fertile, well-drained, neutral to alkaline soil in full sun or partial shade, sheltered from cold, drying winds. Pruning is seldom necessary, but if branches are damaged by frost, cut them back in spring.*

☼ ◐ ◊ ♀H4 ↕↔7m (22ft)

HOHERIA SEXSTYLOSA

Ribbonwood is an evergreen tree or shrub with narrow, glossy, mid-green leaves with toothed margins. It is valued for its graceful shape and abundant clusters of white flowers; attractive to butterflies, they appear in late summer. Grow against a warm, sunny wall in areas with cold winters. The cultivar 'Stardust' is recommended.

CULTIVATION *Best in moderately fertile, well-drained, neutral to alkaline soil in full sun or partial shade, sheltered from cold, drying winds. Protect the roots with a thick winter mulch. Pruning is seldom necessary.*

☼ ◐ ◊ ♀H3 ↕ 8m (25ft) ↔6m (20ft)

HOSTAS

Hostas are evergreen perennials grown principally for their dense mounds of large, overlapping, lance- to heart-shaped leaves. A wide choice of foliage colour is available, from the cloudy blue-green of 'Halcyon' to the bright yellow-green 'Golden Tiara'. Many have leaves marked with yellow or white around the edges; *H. fortunei* var. *albopicta* has bold, central splashes of creamy-yellow. Upright clusters of funnel-shaped flowers, varying from white through lavender-blue to purple, are borne on tall stems in summer. Hostas are effective at the front of a mixed border, in containers or as ground cover under deciduous trees.

CULTIVATION *Grow in well-drained but reliably moist, fertile soil, in full sun or partial shade with shelter from cold winds. Yellow-leaved hostas colour best in full sun with shade at midday. Mulch in spring to conserve moisture throughout the summer.*

☼ ◑ 💧 🏆H4

1 ↕ 50cm (20in) ↔ 1m (3ft)

2 ↕ 55cm (22in) ↔ 1m (3ft)

3 ↕ 55cm (22in) ↔ 1m (3ft)

4 ↕ 55cm (22in) ↔ 1m (3ft)

5 ↕ 60cm (24in) ↔ 1m (3ft)

6 ↕ 30cm (12in) ↔ 50cm (20in)

1 *Hosta crispula* **2** *H. fortunei* var. *albopicta* **3** *H. fortunei* var. *aureomarginata*
4 *H.* 'Francee' **5** *H.* 'Frances Williams' **6** *H.* 'Golden Tiara'

7 ↕1m (3ft) ↔ 75cm (30in)

8 ↕45cm (18in) ↔ 75cm (30in)

9 ↕45cm (18in) ↔ 1m (3ft)

10 ↕60cm (24in) ↔ 1.2m (4ft)

11 ↕45cm (18in) ↔ 75cm (30in)

12 ↕1m (3ft) ↔ 1.2m (4ft)

7 *Hosta* 'Honeybells' **8** *H. lancifolia* **9** *H.* 'Love Pat' **10** *H.* 'Royal Standard'
11 *H.* 'Shade Fanfare' **12** *H. sieboldiana* var. *elegans*

13 ↕75cm (30in) ↔ 1.2m (4ft)

14 ↕35–40cm (14–16in) ↔ 70cm (28in)

15 ↕1m (3ft) ↔ 45cm (18in)

16 ↕45cm (18in) ↔ 70cm (28in)

18 ↕5cm (2in) ↔ 25cm (10in)

17 ↕50cm (20in) ↔ 1m (3ft)

19 ↕75cm (30in) ↔ 1m (3ft)

13 *H.* 'Sum and Substance' **14** *H.* Tardiana Group 'Halcyon' **15** *H. undulata* var. *undulata* **16** *H. undulata* var. *univittata* **17** *H. ventricosa* **18** *H. venusta* **19** *H.* 'Wide Brim'

HUMULUS LUPULUS 'AUREUS'

The golden hop is a twining, perennial climber grown for its attractively lobed, bright golden-yellow foliage. Hanging clusters of papery, cone-like, greenish-yellow flowers appear in autumn. Train over a fence or trellis, or up into a small tree. The flowers dry well for garlands and swags.

CULTIVATION *Grow in moist but well drained, moderately fertile, humus-rich soil. Tolerates partial shade, but leaf colour is best in full sun. Give the twining stems support. Cut back any dead growth to ground level in early spring.*

☼☀ ◊♦ ♀H4 ↕6m (20ft)

HYACINTHUS ORIENTALIS 'BLUE JACKET'

This navy blue hyacinth is a bulbous perennial bearing dense, upright spikes of fragrant, bell-shaped flowers with purple veins in early spring. Good for spring bedding; specially prepared bulbs can be planted in pots during autumn for an indoor display of early flowers. With the paler 'Delft Blue', one of the best true blue hyacinths.

CULTIVATION *Grow in any well-drained, moderately fertile soil or compost, in sun or partial shade. Protect container-grown bulbs from winter wet.*

☼☀ ◊ ♀H4 ↕20–30cm (8–12in) ↔8cm (3in)

HYACINTHUS ORIENTALIS 'CITY OF HAARLEM'

This primrose-yellow hyacinth is a spring-flowering, bulbous perennial bearing upright spikes of fragrant, bell-shaped flowers. The lance-shaped leaves are bright green and emerge from the base of the plant. Good for spring bedding or in containers; specially prepared bulbs can be planted in autumn for early flowers indoors.

CULTIVATION *Grow in well-drained, fairly fertile soil or compost, in sun or partial shade. Protect container-grown plants from excessive winter wet.*

☼ ◐ 💧 🏆H4 ↕20–30cm (8–12in) ↔8cm (3in)

HYACINTHUS ORIENTALIS 'OSTARA'

This violet hyacinth is a bulbous perennial grown as spring bedding for its dense, upright spikes of bell-shaped, fragrant flowers with dark stripes. Bright green, lance-shaped leaves emerge from the base of the plant. Prepared bulbs can be planted indoors during autumn for winter flowers. Startling planted with salmon-orange 'Gipsy Queen'.

CULTIVATION *Best in well-drained, moderately fertile soil or compost, in sun or partial shade. Protect container-grown plants from excessive winter wet.*

☼ ◐ 💧 🏆H4 ↕20–30cm (8–12in) ↔8cm (3in)

HYACINTHUS ORIENTALIS 'PINK PEARL'

This deep pink hyacinth, bearing dense, upright spikes of fragrant, bell-shaped flowers with paler edges, is a spring-flowering, bulbous perennial. The leaves are narrow and bright green. Excellent in a mixed or herbaceous border; specially prepared bulbs can be planted in pots during autumn for an indoor display of early flowers.

CULTIVATION *Grow in any well-drained, moderately fertile soil or compost, in sun or partial shade. Protect container-grown bulbs from winter wet.*

☼ ◑ ◊ ♈H4 ↕20–30cm (8–12in) ↔8cm (3in)

HYDRANGEA ANOMALA SUBSP. *PETIOLARIS*

The climbing hydrangea, often sold simply as *H. petiolaris*, is a woody-stemmed, deciduous, self-clinging climber, usually grown on shady walls for its large, lacecap-like heads of creamy-white, summer flowers. The mid-green leaves are oval and coarsely toothed. Often slow to establish, and may not survive in areas with cold, frost-prone winters.

CULTIVATION *Grow in any reliably moist, fertile soil in sun or deep shade. Little pruning is required, but as the allotted space is filled, cut back overlong shoots after flowering.*

☼ ● ♦ ♈H3 ↕15m (50ft)

HYDRANGEA ARBORESCENS 'ANNABELLE'

An upright, deciduous shrub bearing large, rounded heads of densely-packed, creamy-white flowers from mid-summer to early autumn. The leaves are broadly oval and pointed. Good on its own or in a shrub border; the flowerheads can be dried for winter decoration. 'Grandiflora' has even larger flowerheads.

CULTIVATION *Grow in moist but well-drained, moderately fertile, humus-rich soil, in sun or partial shade. Keep pruning to a minimum, or cut back hard each spring to a low framework.*

 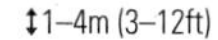 H4 ↕1.5m (5ft) ↔2.5m (8ft)

HYDRANGEA ASPERA VILLOSA GROUP

A group of spreading to upright, deciduous shrubs that can become tree-like with age. In late summer, they produce flattened, lacecap-like heads of small, blue-purple or rich blue flowers, surrounded by larger, lilac-white or rose-lilac flowers. The leaves are lance-shaped and dark green. Excellent in a woodland or wild garden; in cold areas they are best trained against a warm wall.

CULTIVATION *Grow in moist but well-drained, moderately fertile soil that is rich in humus. Site in full sun or semi-shade. Little pruning is necessary.*

H3 ↕1–4m (3–12ft)

HYDRANGEA MACROPHYLLA

Cultivars of the common hydrangea, *H. macrophylla*, are rounded shrubs with oval, mid- to dark green, deciduous leaves. Their large, showy flowerheads, borne from mid- to late summer, are available in two distinct forms: lacecaps, such as 'Veitchii', have flat flowerheads, and mop-head hydrangeas (Hortensias), such as 'Altona', have round flowerheads. Except in white-flowered cultivars, flower colour is directly influenced by soil pH; acid soils produce blue flowers, and alkaline soils give rise to pink flowers. All types of hydrangea are useful for a range of garden sites, and the flowerheads dry well for indoor arrangements.

CULTIVATION *Grow in moist but well-drained, fertile soil, in sun or partial shade with shelter from cold winds. Prune hard in spring to enhance flowering, cutting stems right back to strong pairs of buds. Flower colour can be influenced on neutral soils.*

☼ ◐ ◊ ◆ ♀H3–4

1 ↕1m (3ft) ↔ 1.5m (5ft) **2** ↕2m (6ft) ↔ 2.5m (8ft) **3** ↕↔ 1.5m (5ft) **4** ↕2m (6ft) ↔ 2.5m (8ft) **5** ↕2m (6ft) ↔ 2.5m (8ft)

1 *H. macrophylla* 'Altona' (Mop-head) **2** *H. macrophylla* 'Mariesii Perfecta' (syn. *H. macrophylla* 'Blue Wave') (Lacecap) **3** *H. macrophylla* 'Lanarth White' (Lacecap) **4** *H. macrophylla* 'Générale Vicomtesse de Vibraye' (Mop-head) **5** *H. macrophylla* 'Veitchii' (Lacecap)

HYDRANGEA PANICULATA

Cultivars of *H. paniculata* are fast-growing, upright, deciduous shrubs, with oval, mid- to dark green leaves. They are cultivated for their tall clusters of lacy flowers which usually appear during late summer and early autumn; some cultivars, such as 'Praecox', bloom earlier in the summer. Flowers are mostly creamy-white, with some forms, such as 'Floribunda', becoming pink-tinged as they age. These versatile shrubs are suitable for many different garden uses; as specimen plants, in groups or in containers. The flowerheads look very attractive when dried for indoor decoration.

CULTIVATION *Grow in moist but well-drained, fertile soil. Site in sun or partial shade with shelter from cold, drying winds. Pruning is not essential, but plants flower much better if pruned back annually, in early spring, to the lowest pair of healthy buds on a permanent, woody framework.*

☼◑ ◊♦ ♀H4

1 ↕3–7m (10–22ft) ↔ 2.5m (8ft)

2 ↕3–7m (10–22ft) ↔ 2.5m (8ft)

3 ↕3–7m (10–22ft) ↔ 2.5m (8ft)]

1 *H. paniculata* 'Floribunda' **2** *H. paniculata* 'Grandiflora' **3** *H. paniculata* 'Praecox'

HYDRANGEA QUERCIFOLIA

The oak-leaved hydrangea is a mound-forming, deciduous shrub bearing conical heads of white flowers that fade to pink, from mid-summer to autumn. The deeply lobed, mid-green leaves turn bronze-purple in autumn. Useful in a range of garden sites.

CULTIVATION *Prefers well-drained but moist, moderately fertile soil, in sun or partial shade. Leaves may become yellow in shallow, chalky soil. Keep pruning to a minimum, in spring.*

☼☀ ♁ ♈H3–4 ↕2m (6ft) ↔2.5m (8ft)

HYDRANGEA SERRATA 'BLUEBIRD'

A compact, upright, long-flowering, deciduous shrub bearing flattened heads of tiny, rich blue flowers surrounded by larger, pale blue flowers from summer to autumn. The narrowly oval, pointed, mid-green leaves turn red in autumn. The flowerheads can be dried for indoor arrangements. 'Grayswood' is a similar shrub with mauve flowers.

CULTIVATION *Grow in moist but well-drained, moderately fertile, humus-rich soil, in sun or partial shade. Flowers may turn pink in alkaline soils. Cut back weak, thin shoots in mid-spring.*

☼☀ ♁ ♈H3–4 ↕↔ 1.2m (4ft)

HYDRANGEA SERRATA 'ROSALBA'

An upright, compact, deciduous shrub valued for its flat flowerheads which appear from summer to autumn; these are made up of tiny pink flowers in the centre, surrounded by larger white flowers that become red-marked as they age. The leaves are oval, mid-green and pointed. Ideal as a specimen plant or in a shrub border.

CULTIVATION *Grow in well-drained but moist, moderately fertile, humus-rich soil, in full sun or partial shade. Flowers may turn blue on acid soils. Very little pruning is needed.*

☼◐ 💧 🏆H3–4 ↕↔ 1.2m (4ft)

HYPERICUM 'HIDCOTE'

This dense, evergreen or semi-evergreen shrub produces abundant clusters of large, cupped, golden-yellow flowers which open from mid-summer to early autumn. The leaves are dark green and lance-shaped. Suitable for a shrub border; for a taller but narrower shrub, to 2m (6ft) high but otherwise very similar, look for 'Rowallane'.

CULTIVATION *Grow in well-drained but moist, moderately fertile soil, in sun or partial shade. Deadhead regularly, and trim annually in spring to increase the flowering potential.*

☼◐ 💧💧 🏆H4 ↕1.2m (4ft) ↔1.5m (5ft)

HYPERICUM KOUYTCHENSE

Sometimes known as *H.* 'Sungold', this species of St. John's wort is a rounded, semi-evergreen bush with arching shoots. It has dark blue-green leaves, but its biggest asset is the large clusters of golden-yellow star-shaped flowers borne in profusion during summer and autumn, followed by bright bronze-red fruits. Grow in a shrub or mixed border; a useful rabbit-proof shrub.

CULTIVATION *Grow in moderately fertile, moist but well-drained soil in full sun or partial shade. Prune or trim after flowering, if necessary.*

☼ ◑ ◊◊ ♀H4 ↕1m (3ft) ↔1.5m (5ft)

IBERIS SEMPERVIRENS

A spreading, evergreen subshrub bearing dense, rounded heads of small, unevenly shaped white flowers which are often flushed with pink or lilac. These appear in late spring and early summer, covering the spoon-shaped, dark green leaves. Best in a rock garden or large wall pocket. The recommended cultivar 'Schneeflocke' can be even more floriferous.

CULTIVATION *Grow in well-drained, poor to moderately fertile, neutral to alkaline soil, in full sun. Trim lightly after flowering for neatness.*

☼ ◊ ♀H4 ↕to 30cm (12in) ↔ to 40cm (16in)

ILEX X *ALTACLERENSIS* 'GOLDEN KING' (FEMALE)

A compact, evergreen shrub with glossy, dark green leaves edged in gold. The leaf margins may be smooth or toothed. The flowers are insignificant, but develop into red berries in autumn. Tolerant of pollution and coastal exposure; a good tall windbreak or hedge where winters are not too severe.

CULTIVATION *Grow in moist but well-drained, moderately fertile soil rich in organic matter. For berries, a male holly must grow nearby. A position in full sun is ideal. Trim or prune in early spring, if necessary.*

☼ ◊◊ ♀H4 ↕30m (20ft) ↔4m (12ft)

ILEX X *ALTACLERENSIS* 'LAWSONIANA' (FEMALE)

This dense and bushy holly forms a compact, evergreen tree or shrub. It bears large, usually spineless, oval, bright green leaves, which are splashed with gold and paler green in the centres. Red-brown berries, ripening to red, develop in autumn.

CULTIVATION *Grow in moist but well-drained soil, in sun for best leaf colour. Grow a male holly nearby to ensure a display of berries. Free-standing plants may need some shaping when young. Remove any all-green shoots as seen.*

↕to 6m (20ft) ↔ 5m (15ft)

ENGLISH HOLLIES (*ILEX AQUIFOLIUM*)

Ilex aquifolium, the English holly, has many different cultivars, all upright, evergreen trees or large shrubs, which are usually grown on their own or as spiny hedges. They have purple stems, grey bark and dense, glossy foliage. Most cultivars have multi-coloured, spiny leaves, although those of 'J.C. van Tol' are spineless and dark green. 'Ferox Argentea' has extra-spiny leaves. Male and female flowers are borne on separate plants, so female hollies, like 'Madame Briot', must be near males, such as 'Golden Milkboy', if they are to bear a good crop of berries. Tall specimens make effective windbreaks.

CULTIVATION *Grow in moist, well-drained, fertile, humus-rich soil. Choose a site in full sun for good leaf variegation, but tolerates partial shade. Remove any damaged wood and shape young trees in spring; hedging should be trimmed in late summer. Over-enthusiastic pruning will spoil their form.*

☼ ◑ ◊ 🏆H4

1 ↕ to 25m (80ft) ↔ 8m (25ft)

2 ↕ to 6m (20ft) ↔ 2.5m (8ft)

3 ↕ to 15m (50ft) ↔ 4m (12ft)

1 *Ilex aquifolium* **2** 'Amber' (Female) **3** 'Argentea Marginata' (Female)

4 ↕ to 8m (25ft) ↔ 4m (12ft)

5 ↕ 6m (20ft) ↔ 4m (12ft)

6 ↕ 8m (25ft) ↔ 5m (15ft)

7 ↕ 6m (20ft) ↔ 4m (12ft)

8 ↕ 6m (20ft) ↔ 5m (15ft)

9 ↕ 6m (20ft) ↔ 5m (15ft)

10 ↕ 6m (20ft) ↔ 4m (12ft)

11 ↕ 6m (20ft) ↔ 4m (12ft)

12 ↕ 10m (30ft) ↔ 4m (12ft)

4 'Ferox Argentea' (Male) **5** 'Golden Milkboy' (Male) **6** 'Handsworth New Silver' (Female)
7 'J. C. van Tol' (Female) **8** 'Madame Briot' (Female) **9** 'Pyramidalis' (Female)
10 'Pyramidalis Fructo Luteo' (Female) **11** 'Silver Milkmaid' (Female) **12** 'Silver Queen' (Male)

ILEX CRENATA 'CONVEXA' (FEMALE)

This bushy form of the box-leaved holly is a dense, evergreen shrub with purple-green stems and spineless, oval to elliptic, glossy, mid- to dark green leaves. It bears an abundance of small, black berries in autumn. Lends itself for use as hedging or topiary.

CULTIVATION *Needs moist but well-drained, humus rich soil, in full sun or partial shade. Grow near a male holly for a good crop of berries. Cut out badly placed growth in early spring, and trim shaped plants in summer.*

☼ ◑ ◊ ♈H4 ↕to 2.5m (8ft) ↔2m (6ft)

ILEX X *MESERVEAE* 'BLUE PRINCESS' (FEMALE)

This blue holly is a vigorous, dense, evergreen shrub with oval, softly spiny, very glossy, greenish-blue leaves. White to pinkish-white, late spring flowers are followed by a profusion of glossy red berries in autumn. The dark purplish-green young stems show well when hedging plants are regularly clipped. Dislikes coastal conditions.

CULTIVATION *Grow in moist but well-drained, moderately fertile soil, in full sun or semi-shade. For berries, a male holly will need to be nearby. Prune in late summer to maintain shape.*

☼ ◑ ◊ ♈H4 ↕↔ 3m (10ft)

IMPATIENS SUPER ELFIN SERIES

These busy Lizzies, grown as annual bedding plants, have spreading stems bearing long-lasting summer flowers with flattened faces and slender spurs. These bloom in a range of pastel colours and shades of orange, pink, red and violet. The oval leaves are light green. Excellent summer bedding or container plants for a part-shaded site.

CULTIVATION *Grow in well-drained but moist, humus-rich soil, in partial shade with shelter from wind. Plant out after danger of frost has passed.*

☼ ♦ ♀H3 ↕ to 60cm (24in) ↔ to 25cm (10in)

IMPATIENS TEMPO SERIES

These busy Lizzies, in either single colours or mixed, bear a profusion of flattened, spurred flowers which range in colour from violet and lavender-blue to orange, pink, and red; some have contrasting edges or bi-coloured petals. Leaves are light green and slightly toothed. Grown as an annual bedding plant, it provides long-lasting colour in containers and summer borders.

CULTIVATION *Grow in moist but well-drained, humus-rich soil, in a sheltered site. Tolerates shade. Plant out after any danger of frost has passed.*

☼ ♦ ♀H3 ↕ to 23cm (9in) ↔ to 25cm (10in)

INDIGOFERA HETERANTHA

A medium-sized, spreading shrub grown for its pea-like flowers and elegant foliage. The arching stems carry grey-green leaves made up of many oval to oblong leaflets. Dense, upright clusters of small, purple-pink flowers appear from early summer to autumn. Train against a warm wall in cold climates. *I. ambylantha* is a very similar shrub, also recommended.

CULTIVATION *Grow in well-drained but moist, moderately fertile soil, in full sun. Prune in early spring, cutting back to just above ground level.*

☼ ◊ ♀H3 ↕↔ 2–3m (6–10ft)

IPHEION UNIFLORUM 'WISLEY BLUE'

A vigorous, clump-forming, mainly spring-flowering, bulbous perennial bearing scented, star-shaped, lilac-blue flowers; each petal has a pale base and a dark midrib. Narrow, strap-like, light blue-green leaves are produced in autumn. Useful in a rock garden or for underplanting herbaceous plants. 'Froyle Mill' is another recommended cultivar, its flowers more mauve in colour.

CULTIVATION *Grow in moist but well-drained, moderately fertile, humus-rich soil, in full sun. In colder areas, provide a mulch in winter.*

☼ ◊ ♀H4 ↕15–20cm (6–8in)

IPOMOEA 'HEAVENLY BLUE'

This summer-flowering, twining, fast-growing form of morning glory is grown as a climbing annual. The large, funnel-shaped flowers, azure-blue with pure white throats, appear singly or in clusters of two or three. The heart-shaped, light to mid-green leaves have slender tips. Suitable for a summer border, scrambling among other plants. Seeds are highly toxic if ingested.

CULTIVATION *Grow in well-drained, moderately fertile soil, in sun with shelter from cold, drying winds. Plant out after danger of frost has passed.*

☼ ◊ 🏆H3 ↕to 3–4m (10–12ft)

IPOMOEA INDICA

The blue dawn flower is a vigorous, evergreen climber, perennial in frost-free conditions. Abundant, rich purple-blue, funnel-shaped flowers that often fade to red are borne in clusters of three to five from late spring to autumn. The mid-green leaves are heart-shaped or three-lobed. In mild areas, grow as annuals in a warm conservatory or summer border. The seeds are toxic.

CULTIVATION *Grow in well-drained, faily fertile soil, in sun with shelter from cold, drying winds. Plant out after all danger of frost has passed. Minimum temperature 7°C (45°F).*

☼ ◊ 🏆H1 ↕to 6m (20ft)

IRISES FOR WATER GARDENS

Irises that flourish in reliably moist or wet soils produce swollen, horizontal creeping stems known as rhizomes that lie just below the ground. These produce several new offsets each year, so give plants plenty or room, or divide them regularly. They have strap-shaped leaves and flower in blues, mauves, white or yellow in spring and early summer. The true water irises will grow not only in damp ground but also in shallow water; these include *I. laevigata* and *I. pseudacorus*, very vigorous plants that will soon overwhelm a small pond. Where space is limited, try *I. ensata* or, in moist or even in well-drained soil around the water, plant *I. sibirica* or one of its many attractive cultivars.

CULTIVATION *Grow in deep, acid soil enriched with well-rotted organic matter, in sun or light shade. Best in areas with hot summers. Plant rhizomes in early autumn; divide plants after flowering.*

☼ ◐ 💧💧 🏆H4

1 ↕90cm (36in) ↔ indefinite

2 ↕90cm (36in) ↔ indefinite

3 ↕80cm (32in) ↔ indefinite

4 ↕80cm (32in) ↔ indefinite

5 ↕0.9–1.5m (3–5ft) ↔ indefinite

6 ↕0.9–1.5m (3–5ft) ↔ indefinite

1 *Iris ensata* 'Flying Tiger' **2** *I. ensata* 'Variegata' **3** *I. laevigata* **4** *I. laevigata* 'Variegata' **5** *I. pseudacorus* **6** *I. pseudacorus* 'Variegata'

7 ↕ 1m (3ft) ↔ indefinite
8 ↕ 20–80cm (8–32in) ↔ indefinite
9 ↕ 1m (3ft) ↔ indefinite
10 ↕ 80cm (32in) ↔ indefinite
11 ↕ 80cm (32in) ↔ indefinite
12 ↕ 1m (3ft) ↔ indefinite
13 ↕ 1m (3ft) ↔ indefinite
14 ↕ 80cm (32in) ↔ indefinite
15 ↕ 1m (3ft) ↔ indefinite
16 ↕ 80cm (32in) ↔ indefinite
17 ↕ to 1m (3ft) ↔ indefinite
18 ↕ to 1m (3ft) ↔ indefinite

7 *I.sibirica* 'Annemarie Troeger' **8** *I.versicolor* **9** *I. sibirica* 'Crème Chantilly' **10** *I.sibirica* 'Dreaming Yellow' **11** *I.sibirica* 'Harpswell Happiness' **12** *I.sibirica* 'Mikiko' **13** *I.sibirica* 'Oban' **14** *I.sibirica* 'Perfect Vision' **15** *I.sibirica* 'Roisin' **16** *I.sibirica* 'Smudger's Gift' **17** *I.sibirica* 'Uber den Wolken' **18** *I.sibirica* 'Zakopane'

IRIS BUCHARICA

A fast-growing, spring-flowering, bulbous perennial that carries up to six golden-yellow to white flowers on each stem. The glossy, strap-like leaves die back after flowering. The most commonly available form of this iris has yellow and white flowers.

CULTIVATION *Grow in rich but well-drained, neutral to slightly alkaline soil, in full sun. Water moderately when in growth; after flowering, maintain a period of dry dormancy.*

H3–4 ↕20–40cm (8–16in) ↔ 12cm (5in)

IRIS CONFUSA

This freely spreading, rhizomatous perennial with bamboo-like foliage produces a succession of up to 30 short-lived flowers on each stem during spring. They are white with yellow crests surrounded by purple or yellow spots. The leaves are arranged in fans at the base of the plant. Suitable for a sheltered, mixed or herbaceous border, but may not survive in areas with cold winters.

CULTIVATION *Grow in moist but well-drained, rich soil, in sun or semi-shade. Water moderately when in growth. Keep tidy-looking by removing flowered stems.*

H3 ↕1m (3ft) or more ↔ indefinite

IRIS DELAVAYI

This rhizomatous, deciduous perennial bears three-branched flower stems in summer, each one topped by two light to dark purple-blue flowers. The rounded fall petals have white and yellow flecks. The foliage is grey-green. A handsome perennial with its tall stems, it is easily grown in moist ground.

CULTIVATION *Grow in moist soil in full sun or partial shade. Lift and divide congested clumps after flowering.*

☼ ◑ ♦ ♀H4 ↕to 1.5m (5ft) ↔indefinite

IRIS DOUGLASIANA

A robust, rhizomatous perennial with branched flower stems that each bear two or three white, cream, blue, lavender-blue or red-purple flowers in late spring and early summer. The stiff, glossy, dark green leaves are often red at the bases. A good display plant for a raised bed or trough.

CULTIVATION *Grow in well-drained, neutral to slightly acid laom. SIte in full sun for the best flowers, or light shade. Does not transplant well, so do not lift and divide unnecessarily.*

☼ ◑ ◊ ♀H4 ↕15–70cm (6–28in) ↔indefinite

IRIS FOETIDISSIMA 'VARIEGATA'

The stinking iris is not as unpleasant as it sounds, although the evergreen, silvery leaves, with white stripes in this cultivar, do have a dreadful scent if crushed. A vigorous rhizomatous perennial, it bears yellow-tinged, dull purple flowers in early summer, followed by seed capsules which split in autumn to display decorative scarlet, yellow or, rarely, white seeds. A useful plant for dry shade.

CULTIVATION *Prefers well-drained, neutral to slightly acid loam in shade. Divide congested clumps in autumn.*

H4 ↕30cm–90cm (12–36in) ↔indefinite

IRIS FORRESTII

An elegant, early summer-flowering rhizomatous perennial with slender flower stems that each carry one or two scented, pale yellow flowers with brown markings. The very narrow glossy leaves are mid-green above and grey-green below. Easy to grow in an open border.

CULTIVATION *Grow in moist but well-drained, neutral to slightly acid loam. Position in full sun or partial shade.*

H4 ↕35–40cm (14–16in) ↔indefinite

IRIS GRAMINEA

A deciduous, rhizomatous perennial bearing bright green, strap-like leaves. From late spring, rich purple-violet flowers, with fall petals tipped white and violet-veined, are borne either singly or in pairs; they are often hidden among the leaves. The flowers have a fruity fragrance.

CULTIVATION *Grow in moist but well-drained, neutral to slightly acid loam. Choose a site in full sun or semi-shade. Does not respond well to transplanting.*

☼☀ 🌢🌢 H4 ↕20–40cm (8–16in) ↔ indefinite

IRIS 'KATHARINE HODGKIN'

This very vigorous, tiny but robust, deciduous, bulbous perennial bears delicately patterned, pale blue and yellow flowers, with darker blue and gold markings, in late winter and early spring. The pale to mid-green leaves grow after the flowers have faded. Excellent in a rock garden or at the front of a border, where it will spread slowly to form a group.

CULTIVATION *Grow in well-drained, neutral to slightly alkaline soil, in an open site in full sun.*

☼ 🌢 H4 ↕12cm (5in) when flowering ↔ 5–8cm (2–3in)

BEARDED IRISES (*IRIS*)

These upright, rhizomatous perennials send up fans of sword-shaped, usually broad leaves and simple or branched stems. The flowers are produced in a wide range of colours, with well-developed, often frilly fall and standard petals, and a "beard" of white or coloured hairs in the centre of each fall petal. These are the most widely cultivated group of irises for garden display, usually producing several flowers per stem from spring into early summer, sometimes again later in the season. Taller irises suit a mixed border, and smaller ones may be grown in a rock garden, raised bed, or trough.

CULTIVATION *Grow in well-drained, moderately fertile, neutral to slightly acid soil in full sun. Plant rhizomes in late summer or early autumn, thinly covered with soil. They must not be shaded by other plants. Do not mulch. Divide large or congested clumps in early autumn.*

☼ ◊ ♕H4

1 ↕70cm (28in) or more ↔ 60cm (24in)

2 ↕70cm (28in) or more ↔ 60cm (24in)

3 ↕55m (22in) ↔ 45–60cm (18–24in)

4 ↕70cm (28in) or more ↔ 60cm (24in)

5 ↕↔ 30cm (12in)

1 *Iris* 'Apricorange' **2** *I.* 'Breakers' **3** *I.* 'Brown Lasso' **4** *I.* 'Early Light' **5** *I.* 'Eyebright'

6 ↕ to 70cm (28in) ↔ to 60cm (24in)

7 ↕ 20–40cm (8–16in) ↔ 30cm (12in)

8 ↕ to 70cm (28in) ↔ to 60cm (24in)

9 ↕ to 70cm (28in) ↔ to 60cm (24in)

10 ↕ 70cm (28in) or more ↔ 60cm (24in)

11 ↕ to 70cm (28in) ↔ to 60cm (24in)

12 ↕ 70cm (28in) or more ↔ 60cm (24in)

13 ↕ to 70cm (28in) ↔ to 60cm (24in)

14 ↕ 70cm (28in) or more ↔ 60cm (24in)

15 ↕ 85cm (33in) ↔ 60cm (24in)

6 *I.* 'Happy Mood' **7** *I.* 'Honington' **8** *I.* 'Katie-Koo' **9** *I.* 'Maui Moonlight' **10** *I.* 'Meg's Mantle' **11** *I.* 'Miss Carla' **12** *I.* 'Nicola Jane' **13** *I.* 'Orinoco Flow' **14** *I.* 'Paradise' **15** *I.* 'Paradise Bird'

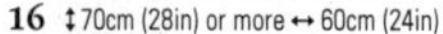

16 ↕70cm (28in) or more ↔ 60cm (24in)

17 ↕40–70cm (16–28in) ↔ 45–60cm (18–24in)

18 ↕70cm (28in) or more ↔ 60cm (24in)

19 ↕to 70cm (28in) ↔ to 60cm (24in)

16 *I.* 'Phil Keen' **17** *I.* 'Pink Parchment' **18** *I.* 'Precious Heather' **19** *I.* 'Quark'

20 ↕25cm (10in) ↔ 30cm (12in)

21 ↕to 70cm (28in) ↔ to 60cm (24in)

22 ↕to 70cm (28in) ↔ to 60cm (24in)

23 ↕to 70cm (28in) ↔ to 60cm (24in)

24 ↕70cm (28in) or more ↔ 60cm (24in)

MORE CHOICES

'Arctic Fancy' White, 50cm (20in) tall.

'Blue-Eyed Brunette' Brown flowers with a lilac blaze, to 90cm (36in) tall

'Bromyard' Blue-grey and ochre flowers, 28cm (11in) tall.

'Stepping Out' White with blue petal edges, 1m (3ft) tall.

25 ↕to 70cm (28in) ↔ to 60cm (24in)

26 ↕90cm (36in) ↔ 60cm (24in)

20 *I.* 'Rain Dance' **21** *I.* 'Sherbet Lemon' **22** *I.* 'Sparkling Lemonade' **23** *I.* 'Sunny Dawn' **24** *I.* 'Sun Miracle' **25** *I.* 'Templecloud' **26** *I.* 'Vanity'

IRIS LACUSTRIS

This dwarf, deciduous, rhizomatous perennial bears small flowers in late spring. These are purple-blue to sky-blue with gold crests and a white patch on each of the the fall petals; they arise from basal fans of narrow leaves. Suitable for growing in a rock garden or trough.

CULTIVATION *Grow in reliably moist, lime-free soil that is rich in humus, in sun or partial shade. Water moderately when in growth.*

☼◑ 💧 ♈H4 ↕10cm (4in) ↔ indefinite

IRIS PALLIDA 'VARIEGATA'

This semi-evergreen, rhizomatous perennial is probably the most versatile and attractive variegated iris. The strap-like, bright green leaves are clearly striped with golden-yellow. (For silver-striped leaves, look for 'Argentea Variegata'.) The large, scented, soft blue flowers with yellow beards are borne in clusters of two to six on branched stems in late spring and early summer. Grow in a mixed or herbaceous border.

CULTIVATION *Best in well-drained, fertile, slightly alkaline soil, in sun. Water moderately when in growth.*

☼ ◊ ♈H4 ↕to 1.2m (4ft) ↔ indefinite

IRIS SETOSA

This rhizomatous perennial flowers in late spring and early summer. Each flowering stem bears several beautiful, blue or blue-purple flowers above the narrow, mid-green leaves. Easily grown in moist soil.

CULTIVATION *Grow in moist, neutral to slightly acid soil in full sun or partial shade. Lift and divide congested clumps after flowering.*

☼ ◐ 💧 🏆H4 ↕15–90cm (6–36in) ↔indefinite

IRIS UNGUICULARIS

A fast-growing, evergreen, rhizomatous perennial, sometimes called *I. stylosa,* with short flower stems bearing large, fragrant blooms from late winter (sometimes even earlier) to early spring. The pale lavender to deep violet petals have contrasting veins and a band of yellow on each fall petal. The leaves are grass-like and mid-green. Ideal for the base of a sunny wall.

CULTIVATION *Grow in sharply drained, neutral to alkaline soil. Choose a warm, sheltered site in full sun. Does not like to be disturbed. Keep tidy by removing dead leaves in late summer and spring.*

☼ ◊ 🏆H4 ↕30cm (12in) ↔indefinite

IRIS VARIEGATA

This slender and robust, deciduous, rhizomatous perennial bears three to six flowers on each branched stem from mid-summer. The striking flowers are pale yellow with brown or violet veins on the fall petals; there are many other colour combinations available. The deep green leaves are strongly ribbed.

CULTIVATION *Grow in well-drained, neutral to alkaline soil, in sun or light shade. Avoid mulching with organic matter, which may encourage rot.*

☼ ◐ ♢ ♀H4 ↕20–45cm (8–18in) ↔ indefinite

ITEA ILICIFOLIA

An evergreen shrub bearing upright at first, then spreading, arching shoots. The oval, holly-like leaves are sharply toothed. Tiny, greenish-white flowers are borne in long, catkin-like clusters from mid-summer to early autumn. Needs a sheltered position in cold areas.

CULTIVATION *Grow in well-drained but moist, fertile soil, preferably against a warm wall in full sun. Protect with a winter mulch when young.*

☼ ♢ ♀H3 ↕3–5m (10–15ft) ↔3m (10ft)

JASMINUM MESNYI

The primrose jasmine is a half hardy, scrambling, evergreen shrub with large, usually semi-double, bright yellow flowers. These appear singly or in small clusters during spring and summer, amid the glossy, dark green leaves which are divided into three oblong to lance-shaped leaflets. Will climb if tied to a support, but may not survive in frosty areas.

CULTIVATION *Grow in any well-drained, fertile soil, in full sun or partial shade. Cut back flowered shoots in summer to encourage strong growth from the base.*

☼ ◑ ◊ 🏆H2–3 ↕to 3m (10ft) ↔1–2m (3–6ft)

JASMINUM NUDIFLORUM

Winter jasmine is a lax, mound-forming, deciduous shrub with slender, arching stems. Small, tubular yellow flowers are borne singly on the leafless, green shoots in late winter. The dark green leaves, which develop after the flowers, are divided into three leaflets. Tie in a framework of stems against a wall, or let it sprawl unsupported.

CULTIVATION *Grow in well-drained, fertile soil. Tolerates semi-shade, but flowers best in sun. Encourage strong growth by cutting back flowered shoots.*

☼ ◑ ◊ 🏆H4 ↕↔ to 3m (10ft)

JASMINUM OFFICINALE 'ARGENTEOVARIEGATUM'

This variegated form of the common jasmine, *J. officinale*, is a vigorous, deciduous or semi-deciduous, woody climber. The grey-green, cream-edged leaves are made up of 5–9 sharply pointed leaflets. Clusters of fragrant white flowers open from summer to early autumn. If tied in initially, it will twine over supports, such as a trellis or an arch. May not survive in cold, exposed areas.

CULTIVATION *Grow in well-drained, fertile soil. Tolerates shade, but flowers best in full sun. Thin out crowded growth after flowering.*

☼ ◑ ◊ ♡H4 ↕to 12m (40ft)

JASMINUM POLYANTHUM

A fast-growing, woody-stemmed, twining, evergreen climber that bears, in very sheltered conditions, abundant clusters of small, strongly fragrant white flowers. These open from pink buds in late spring and early summer, amid the deep green, divided leaves. Allow to climb over a trellis, fence, arch or large shrub, in a frost-free position.

CULTIVATION *Grow in well-drained, moderately fertile soil, against a warm, sunny wall. Thin out overcrowded growth after flowering.*

☼ ◊ ♡H1–2 ↕to 3m (12ft)

JUNIPERUS COMMUNIS 'COMPRESSA'

This slow-growing, spindle-shaped, dwarf form of the common juniper bears deep to blue-green, aromatic, evergreen scale-like leaves, borne in whorls of three along the stems. Small, oval or spherical fruits remain on the plant for three years, ripening from green through to cloudy-blue to black. 'Hibernica' is another recommended narrowly upright juniper, slightly faster-growing.

CULTIVATION *Grow in any well-drained soil, preferably in full sun or light dappled shade. No pruning is needed.*

☼☀ ◊ ♈H4 ↕to 80cm (32in) ↔45cm (18in)

JUNIPERUS X *PFITZERIANA* 'PFITZERIANA'

This spreading, dense, evergreen shrub has ascending branches of grey-green foliage which droop at the tips; it eventually forms a flat-topped, tiered bush. The flattened, scale-like leaves are borne in whorls of three. Spherical fruits are at first dark purple, becoming paler as they age. Looks well as a specimen plant or in a large rock garden.

CULTIVATION *Grow in any well-drained soil, preferably in full sun or light dappled shade. Keep pruning to a minimum, in late autumn if necessary.*

☼☀ ◊ ♈H4 ↕1.2m (4ft) ↔3m (10ft)

JUNIPERUS PROCUMBENS 'NANA'

A compact, mat-forming conifer that is excellent for ground cover in a wide range of situations. The needle-like, aromatic, yellow-green or light green leaves are carried in groups of three. Bears berry-like, brown to black, fleshy fruits which take two or three years to ripen.

CULTIVATION *Grow in any well-drained soil, including sandy, dry, or chalky conditions. Site in full sun or dappled shade. No pruning is required.*

☼☀ ◊ H4 ↕15–20cm (6–8in) ↔75cm (30in)

JUNIPERUS SQUAMATA 'BLUE STAR'

This conifer is a low-growing, dense, compact, rounded bush with rust-coloured, flaky bark. The silvery-blue leaves are sharply pointed, and grouped in whorls of three. The ripe fruits are oval and black. Useful as ground cover or in a rock garden.

CULTIVATION *Grow in any well-drained soil, in full sun or dappled shade. Very little pruning is required.*

☼☀ ◊ H4 ↕to 40cm (16in) ↔to 1m (3ft)

KALMIA ANGUSTIFOLIA

The sheep laurel is a tough, rabbit-proof shrub grown for its spectacular, rounded clusters of small flowers, usually pink to deep red, but occasionally white. They appear in early summer amid the dark green leaves. Useful for a shrub border or rockery; naturally mound-forming, it tolerates trimming to a neat shape if necessary.

CULTIVATION *Choose a part-shaded site in moist, acid soil, rich in organic matter. Grow in full sun only where the soil remains reliably moist. Mulch in spring with leaf mould or pine needles. Trim or prune hard after flowering.*

◐ 💧 🏆H4 ↕60cm (24in) ↔1.5m (5ft)

KALMIA LATIFOLIA

The calico bush is a dense evergreen shrub producing large clusters of flowers from late spring to mid-summer. These are cup-shaped, pink or occasionally white, and open from distinctively crimped buds. The oval leaves are glossy and dark green. An excellent specimen shrub for woodland gardens, but flowers best in full sun. 'Ostbo Red' is a recommended cultivar.

CULTIVATION *Grow in moist, humus-rich, acid soil, in sun or partial shade. Mulch each spring with pine needles or leaf mould. Requires very little pruning, although deadheading is worthwhile.*

☼◐ 💧 🏆H4 ↕↔ 3m (10ft)

KERRIA JAPONICA 'GOLDEN GUINEA'

This vigorous, suckering, deciduous shrub forms clumps of arching, cane-like shoots which arise from ground level each year. Large, single yellow flowers are borne in mid- and late spring along the previous year's growth. ('Flore Pleno' has double flowers.) The bright green leaves are oval and sharply toothed.

CULTIVATION *Grow in well-drained, fertile soil, in full sun or partial shade. Cut flowered canes back to different levels to obtain flowers at different heights. Chop out unwanted canes and suckers with a spade to restrict spread.*

☼ ◑ ◊ 🏆H4 ↕2m (6ft) ↔2.5m (8ft)

KIRENGESHOMA PALMATA

A handsome, upright perennial with broad, lobed, pale green leaves. In late summer and early autumn, these are topped by loose clusters of nodding, pale yellow flowers, which are sometimes called "yellow wax bells". Brings a gentle elegance to a shady border, poolside or woodland garden.

CULTIVATION *Thrives in moist, lime-free soil, enriched with leaf mould, in partial shade sheltered from wind. If necessary, divide clumps in spring.*

◑ ◆ 🏆H4 ↕60–120cm (24–48in) ↔75cm (30in)

KNIPHOFIA 'BEES' SUNSET'

This red-hot poker is a deciduous perennial grown for its elegant spikes of soft yellowish-orange flowers. These appear through summer above the clumps of arching, grass-like leaves. Very attractive to bees.

CULTIVATION *Grow in deep, fertile, moist but well-drained soil, ideally sandy but enriched with organic matter. Choose a site in full sun or partial shade. Mulch young plants for their first winter, and divide mature, crowded clumps in late spring.*

☼☀ ◊◊ ♈H4 ↕90cm (36in) ↔60cm (24in)

KNIPHOFIA CAULESCENS

This stately evergreen perennial bears tall spikes of flowers from late summer to mid-autumn; coral-red, they fade upwards as they age to pale yellow, giving them their common name of red-hot pokers. They are carried well above basal rosettes of arching, grass-like, blue-green leaves. Good for a herbaceous border; tolerant of coastal exposure.

CULTIVATION *Grow in deep, fertile, moist but well-drained soil, preferably sandy and enriched with organic matter. Position in full sun or partial shade. Mulch young plants for their first winter. Divide large clumps in late spring.*

☼☀ ◊◊ ♈H3–4 ↕to 1.2m (4ft) ↔60cm (24in)

KNIPHOFIA 'LITTLE MAID'

This clump-forming, deciduous perennial has tall heads of tubular flowers which appear from late summer to early autumn. They are pale green in bud, opening to pale buff-yellow, then fading to ivory. The leaves are narrow and grass-like. Good for late displays in a mixed or herbaceous border.

CULTIVATION *Grow in well-drained, deep, fertile, humus-rich soil, in full sun. Keep moist when in growth. In their first winter and in cold areas, provide a mulch of straw or leaves.*

☼ ◊ 🏆H3–4 ↕60cm (24in) ↔45cm (18in)

KNIPHOFIA 'ROYAL STANDARD'

A clump-forming, herbaceous perennial of classic red-hot poker appearance that bears tall, conical flowerheads from mid- to late summer. The bright yellow, tubular flowers open from red buds, starting at the base and moving upwards. The arching, grass-like leaves are deciduous, dying back in winter.

CULTIVATION *Grow in deep, moist but well-drained, humus-rich soil, in sun. Water freely when in growth. Provide a mulch in cold areas, especially for young plants in their first winter.*

☼ ◊◆ 🏆H4 ↕1m (3ft) ↔60cm (24in)

KNIPHOFIA TRIANGULARIS

A clump-forming, deciduous perennial grown for its early to mid-autumn display of long, reddish-orange, tubular flowers. These are borne in dense, spike-like heads, and become slightly yellow around the mouths. The leaves are narrow, grass-like and arching. A good specimen for waterside planting.

CULTIVATION *Best in deep, moist but well-drained, humus-rich, fertile soil, in full sun. Water freely when in growth. Mulch in cold areas, especially young plants in their first winter.*

☼ ◊◆ 🏆H3–4 ↕60–90cm (24–36in) ↔45cm (18in)

KOLKWITZIA AMABILIS 'PINK CLOUD'

The beauty bush is a fast-growing, suckering, deciduous shrub with an arching habit. Dense clusters of bell-shaped pink flowers with yellow-flushed throats appear in abundance from late spring to early summer. The leaves are dark green and broadly oval. Excellent for a shrub border or as a specimen plant.

CULTIVATION *Grow in any well-drained, fertile soil, in full sun. Let the arching habit of young plants develop without pruning, then thin out the stems each year after flowering, to maintain vigour.*

☼ ◊ 🏆H4 ↕3m (10ft) ↔4m (12ft)

LABURNUM x *WATERERI* 'VOSSII'

This spreading, deciduous tree bears long, hanging clusters of golden-yellow, pea-like flowers in late spring and early summer. The dark green leaves are made up of three oval leaflets. A fine specimen tree for small gardens; it can also be trained on an arch, pergola or tunnel framework. All parts are toxic if eaten.

CULTIVATION *Grow in well-drained, moderately fertile soil, in full sun. Cut back badly placed growth in winter or early spring. Remove any suckers or buds at the base of the trunk.*

☼ ◊ ♈H4 ↕↔ 8m (25ft)

LAGURUS OVATUS

Hare's tail is an annual grass that bears fluffy, oval flowerheads in summer. These are pale green, often purple-tinged, and fade to a pale, creamy-buff. The flat, narrow leaves are pale green. Effective in a border; the flowers can be cut for indoor arrangements; for drying, pick the heads before they are fully mature.

CULTIVATION *Best in light, well-drained, moderately fertile, ideally sandy soil. Choose a position in full sun.*

☼ ◊ ♈H3 ↕to 50cm (20in) ↔30cm (12in)

LAMIUM MACULATUM 'WHITE NANCY'

This colourful dead-nettle is a semi-evergreen perennial which spreads to form mats; this makes it effective as ground cover between shrubs. Spikes of pure white, two-lipped flowers are produced in summer above triangular to oval, silver leaves which are edged with green.

CULTIVATION *Grow in moist but well-drained soil, in partial or deep shade. Can be invasive, so position away from other small plants, and dig up invasive roots or shoots to limit spread.*

☀ ☀ 💧💧 🏆H4 ↕to 15cm (6in) ↔to 1m (3ft) or more

LAPAGERIA ROSEA

The Chilean bell flower is a long-lived, twining, evergreen climber. From summer to late autumn, it produces large, narrowly bell-shaped, waxy red flowers which are borne either singly or in small clusters. The leaves are oval and dark green. In cold areas, it needs the protection of a warm, but partially shaded, wall. 'Nash Court' has soft pink flowers.

CULTIVATION *Grow in well-drained, moderately fertile soil, preferably in partial shade. In frost-prone areas, shelter from wind and provide a winter mulch. Keep pruning to a minimum, removing damaged growth in spring.*

☀ 💧 🏆H3 ↕5m (15ft)

LATHYRUS LATIFOLIUS

The everlasting or perennial pea is a tendril-climbing, herbaceous perennial with winged stems, ideal for growing through shrubs or over a bank. Clusters of pea-like, pink-purple flowers appear during summer and early autumn, amid the deciduous, blue-green leaves which are divided into two oblong leaflets. The seeds are not edible. For white flowers, choose 'Albus' or 'White Pearl'.

CULTIVATION *Grow in well-drained, fertile, humus-rich soil, in sun or semi-shade. Cut back to ground level in spring and pinch out shoot tips to encourage bushiness. Resents disturbance.*

☼☀ ◊ ♈H4 ↕2m (6ft) or more

LATHYRUS VERNUS

Spring vetchling is a dense, clump-forming, herbaceous perennial with upright stems. Despite its pea-like appearance, it does not climb. In spring, clusters of purplish-blue flowers appear above the mid- to dark green leaves which are divided into several pointed leaflets. Suitable for a rock or woodland garden. The cultivar 'Alboroseus' has pretty bicoloured flowers, in pink and white.

CULTIVATION *Grow in well-drained soil, in full sun or partial shade. Tolerates poor soil, but resents disturbance.*

☼☀ ◊ ♈H4 ↕20–45cm (8–18in) ↔45cm (18in)

SWEET PEAS (*LATHYRUS ODORATUS*)

The many cultivars of *Lathyrus odoratus* are annual climbers cultivated for their long display of beautiful and fragrant flowers which cut well and are available in most colours except yellow. The flowers are arranged in clusters during summer to early autumn. The seeds are not edible. Most look very effective trained on a pyramid of canes or trellis, or scrambling amid shrubs and perennials. Compact cultivars such as 'Patio Mixed' suit containers; some of these are self-supporting. Grow sweet peas, also, in a vegetable garden, because they attract pollinating bees and other beneficial insects.

CULTIVATION *Grow in well-drained, fertile soil; for the best flowers, add well-rotted manure the season before planting. Site in full sun or partial shade. Feed with a balanced fertiliser fortnightly when in growth. Deadhead or cut flowers for the house regularly. Support the climbing stems.*

☼ ◑ ◊ ♦ 🏆H4

1 ↕2–2.5m (6–8ft) 2 ↕2–2.5m (6–8ft) 3 ↕2–2.5m (6–8ft)

4 ↕1m (3ft) 5 ↕2–2.5m (6–8ft) 6 ↕2–2.5m (6–8ft)

1 *Lathyrus odoratus* 'Aunt Jane' **2** *L. odoratus* 'Evening Glow' **3** *L. odoratus* 'Noel Sutton' **4** *L. odoratus* 'Patio Mixed' **5** *L. odoratus* 'Teresa Maureen' **6** *L. odoratus* 'White Supreme'

LAURUS NOBILIS

Bay, or bay laurel, is a conical, evergreen tree grown for its oval, aromatic, leathery, dark green leaves which are used in cooking. Clusters of small, greenish-yellow flowers appear in spring, followed by black berries in autumn. Effective when trimmed into formal shapes.

CULTIVATION *Grow in well-drained but moist, fertile soil, in sun or semi-shade with shelter from cold, drying winds. Grow male and female plants together for a reliable crop of berries. Prune young plants to shape in spring; trim established plants lightly in summer.*

☼☼ ◊ ♀H4 ↕12m (40ft) ↔10m (30ft)

LAURUS NOBILIS 'AUREA'

This golden-yellow-leaved bay is a conical tree bearing aromatic, evergreen leaves. These are oval and leathery, and can be used in cooking. In spring, clusters of small, greenish-yellow flowers appear, followed in autumn by black berries on female plants. Good for topiary and in containers, where it makes a much smaller plant.

CULTIVATION *Grow in well-drained but moist, fertile soil. Position in full sun or partial shade with shelter from cold winds. Prune young plants to shape in spring; once established, trim lightly in summer to encourage a dense habit.*

☼☼ ◊ ♀H4 ↕12m (40ft) ↔10m (30ft)

LAVANDULA ANGUSTIFOLIA 'HIDCOTE'

This compact lavender with thin, silvery-grey leaves and dark purple flowers is an evergreen shrub useful for edging. Dense spikes of fragrant, tubular flowers, borne at the ends of long, unbranched stalks, appear during mid- to late summer. Like all lavenders, the flowers dry best if cut before they are fully open.

CULTIVATION *Grow in well-drained, fertile soil, in sun. Cut back flower stems in autumn, and trim the foliage lightly with shears at the same time. In frost-prone areas, leave trimming until spring. Do not cut into old wood.*

☼ ◊ 🏆H4 ↕60cm (24in) ↔75cm (30in)

LAVANDULA ANGUSTIFOLIA 'TWICKEL PURPLE'

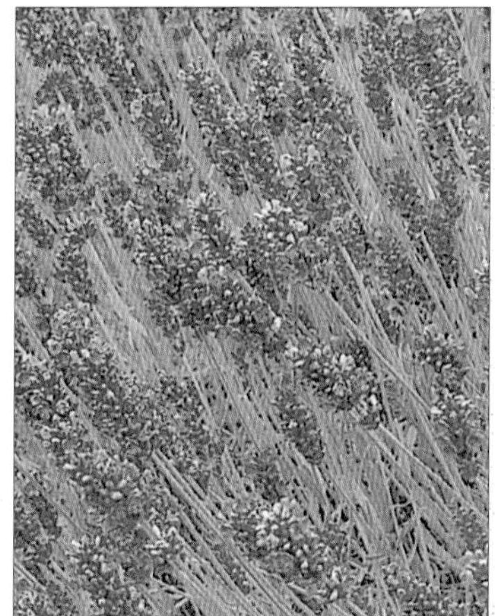

This evergreen shrub is a close relative of 'Hidcote' (above) with a more spreading habit, paler flowers and greener leaves. Dense spikes of fragrant purple flowers are borne in mid-summer above narrowly oblong, grey-green leaves. Suited to a shrub border; like all lavenders, the flower-heads are very attractive to bees.

CULTIVATION *Grow in well-drained, fairly fertile soil, in full sun. Trim in autumn, or delay until spring in frost-prone areas. Do not cut into old wood.*

 ↕60cm (24in) ↔1m (3ft)

LAVANDULA X *INTERMEDIA* DUTCH GROUP

A tall, robust, bushy lavender, with broad, silvery, aromatic leaves and tall, slender spikes of scented lavender-blue flowers. Suited to a sunny shrub border or large rock or scree garden; makes good low hedging if regularly trimmed. 'Grappenhall' is very similar, but not as hardy.

CULTIVATION *Grow in well-drained, fertile soil, in sun. Cut back flower stems in autumn, and trim the foliage lightly with shears at the same time. In frost-prone areas, leave trimming until spring. Do not cut into old wood.*

☼ ◊ ♈H4 ↕↔1.2m (4ft)

LAVANDULA STOECHAS SUBSP. *PEDUNCULATA*

French lavender is a compact, evergreen shrub which blooms from late spring to summer. Dense spikes of tiny, fragrant, dark purple flowers, each spike topped by distinctive, rose-purple bracts, are carried on long stalks well above the narrow, woolly, silvery-grey leaves. Effective in a sheltered shrub border or rock garden. May suffer in cold areas.

CULTIVATION *Grow in well-drained, fairly fertile soil, in sun. Trim back in spring or, in frost-free climates, after flowering. Avoid cutting into old wood.*

☼ ◊ ♈H3 ↕↔60cm (24in)

SHRUBBY LAVATERAS (*LAVATERA*)

These upright and showy, flowering shrubs with stiff stems and sage-green leaves usually bloom from mid-summer to autumn in pinks and purples. Although they are short-lived, they grow quickly on any well-drained soil, including thin, dry soil, which makes them a welcome addition to any garden where quick results are desired. They also perform well in coastal areas, being able to tolerate salt-laden winds, but the shrubs will need staking if grown in a site exposed to wind. In areas prone to severe frost, grow against a warm, sunny wall.

CULTIVATION *Grow in any poor to moderately fertile, well-drained to dry soil in full sun. Shelter from cold, drying winds in frost-prone areas. After cold winters, frost-damaged plants are best pruned right down to the base in spring to encourage new growth and tall, vigorous stems. In milder areas, trim to shape after flowering.*

☼ ◊ ♀H3–4

1 ↕↔ 2m (6ft) 2 ↕↔ 2m (6ft) 3 ↕↔ 2m (6ft) 4 ↕↔ 2m (6ft) 5 ↕↔ 2m (6ft)

1 *Lavatera* 'Barnsley' **2** *L.* 'Bredon Springs' **3** *L.* 'Burgundy Wine' **4** *L.* 'Candy Floss' **5** *L.* 'Rosea'

ANNUAL LAVATERAS (*LAVATERA*)

These sturdy, bushy annuals are an excellent choice for planting in groups in a herbaceous border or for summer bedding, flowering continuously from mid-summer into autumn. They are softly hairy or downy plants with shallowly lobed, soft green leaves; the open funnel-shaped, pink, reddish-pink, or purple blooms are nicely complemented by the foliage. Although annual lavateras are only a temporary visitor to the garden, with their cottage-garden appearance they seem to look as if they have been around for years. Excellent for a dry, sunny site and as a source of cut flowers.

CULTIVATION *Grow in light, moderately fertile, well-drained soil in full sun. Young plants need a regular supply of water until they become established; after this, the plants are fairly drought-tolerant. Watch out for aphids, which often attack young growth.*

☼ ◊ ♀H3

1 ↕ to 60cm (24in) ↔ 45cm (18in)

2 ↕ to 60cm (24in) ↔ 45cm (18in)

3 ↕ to 60cm (24in) ↔ 45cm (18in)

4 ↕ to 75cm (30in) ↔ 45cm (18in)

5 ↕ to 60cm (24in) ↔ 45cm (18in)

6 ↕ to 75cm (30in) ↔ 45cm (18in)

1 *Lavatera* 'Beauty Formula Mixture' **2** *L.* 'Pink Beauty' **3** *L.* 'Salmon Beauty' **4** *L.* 'Silver Cup' **5** *L.* 'White Beauty' **6** *L.* 'White Cherub'

LEIOPHYLLUM BUXIFOLIUM

Sand myrtle is an upright to mat-forming, evergreen perennial, grown for its glossy, dark green foliage and for its abundance of star-shaped, pinkish-white flowers in late spring and early summer. The leaves tint bronze in winter. Good, free-flowering underplanting for a shrub border or woodland garden.

CULTIVATION *Grow in moist but well-drained, acid soil rich in organic matter. Choose a site in partial or deep shade with protection from cold, drying winds. Trim after flowering; it may spread widely if left unattended.*

◐● ◊◊ ♈H4 ↕30–60cm (12–24in) ↔60cm (24in) or more

LEPTOSPERMUM RUPESTRE

This low-growing, evergreen shrub with dense foliage bears star-shaped white flowers from late spring to summer. The small, aromatic leaves are glossy, elliptic and dark green. Native to coastal areas of Tasmania, it is useful in seaside gardens provided that the climate is mild. May be sold as *L. humifusum*.

CULTIVATION *Grow in well-drained, fertile soil, in full sun or partial shade. Trim young growth in spring to promote bushiness, but do not cut into old wood.*

☼◐ ◊ ♈H3 ↕0.3–1.5m (1–5ft) ↔1–1.5m (3–5ft)

LEPTOSPERMUM SCOPARIUM 'KIWI'

A compact shrub with arching shoots bearing an abundance of small, flat, dark crimson flowers during late spring and early summer. The small, aromatic leaves are flushed with purple when young, maturing to mid- or dark green. Suitable for a large rock garden, and attractive in an alpine house display; may not survive cold winters. 'Nicholsii Nanum' is similarly compact.

CULTIVATION *Grow in well-drained, moderately fertile soil, in full sun or part-shade. Trim new growth in spring for bushiness; do not cut into old wood.*

H3 ↕↔ 1m (3ft)

LEUCANTHEMUM X *SUPERBUM* 'WIRRAL SUPREME'

A robust, clump-forming, daisy-flowered perennial with, from early summer to early autumn, dense, double white flowerheads. These are carried singly at the end of long stems, above lance-shaped, toothed, dark green leaves. Good for cut flowers. 'Aglaia' and 'T.E. Killin' are other recommended cultivars.

CULTIVATION *Grow in moist but well-drained, moderately fertile soil, in full sun or partial shade. May need staking.*

H4 ↕ 90cm (36in) ↔ 75cm (30in)

LEUCOJUM AESTIVUM 'GRAVETYE GIANT'

This robust cultivar of summer snowflake is a spring-flowering, bulbous perennial with upright, strap-shaped, dark green leaves, to 40cm (16in) tall. The faintly chocolate-scented, drooping, bell-shaped white flowers with green petal-tips are borne in clusters. Good planted near water, or for naturalizing in grass.

CULTIVATION *Grow in reliably moist, humus-rich soil, preferably near water. Choose a position in partial shade*

☀ 💧 🏆H4 ↕1m (3ft) ↔ 8cm (3in)

LEUCOJUM AUTUMNALE

A slender, late-summer-flowering, bulbous perennial bearing stems of two to four drooping, bell-shaped white flowers, tinged red at the petal bases. Narrow, upright, grass-like leaves appear at the same time as or just after the flowers. Suitable for a rock garden.

CULTIVATION *Grow in any moist but well-drained soil. Choose a position in full sun. Divide and replant bulbs once the leaves have died down.*

☼ 💧 🏆H4 ↕10–15cm (4–6in) ↔5cm (2in)

LEWISIA COTYLEDON

This evergreen perennial produces tight clusters of open funnel-shaped, usually pinkish-purple flowers; they may be white, cream, yellow or apricot. These are borne on long stems from spring to summer. The dark green, fleshy, lance-shaped leaves are arranged in basal rosettes. Suitable for growing in wall crevices. The Sunset Group are recommended garden forms.

CULTIVATION *Grow in sharply drained, fairly fertile, humus-rich, neutral to acid soil. Choose a site in light shade where plants will be sheltered from winter wet.*

◑ ◊ ♀H4 ↕15–30 cm (6–12in) ↔20–40cm (8–16in)

LEWISIA TWEEDYI

An evergreen perennial with upright to arching stems that bear one to four open funnel-shaped, white to peach-pink flowers in spring and early summer. Lance-shaped, fleshy, deep green leaves flushed with purple are arranged in rosettes at the base of the plant. Good in a rock garden, or in an alpine house in climates with frosty winters.

CULTIVATION *Grow in sharply drained, humus-rich, fairly fertile, neutral to acid soil, in light shade. Protect plants in winter from excessive winter wet.*

◑ ◊ ♀H2 ↕20cm (8in) ↔30cm (12in)

LIGULARIA 'GREGYNOG GOLD'

A large, robust, clump-forming perennial with pyramidal spikes of daisy-like, golden-orange, brown-centred flowers from late summer to early autumn. These are carried on upright stems above the large, rounded, mid-green leaves. Excellent by water; it naturalizes readily in moist soils. Dark-leaved 'Desdemona', only 1m (3ft) tall, and 'The Rocket' are other recommended ligularias.

CULTIVATION *Grow in reliably moist, deep, moderately fertile soil. Position in full sun with some midday shade, and shelter from strong winds.*

☼◑ 💧 🏆H4 ↕to 2m (6ft) ↔1m (3ft)

LIGUSTRUM LUCIDUM

The Chinese privet is a vigorous, conical, evergreen shrub with glossy dark green, oval leaves. Loose clusters of small white flowers are produced in late summer and early autumn, followed by oval, blue-black fruits. Good as hedging, but also makes a useful, well-shaped plant for a shrub border.

CULTIVATION *Best in well-drained soil, in full sun or partial shade. Cut out any unwanted growth in winter.*

☼◑ ♢ 🏆H4 ↕↔ 10m (30ft)

LIGUSTRUM LUCIDUM 'EXCELSUM SUPERBUM'

This variegated Chinese privet, with yellow-margined, bright green leaves, is a fast-growing, conical, evergreen shrub. Loose clusters of small, creamy-white flowers appear in late summer and early autumn, followed by oval, blue-black fruits.

CULTIVATION *Grow in any well-drained soil, in full sun for the best leaf colour. Remove unwanted growth in winter. Remove any shoots that have plain green leaves as soon as seen.*

☼ ◇ ♈H4 ↕↔ 10m (30ft)

LILIUM CANDIDUM

The Madonna lily is an upright, bulbous perennial, bearing sprays of up to 20 highly fragrant, trumpet-shaped flowers on each stiff stem in mid-summer. The flowers have pure white petals with tinted yellow bases, and yellow anthers. The lance-shaped, glossy bright green leaves that appear after the flowers usually last over winter.

CULTIVATION *Grow in well-drained, neutral to alkaline soil that is rich in well-rotted organic matter. Tolerates drier soil than most lilies. Position in full sun with the base in shade.*

☼ ◇ ♈H4 ↕1–2m (3–6ft)

LILIUM FORMOSANUM VAR. *PRICEI*

An elegant, clump-forming perennial bearing very fragrant, slender, trumpet-shaped flowers. These are borne singly or in clusters of up to three during summer. The flowers have curved petal tips, white insides and strongly purple-flushed outsides. Most of the oblong, dark green leaves grow at the base of the stem. Lovely for an unheated greenhouse or conservatory.

CULTIVATION *Grow in moist, neutral to acid, humus-rich soil or compost, in sun with the base in shade. Protect from excessively hot sun.*

☼ ◐ 🏆H4 ↕0.6–1.5m (2–5ft)

LILIUM HENRYI

A fast-growing, clump-forming perennial that bears a profusion of slightly scented, "turkscap" flowers (with reflexed, or backward-bending petals) in late summer. These are deep orange with brown spots and red anthers, carried on purple-marked green stems above lance-shaped leaves. Excellent for a wild garden or woodland planting.

CULTIVATION *Grow in well-drained, neutral to alkaline soil with added leaf mould or well-rotted organic matter. Choose a position in partial shade.*

☀ ◇ 🏆H4 ↕1–3m (3–10ft)

LILIUM LONGIFLORUM

The Easter lily is a fast-growing perennial carrying short clusters of one to six pure white, strongly fragrant, trumpet-shaped flowers with yellow anthers. They appear during mid-summer above scattered, lance-shaped, deep green leaves. One of the smaller, less hardy lilies, it grows well in containers and under glass.

CULTIVATION *Best in well-drained soil or compost with added organic matter, in partial shade. Tolerates lime.*

H2–3 ↕40–100cm (16–39in)

LILIUM MARTAGON VAR. *ALBUM*

A clump-forming, vigorous perennial bearing sprays of up to 50 small, nodding, glossy white, "turkscap" flowers with strongly curled petals. The leaves are elliptic to lance-shaped, mostly borne in dense whorls. Unlike most lilies, it has an unpleasant smell and is better sited in a border or wild garden, for which it is well-suited. Looks good planted with var. *cattaniae*, with its deep maroon flowers.

CULTIVATION *Best in almost any well-drained soil, in full sun or partial shade. Water freely when in growth.*

H4 ↕1–2m (3–6 ft)

LILIUM MONADELPHUM

This stout, clump-forming perennial, also known as *L. szovitsianum*, bears up to 30 large, fragrant, trumpet-shaped flowers on each stiff stem in early summer. The blooms are pale yellow, flushed brown-purple on the outsides and flecked purple-maroon on the insides. The scattered, bright green leaves are narrowly oval. Good for containers, as it tolerates drier conditions than most lilies.

CULTIVATION *Grow in any well-drained soil, in full sun. Tolerates fairly heavy soils and lime.*

☼ ◊ H4 ↕1–1.5m (3–5ft)

LILIUM PINK PERFECTION GROUP

These stout-stemmed lilies bear clusters of large, scented, trumpet-shaped flowers with curled petals in mid-summer. Flower colours range from deep purplish-red to purple-pink, all with bright orange anthers. The mid-green leaves are strap-like. Excellent for cutting.

CULTIVATION *Grow in well-drained soil that is enriched with leaf mould or well-rotted organic matter. Choose a position in full sun with the base in shade.*

☼ ◊ H4 ↕1.5–2m (5–6ft)

LILIUM PYRENAICUM

A relatively short, bulbous lily that produces up to 12 nodding, green-yellow or yellow, purple-flecked flowers per stem, in early to mid-summer. Their petals are strongly curved back. The green stems are sometimes spotted with purple, and the lance-shaped, bright green leaves often have silver edges. Not a good lily for patio planting, since its scent is unpleasant.

CULTIVATION *Grow in well-drained, neutral to alkaline soil with added leaf mould or well-rotted organic matter. Position in full sun or partial shade.*

☼ ◑ ♢ H4 ↕30–100cm (12–39in)

LILIUM REGALE

The regal lily is a robust, bulbous perennial with very fragrant, trumpet-shaped flowers opening during mid-summer. They can be borne in sprays of up to 25, and are white, flushed with purple or purplish-brown on the outsides. The narrow leaves are numerous and glossy dark green. A bold statement in a mixed border. Suitable for growing in pots.

CULTIVATION *Grow in well-drained soil enriched with organic matter. Dislikes very alkaline conditions. Position in full sun or partial shade.*

☼ ◑ ♢ H4 ↕0.6m–2m (2–6ft)

LIMNANTHES DOUGLASII

The poached egg plant is an upright to spreading annual which produces a profusion of white-edged, yellow flowers from summer to autumn. The deeply toothed, glossy, bright yellow-green leaves are carried on slender stems. Good for brightening up a rock garden or path edging, and attractive to hoverflies which help control aphids.

CULTIVATION *Grow in moist but well-drained, fertile soil, in full sun. Sow seed outdoors during spring or autumn. After flowering, it self-seeds freely.*

☼ ◊ ♀H4 ↕↔ to 15cm (6in) or more

LIMONIUM SINUATUM 'FOREVER GOLD'

An upright perennial bearing tightly packed clusters of bright yellow flowers from summer to early autumn. The stiff stems have narrow wings. Most of the dark green leaves are arranged in rosettes around the base of the plant. Suitable for a sunny border or in a gravel garden; the flowers dry well for indoor arrangements. Like other statice, this is usually grown as an annual since it may not survive cold winters.

CULTIVATION *Grow in well-drained, preferably sandy soil, in full sun. Tolerates dry and stony conditions.*

☼ ◊ ♀H3 ↕to 60cm (24in) ↔ 30cm (12in)

LINUM 'GEMMELL'S HYBRID'

A dome-forming, semi-evergreen, perennial flax bearing abundant clusters of yellow, broadly funnel-shaped flowers which open for many weeks throughout summer. The leaves are oval and blue-green. Suitable for a rock garden.

CULTIVATION *Grow in light, moderately fertile, humus-rich soil, with protection from winter wet. Position in full sun.*

☼ ◊ ♀H4 ↕15cm (6in) ↔ to 20cm (8in)

LIRIODENDRON TULIPIFERA

The stately tulip tree has a broadly columnar habit, spreading with age. The deciduous, squarish, lobed leaves are dark green, turning butter-yellow in autumn. Tulip-shaped, pale green flowers, tinged orange at the base, appear in summer and are followed by cone-like fruits in autumn. An excellent specimen tree for a large garden.

CULTIVATION *Grow in moist but well-drained, moderately fertile, preferably slightly acid soil. Choose a site in full sun or partial shade. Keep pruning of established specimens to a minimum.*

☼ ◐ ◊ ♦ ♀H4 ↕30m (100ft) ↔ 15m (50ft)

LIRIOPE MUSCARI

This stout, evergreen perennial forms dense clumps of dark green, strap-like leaves. Spikes of small, violet-purple flowers open in autumn amid the foliage, and may be followed by black berries. Good in a woodland border, or use as drought-tolerant ground cover for shady areas.

CULTIVATION *Grow in light, moist but well-drained, moderately fertile soil. Prefers slightly acid conditions. Position in partial or full shade with shelter from cold, drying winds. Tolerates drought.*

◐● ◊◆ 🏆H4 ↕30cm (12in) ↔45cm (18in)

LITHODORA DIFFUSA 'HEAVENLY BLUE'

A spreading, evergreen shrub, sometimes sold as *Lithospermum* 'Heavenly Blue', that grows flat along the ground. Deep azure-blue, funnel-shaped flowers are borne in profusion over long periods from late spring into summer. The leaves are elliptic, dark green and hairy. Suits an open position in a rock garden or raised bed.

CULTIVATION *Grow in well-drained, humus-rich, acid soil, in full sun. Trim lightly after flowering.*

☼ ◊ 🏆H4 ↕15cm (6in) ↔60cm (24in) or more

LOBELIA 'CRYSTAL PALACE'

A compact, bushy perennial that is almost always grown as an annual, with vibrant clusters of two-lipped, dark blue flowers during summer to autumn. The tiny leaves are dark green and bronzed. Useful for edging and to spill over the edges of containers.

CULTIVATION *Grow in deep, fertile soil or compost that is reliably moist, in full sun or partial shade. Plant out, after the risk of frost has passed, in spring.*

☼ ◐ 💧 ♈H3 ↕to 10cm (4in) ↔10–15cm (4–6in)

LOBELIA 'QUEEN VICTORIA'

This short-lived, clump-forming perennial bears almost luminous spikes of vivid red, two-lipped flowers from late summer to mid-autumn. Both the leaves and stems are deep purple-red. Effective in a mixed border or waterside planting. May not survive in cold areas. 'Bee's Flame' is a very similar plant, enjoying the same conditions.

CULTIVATION *Grow in deep, reliably moist, fertile soil, in full sun. Short-lived, but can be easily propagated by division in spring.*

☼ 💧 ♈H3 ↕1m (3ft) ↔30cm (12in)

LONICERA X *ITALICA*

This vigorous, deciduous, woody-stemmed honeysuckle is a free-flowering, twining or scrambling climber. In summer and early autumn, it bears large whorls of tubular, very fragrant, soft flesh-pink flowers, flushed red-purple with yellow insides; red berries follow later in the season. The leaves are oval and dark green. Train onto a wall, or up into a small tree.

CULTIVATION *Grow in moist but well-drained, fertile, humus-rich soil, in full sun or partial shade. Once established, cut back shoots by up to one-third after they have flowered.*

☼ ◐ ◊ ◆ 🏆H4 ↕7m (22ft)

LONICERA NITIDA 'BAGGESEN'S GOLD'

A dense, evergreen shrub bearing tiny, oval, bright yellow leaves on arching stems. Inconspicuous yellow-green flowers are produced in spring, occasionally followed by small, blue-purple berries. Excellent for hedging or topiary in urban gardens, as it is pollution-tolerant.

CULTIVATION *Grow in any well drained soil, in full sun or partial shade. Trim hedges at least 3 times a year, between spring and autumn. Plants that become bare at the base will put out renewed growth if cut back hard.*

 🏆H4 ↕↔ 1.5m (5ft)

LONICERA PERICLYMENUM 'GRAHAM THOMAS'

This long-flowering form of English honeysuckle is a woody, deciduous, twining climber bearing abundant, very fragrant, tubular white flowers. These mature to yellow over a long period in summer, without the red flecking seen on *L. periclymenum* 'Belgica'. The leaves are mid-green and oval in shape.

CULTIVATION *Grow in moist but well-drained, fertile, humus-rich soil. Thrives in full sun, but prefers shade at the base. Once established, cut back shoots by up to one-third after flowering.*

☼☀ ◊◊ ♈H4 ↕7m (22ft)

LONICERA PERICLYMENUM 'SEROTINA'

The late Dutch honeysuckle is a fast-growing, deciduous, twining climber with very fragrant, rich red-purple flowers which appear in abundance during mid- and late summer. These may be followed by red berries. The leaves are oval and mid-green. If given plenty of space, it scrambles naturally with little pruning needed.

CULTIVATION *Grow in well-drained but moist, humus-rich, fertile soil, in sun with shade at the base. To keep trained specimens within bounds, prune shoots back by one-third after flowering.*

☼☀ ◊◊ ♈H4 ↕7m (22ft)

LONICERA X *PURPUSII* 'WINTER BEAUTY'

This dense, semi-evergreen shrub makes an excellent winter-flowering hedge with an appealing scent. The clusters of small, highly fragrant, creamy-white flowers, borne during winter to early spring, are carried on red-purple shoots. The leaves are dark green, and the flowers are occasionally followed by red berries.

CULTIVATION *Grow in any well-drained soil, in full sun for the best flowers, or partial shade. Prune after flowering to remove dead wood or to restrict size.*

☼ ◐ ◇ 🏆H4 ↕2m (6ft) ↔2.5m (8ft)

LONICERA X *TELLMANNIANA*

A twining, deciduous, woody-stemmed climber that bears clusters of coppery-orange, tubular flowers which open from late spring to mid-summer. The deep green, elliptic leaves have blue-white undersides. Train onto a wall or fence, or up into a large shrub. Summer-flowering *L. tragophylla*, one of its parents, is similar and is also recommended.

CULTIVATION *Grow in moist but well-drained, fertile, humus-rich soil. Will tolerate full sun, but produces better flowers in a more shaded position. After flowering, trim shoots by one-third.*

☼ ◐ ◇◆ 🏆H4 ↕5m (15ft)

LOTUS BERTHELOTII

Known as parrot's beak for its hooked flowers, this trailing, evergreen subshrub bears narrowly cut silvery leaves on long stems. Striking orange-red to scarlet flowers appear in profusion throughout spring and early summer. Will not survive winter cold; it should be planted out in containers during summer, and brought under glass for winter. *L. maculatus* is similar and can be used in the same way.

CULTIVATION *Grow in well-drained, moderately fertile soil or compost, in full sun. Cut out some older stems after flowering to encourage new growth. Minimum temperature 2°C (35°F).*

☼ ◊ H1+3 ↕20cm (8in) ↔indefinite

LUPINUS ARBOREUS

The tree lupin is a fast-growing, sprawling, semi-evergreen shrub grown for its spikes of fragrant, clear yellow flowers which open through the summer. The divided leaves are bright and grey-green. Native to scrub of coastal California, it is tolerant of seaside conditions. May not survive in cold, frosty winters.

CULTIVATION *Grow in light or sandy, well-drained, fairly fertile, slightly acid soil, in full sun. Cut off seedheads to prevent self-seeding and trim after flowering to keep compact.*

☼ ◊ H4 ↕↔ 2m (6ft)

LYCHNIS CHALCEDONICA

The Jerusalem cross is a clump-forming perennial which produces slightly domed, brilliant red flower-heads in early to mid-summer. These are borne on upright stems above the oval, mid-green, basal leaves. The flowers are small and cross-shaped. Good for a sunny border or wild garden, but needs some support. It self-seeds freely.

CULTIVATION *Grow in moist but well-drained, fertile, humus-rich soil, in full sun or light dappled shade.*

☼ ◑ ◊♦ 🏆H4 ↕1–1.2m (3–4ft) ↔30cm (12in)

LYSICHITON AMERICANUS

Yellow skunk cabbage is a striking, colourful perennial, which flowers in early spring and is ideal for a waterside planting. Each dense spike of tiny, greenish-yellow flowers is hooded by a bright yellow spathe and has an unpleasant, slightly musky scent. The large, dark green leaves, 50–120cm (20–48in) long, emerge from the base of the plant.

CULTIVATION *Grow in moist, fertile, humus-rich soil. Position in full sun or partial shade, allowing plenty of room for the large leaves to develop.*

☼ ◑ ♦ 🏆H4 ↕1m (3ft) ↔1.2m (4ft)

LYSICHITON CAMTSCHATCENSIS

White skunk cabbage is a bold waterside perennial with a slightly musky scent. In early spring, dense spikes of tiny green flowers emerge, cloaked by pointed white spathes. The large, dark green leaves, up to 1m (3ft) long, grow from the base. Ideal beside a stream or a pool.

CULTIVATION *Grow in moist, waterside conditions, in fertile, humus-rich soil. Position in full sun or partial shade, allowing room for the leaves to develop.*

☼☀ ◊ H4 ↕↔ to 1m (3ft)

LYSIMACHIA CLETHROIDES

This spreading, clump-forming, herbaceous perennial is grown for its tapering spikes of tiny, star-shaped white flowers. These droop when in bud, straightening up as the flowers open in mid- to late summer. The narrow leaves are yellow-green when young, maturing to mid-green with pale undersides. Suitable for naturalizing in a wild woodland or bog garden.

CULTIVATION *Grow in reliably moist soil that is rich in humus, in full sun or partial shade. May need staking.*

☼☀ ◊ H4 ↕1m (3ft) ↔60cm (24in)

LYSIMACHIA NUMMULARIA 'AUREA'

Golden creeping Jenny is a rampant, sprawling, evergreen perennial, which makes an excellent ground cover plant. Its golden-yellow leaves are broadly oval in shape, with heart-shaped bases. The bright yellow, cup-shaped, summer flowers further enhance the foliage colour.

CULTIVATION *Grow in reliably moist soil that is enriched with well-rotted organic matter. Site in full sun or partial shade.*

☼◐ ◑ 🏆H4 ↕to 5cm (2in) ↔indefinite

LYTHRUM SALICARIA 'FEUERKERZE'

This cultivar of purple loosestrife, with more intense rose-red flowers than the species, is a clump-forming, upright perennial. The slender spikes of flowers bloom from mid-summer to early autumn, above the lance-shaped, mid-green foliage. Suited to a moist border or waterside planting. Sometimes sold as 'Firecandle'.

CULTIVATION *Grow in moist, preferably fertile soil, in full sun. Remove flowered stems to prevent self-seeding.*

 ↕1m (3ft) ↔ 45cm (18in)

MACLEAYA X *KEWENSIS* 'KELWAY'S CORAL PLUME'

A clump-forming perennial grown for its foliage and large, graceful plumes of tiny, coral-pink to deep-buff flowers. These open from pink buds from early summer, appearing to float above large, olive-green leaves. Grow among shrubs, or group to form a hazy screen. *M. cordata*, with paler flowers, can be used similarly.

CULTIVATION *Best in moist but well-drained, moderately fertile soil, in sun or light shade. Shelter from cold winds. May be invasive; chop away roots at the margins of the clump to confine.*

☼☀ ◊◊ ♀H4 ↕2.2m (7ft) ↔ 1m (3ft) or more

MAGNOLIA GRANDIFLORA 'EXMOUTH'

This hardy cultivar of the bull bay is a dense, evergreen tree bearing glossy dark green leaves with russet-haired undersides. *M. grandiflora* is quite distinct from other magnolias, since it blooms sporadically from late summer to autumn, producing flowers that are large, very fragrant, cup-shaped and creamy-white.

CULTIVATION *Best in well-drained but moist, humus-rich, acid soil, in full sun or light shade. Tolerates dry, alkaline conditions. Mulch with leaf mould in spring. Keep pruning to a minimum.*

☼☀ ◊◊ ♀H3–4 ↕6–18m (20–60ft) ↔to 15m (50ft)

MAGNOLIA GRANDIFLORA 'GOLIATH'

This cultivar of bull bay is slightly less hardy than 'Exmouth' (see facing page, below), but has noticeably larger flowers, to 30cm (12in) across. It is a dense, conical, evergreen tree with slightly twisted, dark green leaves. The cup-shaped, very fragrant, creamy-white flowers appear from late summer to autumn.

CULTIVATION *Best in well-drained but moist, humus-rich, acid soil, in full sun or light shade. Tolerates dry, alkaline conditions. Mulch with leaf mould in spring. Keep pruning to a minimum.*

☼ ◑ ◊ ♦ 🏆H3–4 ↕6–18m (20–60ft) ↔to 15m (50ft)

MAGNOLIA LILIIFLORA 'NIGRA'

A dense, summer-flowering shrub bearing goblet-shaped, deep purple-red flowers. The deciduous leaves are elliptic and dark green. Plant as a specimen or among other shrubs and trees. Unlike many magnolias, it begins to flower when quite young.

CULTIVATION *Grow in moist but well-drained, rich, acid soil, in sun or semi-shade. Provide a mulch in early spring. Prune young shrubs in mid-summer to encourage a good shape; once mature, very little other pruning is needed.*

☼ ◑ ◊ ♦ 🏆H4 ↕3m (10ft) ↔2.5m (8ft)

SPRING-FLOWERING MAGNOLIAS

Spring-flowering magnolias are handsome deciduous trees and shrubs valued for their elegant habit and beautiful, often fragrant blooms which emerge just before the leaves. Flowers range from the tough but delicate-looking, star-shaped blooms of *M. stellata*, to the exotic, goblet-like blooms of hybrids such as 'Ricki'. Attractive red fruits form in autumn. The architectural branch framework of the bare branches makes an interesting display for a winter garden, especially when grown as free-standing specimens. *M.* x *soulangeana* 'Rustica Rubra' can be trained against a wall.

CULTIVATION *Best in deep, moist but well-drained, humus-rich, neutral to acid soil.* M. wilsonii *tolerates alkaline conditions. Choose a position in full sun or partial shade. After formative pruning when young, restrict pruning to the removal of dead or diseased branches after flowering.*

☼ ◐ ◊ ♦ ♀H3–4, H4

1 ↕10m (30ft) ↔ 8m (25ft)

2 ↕15m (50ft) ↔ 10m (30ft)

3 ↕↔ 10m (30ft)

4 ↕10m (30ft) ↔ 6m (20ft)

1 *M.* x *loebneri* 'Merrill' (H4) **2** *M. campbellii* 'Charles Raffill' (H3–4) **3** *M. denudata* (H3–4) **4** *M.* 'Elizabeth' (H4)

5 ↕↔ 4m (12ft)

6 ↕8m (25ft) ↔ 6m (20ft)

7 ↕10m (30ft) ↔ 5m (15ft)

8 ↕↔ 6m (20ft)

9 ↕9m (28ft) ↔ 6m (20ft)

10 ↕8m (25ft) ↔ 6m (20ft)

11 ↕3m (10ft) ↔ 4m (12ft)

5 *M.* 'Ricki' (H4) **6** *M.* x *loebneri* 'Leonard Messel' (H4) **7** *M. salicifolia* (H3–4)
8 *M.* x *soulangeana* 'Rustica Rubra' (H3–4) **9** *M.* x *kewensis* 'Wada's Memory' (H4)
10 *M. wilsonii* (H4) **11** *M. stellata* (H4)

MAHONIA AQUIFOLIUM 'APOLLO'

Oregon grape is a low-growing, evergreen shrub with dark green leaves; these are divided into several spiny leaflets and turn brownish-purple in winter. Dense clusters of deep golden flowers open in spring, followed by small, blue-black fruits. Can be grown as ground cover.

CULTIVATION *Grow in moist but well-drained, rich, fertile soil, in semi-shade; tolerates sun if the soil remains moist. Every 2 years after flowering, shear ground-cover plants close to the ground.*

☼☀ ◊♦ ♈H4 ↕60cm (24in) ↔1.2m (4ft)

MAHONIA JAPONICA

A dense, upright, winter-flowering, evergreen shrub carrying large, glossy dark green leaves divided into many spiny leaflets. Long, slender spikes of fragrant, soft yellow flowers are borne from late autumn into spring, followed by purple-blue fruits. Suits a shady border or woodland garden.

CULTIVATION *Grow in well-drained but moist, moderately fertile, humus-rich soil. Prefers shade, but will tolerate sun if soil remains moist. Limit pruning to removal of dead wood, after flowering.*

☼☀ ◊♦ ♈H4 ↕2m (6ft) ↔3m (10ft)

MAHONIA X *MEDIA* 'BUCKLAND'

A vigorous, upright, evergreen shrub bearing dense and sharply spiny, dark green foliage. Small, fragrant, bright yellow flowers are produced in arching spikes from late autumn to early spring. A good vandal-resistant shrub for a boundary or front garden.

CULTIVATION *Best in moist but well-drained, fairly fertile, humus-rich soil. Thrives in semi-shade, but will become leggy in deep shade. Little pruning is needed, but over-long stems can be cut back to a low framework after flowering.*

☀ 💧💧 🏆H4 ↕5m (15ft) ↔4m (12ft)

MAHONIA X *MEDIA* 'CHARITY'

A fast-growing, evergreen shrub, very similar to 'Buckland' (above), but has more upright, densely-packed flower spikes. The dark green leaves are spiny, making it useful for barrier or vandal-proof planting. Fragrant yellow flowers are borne from late autumn to spring.

CULTIVATION *Grow in moist but well-drained, moderately fertile, rich soil. Prefers partial shade, and will become leggy in deep shade. After flowering, bare, leggy stems can be pruned hard to promote strong growth from lower down.*

☀ 💧💧 🏆H4 ↕5m (15ft) ↔4m (12ft)

MALUS FLORIBUNDA

The Japanese crab apple is a dense, deciduous tree with a long season of interest. Graceful, arching branches, bearing dark green foliage, flower during mid- to late spring to give a glorious display of pale pink blossom. The flowers are followed by small yellow crab apples; these often persist, providing a valuable source of winter food for garden wildlife.

CULTIVATION *Grow in moist but well-drained, moderately fertile soil, in sun or light shade. Prune to shape in the winter months when young; older specimens require little pruning.*

☼☀ ◊◆ ♈H4 ↕↔ 10m (30ft)

MALUS 'JOHN DOWNIE'

This vigorous, deciduous tree is upright when young, becoming conical with age. Large, cup-shaped white flowers, which open from pale pink buds in late spring, are followed by egg-shaped, orange and red crab apples. The oval leaves are bright green when young, maturing to dark green. An ideal small garden tree.

CULTIVATION *Grow in well-drained but moist, fairly fertile soil. Flowers and fruits are best in full sun, but tolerates some shade. Remove damaged or crossing shoots when dormant, to form a well-spaced crown. Avoid hard pruning of established branches.*

☼☀ ◊◆ ♈H4 ↕10m (30ft) ↔6m (20ft)

MALUS TSCHONOSKII

This upright, deciduous tree with upswept branches produces pink-flushed white blossom in late spring, followed in autumn by red-flushed yellow crab apples. The leaves turn from green to a vibrant gold, then red-purple in autumn. It is taller than many crab apples, but is still a beautiful specimen tree which can be accommodated in small gardens.

CULTIVATION *Grow in well-drained, moderately fertile soil. Best in full sun, but tolerates some shade. Forms a good shape with little or no pruning; does not respond well to hard pruning.*

☼☀ ◊ 🏆H4 ↕12m (40ft) ↔7m (22ft)

MALUS x *ZUMI* 'GOLDEN HORNET'

A broadly pyramidal, deciduous tree bearing a profusion of large, cup-shaped, pink-flushed white flowers which open from deep pink buds in late spring. Small yellow crab apples follow, and persist well into winter. The display of golden fruit is further enhanced when the dark green foliage turns yellow in autumn.

CULTIVATION *Grow in any but water-logged soil, in full sun for best flowers and fruit. To produce a well-spaced crown, remove damaged or crossing shoots on young plants when dormant. Do not prune older specimens.*

☼☀ ◊♦ 🏆H4 ↕10m (30ft) ↔8m (25ft)

MALVA MOSCHATA F. *ALBA*

This white- to very light pink-flowered musk mallow is a bushy, upright perennial suitable for wild-flower gardens or borders. The very attractive and showy flowers are borne in clusters from early to late summer, amid the slightly musk-scented, mid-green foliage.

CULTIVATION *Grow in moist but well-drained, moderately fertile soil, in full sun. Taller plants may need staking. Often short-lived, but self-seeds readily.*

☼ ◊ ♈H4 ↕1m (3ft) ↔60cm (24in)

MATTEUCCIA STRUTHIOPTERIS

The shuttlecock fern forms clumps of upright or gently arching, pale green, deciduous fronds. In summer, smaller, dark brown fronds form at the centre of each clump, which persist until late winter. No flowers are produced. An excellent foliage perennial for a damp, shady border, and woodland or waterside plantings.

CULTIVATION *Grow in moist but well-drained, humus-rich, neutral to acid soil. Chose a site in light dappled shade.*

☀ ◊♦ ♈H4 ↕1–1.5m (3–5ft) ↔45–75cm (18–30in)

MATTHIOLA CINDERELLA SERIES

These woody-based stocks, short-lived perennials grown as annuals, are valued for their dense spikes of sweet-scented, double flowers, in a range of colours from white through pink to dark blue-purple. 'Cinderella White' and 'Cinderella Purple' are recommended single colours. An attractive addition to a summer border; the flowers cut well.

CULTIVATION *Grow in well-drained but moist, fertile, neutral to acid soil, in a sheltered, sunny site. Plant out after the danger of frost has passed.*

☼ ◊ 🏆H3 ↕20–25cm (8–10in) ↔to 25cm (10in)

MECONOPSIS BETONICIFOLIA

The Tibetan blue poppy is a clump-forming perennial bearing upright stems of large, saucer-shaped flowers which are clear blue or often purple-blue or white. These appear in early summer, above the oval and bluish-green leaves. Naturalizes well in a woodland garden.

CULTIVATION *Best in moist but well-drained, rich, acid soil. Site in partial shade with shelter from cold winds. May be short-lived, especially in hot or dry conditions. Divide clumps after flowering to maintain vigour.*

◑ ◊♦ 🏆H4 ↕1.2m (4ft) ↔45cm (18in)

MECONOPSIS GRANDIS

The Himalayan blue poppy is an upright, clump-forming perennial, similar to *M. betonicifolia* (see previous page, bottom), but with larger, less clustered, rich blue to purplish-red flowers. These are carried above the mid- to dark green foliage in early summer. Appealing when grown in large groups in a woodland setting.

CULTIVATION *Grow in moist, leafy, acid soil. Position in partial shade with shelter from cold, drying winds. Mulch generously and water in dry spells; may fail to flower if soil becomes too dry*

☀ 💧 ♀H4 ↕1–1.2m (3–4ft) ↔60cm (24in)

MELIANTHUS MAJOR

The honey bush is an excellent foliage shrub of upright to spreading habit. From late spring to mid-summer, spikes of blood-red flowers may appear above the grey-green to bright blue-green, divided leaves. In cold areas, treat as a herbaceous perennial, but in milder climates it is ideal for a coastal garden.

CULTIVATION *Grow in moist but well-drained, fertile soil. Choose a sunny site protected from cold, drying winds and winter wet. In cold areas, cut back the last year's growth in spring; in warm areas, cut out flowered stems in autumn.*

☀ 💧 ♀H3 ↕2–3m (6–10ft) ↔1–3m (3–10ft)

MIMULUS AURANTIACUS

This domed or sprawling, evergreen shrub bears open trumpet-shaped, yellow, orange or dark red flowers from late summer to autumn. The rich green leaves are lance-shaped and toothed. Grow against a warm, sunny wall; in regions with cold winters it may be better off in a cool conservatory.

CULTIVATION *Best in well-drained, humus-rich soil or compost, in full sun. Often short-lived, but easily propagated by cuttings in mid-summer.*

☼ ◊ 🏆H2–3 ↕↔ 1m (3ft)

MIMULUS CARDINALIS

The scarlet monkey flower is a creeping perennial with tubular, scarlet, sometimes yellow-marked flowers. They appear throughout summer amid the oval, light green leaves, on hairy stems. Good for adding colour to a warm border. May not survive in cold climates.

CULTIVATION *Grow any well-drained, fertile, humus-rich soil; tolerates quite dry conditions. Position in sun or light dappled shade. May be short-lived, but easily propagated by division in spring.*

☼ ◑ ◊ 🏆H3 ↕1m (3ft) ↔60cm (24in)

MOLINIA CAERULEA 'VARIEGATA'

The purple moor grass is a tufted perennial forming clumps of cream-striped, narrow, dark green leaves. Dense, purple flowering spikes are produced over a long period from spring to autumn on tall, ochre-tinted stems. A good structural plant for a border or a woodland garden.

CULTIVATION *Grow in any moist but well-drained, preferably acid to neutral soil, in full sun or partial shade.*

☼ ◑ ◊ ♀H4 ↕to 60cm (24in) ↔40cm (16in)

MONARDA 'CAMBRIDGE SCARLET'

This bergamot hybrid is a clump-forming perennial bearing shaggy heads of rich scarlet-red, tubular flowers from mid-summer to early autumn. They appear in profusion above the aromatic leaves, and are very attractive to bees; a common name for Monarda is "bee-balm". A colourful addition to any mixed or herbaceous border.

CULTIVATION *Prefers moist but well-drained, moderately fertile, humus-rich soil, in full sun or light dappled shade. Keep moist in summer, but protect from excessive wet in winter.*

☼ ◑ ◊ ♦ ♀H4 ↕1m (3ft) ↔45cm (18in)

MONARDA 'CROFTWAY PINK'

A clump-forming, herbaceous perennial with shaggy heads of tubular pink flowers carried above the small, aromaticsmall, light green leaves from mid-summer to early autumn. Suits mixed or herbaceous borders; the flowers are attractive to bees. 'Beauty of Cobham' is very similar, but with mauve outer bracts surrounding the pink flowers.

CULTIVATION *Grow in well-drained, fairly fertile, humus-rich soil that is reliably moist in summer. Site in full sun or light shade, with protection from excessive winter wet.*

☼☀ ◆ 𝕐H4 ↕1m (3ft) ↔45cm (18in)

MUSCARI ARMENIACUM

A vigorous, bulbous perennial that bears dense spikes of tubular, rich blue flowers in early spring. The mid-green leaves are strap-like, and begin to appear in autumn. Plant massed together in borders or allow to spread and naturalize in grass, although it can be invasive.

CULTIVATION *Grow in moist but well-drained, fairly fertile soil, in full sun. Divide clumps of bulbs in summer.*

☼ ◇ 𝕐H4 ↕20cm (8in) ↔5cm (2in)

MUSCARI AUCHERI

This bulbous perennial, less invasive than *M. armeniacum* (see previous page, bottom), bears dense spikes of small, bright blue spring flowers. The flower spikes, often topped with paler blue flowers, are carried above the basal clumps of strap-shaped, mid-green leaves. Suitable for a rock garden. Sometimes known as *M. tubergianum*.

CULTIVATION *Grow in moist but well-drained, moderately fertile soil. Choose a position in full sun.*

☼ ◊ 🏆H4 ↕10–15cm (4–6in) ↔5cm (2in)

MYRTUS COMMUNIS

The common myrtle is a rounded shrub with dense, evergreen foliage. From mid-summer to early autumn, amid the small, glossy dark green, aromatic leaves, it bears great numbers of fragrant white flowers with prominent tufts of stamens. Purple-black berries appear later in the season. Can be grown as an informal hedge or a specimen shrub. In areas with cold winters, grow against a warm, sunny wall.

CULTIVATION *Best in well-drained but moist, fertile soil, in full sun. Protect from cold, drying winds. Trim in spring; tolerates close clipping.*

☼ ◊ 🏆H3 ↕↔ 3m (10ft)

MYRTUS COMMUNIS SUBSP. *TARENTINA*

This dense, evergreen shrub is more compact and rounded than the species (see facing page, below), with smaller leaves and pink-tinted cream flowers. These appear during mid-spring to early autumn and are followed by white berries. Grow in a border or as an informal hedge; may not survive in cold climates.

CULTIVATION *Grow in moist but well-drained, moderately fertile soil. Choose a site in full sun with shelter from cold, drying winds. Trim back in spring; tolerates close clipping.*

☼ ◊ H3 ↕↔ 1.5m (5ft)

NANDINA DOMESTICA

Heavenly bamboo is an upright, evergreen or semi-evergreen shrub with fine spring and autumn colour. The divided leaves are red when young, maturing to green, then flushing red again in late autumn. Conical clusters of small white flowers with yellow centres appear in mid-summer, followed, in warm climates, by long-lasting, bright red fruits. May not survive cold winters.

CULTIVATION *Grow in moist but well-drained soil, in full sun. Cut back on planting, then prune in mid-spring to keep the plant tidy. Deadhead regularly.*

☼ ♦ H3 ↕2m (6ft) ↔1.5m (5ft)

SMALL DAFFODILS (*NARCISSUS*)

Small and miniature daffodils make good spring-flowering, bulbous perennials for both indoor and outdoor displays. They are cultivated for their elegant, mostly yellow or white flowers, of which there is great variety in shape. The blooms are carried either singly or in clusters above basal clumps of long, strap-shaped leaves on upright, leafless stems. All small daffodils are suitable for a rock garden, and they can look effective when massed together to form drifts. Because of their manageable size, these daffodils are useful for indoor displays. Some, such as *N. bulbocodium*, will naturalize well in short, fine grass.

CULTIVATION *Best in well-drained, fertile soil that is moist during growth, preferably in sun. Feed with a balanced fertiliser after flowering to encourage good flowers the following year. Deadhead as flowers fade, and allow the leaves to die down naturally; do not tie them into bunches.*

☼ 💧 🏆H3, H3–4, H4

1 ↕35cm (14in) ↔ 8cm (3in)

2 ↕10–15cm (4–6in) ↔ 5–8cm (2–3in)

3 ↕30cm (12in) ↔ 8cm (3in)

4 ↕15–20cm (6–8in) ↔ 5–8cm (2–3in)

5 ↕30cm (12in) ↔ 8cm (3in)

6 ↕30cm (12in) ↔ 8cm (3in)

1 *N.* 'Avalanche' (H3) **2** *N. bulbocodium* (H3–4) **3** *N.* 'Charity May' (H4)
4 *N. cyclamineus* (H4) **5** *N.* 'Dove Wings' (H4) **6** *N.* 'February Gold' (H4)

7 ↕17cm (7in) ↔ 5–8cm (2–3in)

8 ↕20cm (8in) ↔ 8cm (3in)

9 ↕20cm (8in) ↔ 8cm (3in)

10 ↕17cm (7in) ↔ 5–8cm (2–3in)

11 ↕10–15cm (4–6in) ↔ 5–8cm (2–3in)

12 ↕15cm (6in) ↔ 5–8cm (2–3in)

13 ↕10–25cm (4–10in) ↔ 5–8cm (2–3in)

7 *N.* 'Hawera' (H4) **8** *N.* 'Jack Snipe' (H4) **9** *N.* 'Jetfire' (H4) **10** *N.* 'Jumblie' (H4) **11** *N. minor* (H4) **12** *N.* 'Tête-à-tête' (H4) **13** *N. triandrus* (H3)

LARGE DAFFODILS (*NARCISSUS*)

Large daffodils are tall, bulbous perennials, easily cultivated for their showy, mostly white or yellow flowers in spring. These are borne singly or in clusters on upright, leafless stems above long, strap-shaped, mid-green foliage arising from the bulb. A great diversity of elegant flower shapes is available, those illustrated all are excellent for cutting; 'Sweetness' has blooms that last particularly well when cut. Most look very effective flowering in large groups between shrubs or in a border. Some naturalize easily in grass, or under deciduous trees and shrubs in a woodland garden.

CULTIVATION *Best in well-drained, fertile soil, preferably in full sun. Keep soil reliably moist during the growing season, and feed with a balanced fertiliser after flowering to ensure good blooms the following year. Deadhead as flowers fade, and allow leaves to die down naturally.*

☼ ◊ ♈H4

1 ↕45cm (18in) ↔ 15cm (6in)

2 ↕40cm (16in) ↔ 12cm (5in)

3 ↕40cm (16in) ↔ 12cm (5in)

4 ↕40cm (16in) ↔ 15cm (6in)

1 *N.* 'Actaea' **2** *N.* 'Empress of Ireland' **3** *N.* 'Ceylon' **4** *N.* 'Cheerfulness'

5 ↕40cm (16in) ↔ 15cm (6in)
6 ↕45cm (18in) ↔ 15cm (6in)
7 ↕35cm (14in) ↔ 15cm (6in)
8 ↕45cm (18in) ↔ 15cm (6in)
9 ↕40cm (16in) ↔ 15cm (6in)
10 ↕40cm (16in) ↔ 8cm (3in)
11 ↕45cm (18in) ↔ 15cm (6in)
12 ↕45cm (18in) ↔ 15cm (6in)
13 ↕40cm (16in) ↔ 8cm (3in)
14 ↕45cm (18in) ↔ 15cm (6in)

5 *N.* 'Ice Follies' **6** *N.* 'Kingscourt' **7** *N.* 'Merlin' **8** *N.* 'Mount Hood' **9** *N.* 'Passionale' **10** *N.* 'Suzy' **11** *N.* 'Saint Keverne' **12** *N.* 'Tahiti' **13** *N.* 'Sweetness' **14** *N.* 'Yellow Cheerfulness'

NERINE BOWDENII

This robust perennial is one of the best late flowering bulbs. In autumn, it bears open sprays of five to ten trumpet-shaped, faintly scented, bright pink flowers with curled, wavy-edged petals. The strap-like, fresh green leaves appear after the flowers, at the base of the plant. Despite its exotic appearance, it will survive winter cold with protection at the base of a warm, sunny wall. The flowers are good for cutting.

CULTIVATION *Grow in well-drained soil, in a sunny, sheltered position. Provide a deep, dry mulch in winter.*

☼ ♢ ♀H3–4 ↕45cm (18in) ↔12–15cm (5–6in)

NICOTIANA 'LIME GREEN'

This striking tobacco plant is an upright, free-flowering, bushy annual, ideal for a summer border. It bears loose clusters of night-scented, yellow-green flowers, with long throats and flattened faces, from mid- to late summer. The leaves are mid-green and oblong. Also good on patios, where the evening scent can be appreciated. For tobacco flowers in mixed colours that include lime green, the Domino Series is recommended.

CULTIVATION *Grow in moist but well-drained, fertile soil. Choose a position in full sun or partial shade.*

☼ ◑ ♢ ♀H3 ↕60cm (24in) ↔25cm (10in)

NICOTIANA SYLVESTRIS

A vigorous, upright perennial that is grown as an annual in cold climates. Throughout summer, open clusters of long-tubed, flat-faced, sweetly scented white flowers are borne above the dark green, sticky leaves. Although plants are tall (and may need staking), try to position where their scent can be appreciated.

CULTIVATION *Grow in any well-drained, moderately fertile soil, in sun or partial shade; the flowers close in full sun. Given a dry winter mulch, it can survive occasional frosts, resprouting from the base in spring.*

NIGELLA 'MISS JEKYLL'

This tall, slender annual bears pretty, sky-blue flowers during summer, which are surrounded by a feathery "ruff" of bright green foliage. These are followed later in the season by attractive, inflated seed pods. The blooms last well when cut, and the seed pods can be dried for indoor flower arrangements. There is a white version, also recommended: 'Miss Jekyll Alba'.

CULTIVATION *Grow in well-drained soil, in full sun. Like all love-in-a-mists, it self-seeds freely, although seedlings may differ from the parent.*

☼ ◊ ♈H4 ↕to 45cm (18in) ↔to 23cm (9in)

WATER LILIES (*NYMPHAEA*)

These aquatic perennials are cultivated for their showy, sometimes fragrant summer flowers and rounded, floating leaves. The shade cast by the leaves is useful in reducing growth of pond algae. Flowers are mostly white, yellow, pink, or red with yellow stamens in the centres. Spread varies greatly, so choose carefully to match the size of your water feature. Planting depths may be between 45cm and 1m (18–30in), though 'Pygmaea Helvola' can grow in water no deeper than 15cm (6in). Using aquatic plastic mesh planting baskets makes lifting and dividing much easier.

CULTIVATION *Grow in still water in full sun. Plant in aquatic compost with rhizomes just below the surface, anchored with a layer of grit. Stand young plants on stacks of bricks so that shoot tips reach the water surface; lower plants as stems lengthen. Remove yellow leaves regularly. Divide in spring, giving plants an aquatic fertilizer.*

☼ ♈H4

1 ↔ 1.2–1.5m (4–5ft)
2 ↔ 1.5–2.5m (5–8ft)
3 ↔ 0.9–1.2m (3–4ft)
4 ↔ 0.9–1.2m (3–4ft)
5 ↔ 1.2–1.5m (4–5ft)
6 ↔ 25–40cm (10–16in)

1 *Nymphaea* 'Escarboucle' **2** *N.* 'Gladstoneana' **3** *N.* 'Gonnère' **4** *N.* 'James Brydon' **5** *N.* 'Marliacea Chromatella' **6** *N.* 'Pygmaea Helvola'

NYSSA SINENSIS

The Chinese tupelo is a deciduous tree, conical in form, grown for its lovely foliage. The elegant leaves are bronze when young, maturing to dark green, then becoming brilliant shades of orange, red and yellow in autumn before they fall. The flowers are inconspicuous. Ideal as a specimen tree near water.

CULTIVATION *Grow in fertile, reliably moist but well-drained, neutral to acid soil in sun or partial shade, with shelter from cold, drying winds. Thin out crowded branches in late winter.*

☼ ◑ 🌢🌢 ♈H4 ↕↔10m (30ft)

NYSSA SYLVATICA

A taller tree than *N. sinensis* (above), the tupelo is similar in form, with drooping lower branches. Its dark green leaves change to a glorious display of orange, yellow, or red in autumn. A brilliant tree for autumn colour, with brownish-grey bark that breaks up into large pieces on mature specimens. It tolerates acid soil.

CULTIVATION *Grow in moist but well-drained, fertile, neutral to acid soil in sun or partial shade. Shelter from cold, drying winds, and prune in late winter, if necessary.*

☼ ◑ 🌢🌢 ♈H4 ↕20m (70ft) ↔10m (30ft)

OENOTHERA FRUTICOSA 'FYRVERKERI'

An upright, clump-forming perennial carrying clusters of short-lived, cup-shaped, bright yellow flowers which open in succession from late spring to late summer. The lance-shaped leaves are flushed red-purple when young, contrasting beautifully with the red stems. It associates well with bronze- or copper-leaved plants. *O. macrocarpa* subsp. *glauca* is quite similar, with slightly paler flowers.

CULTIVATION *Grow in sandy, well-drained soil that is well-fertilized. Choose a site in full sun.*

☼ ♢ ♀H4 ↕30–100cm (12–39in) ↔30cm (12in)

OENOTHERA MACROCARPA

This vigorous perennial has similar flowers to *O. fruticosa* 'Fyrverkeri' (above), but its trailing habit makes this plant more suitable for border edging. The golden-yellow blooms appear from late spring to early autumn, amid the lance-shaped, mid-green leaves. Can also be used in a scree bed or rock garden. Also known as *O. missouriensis*.

CULTIVATION *Grow in well-drained, poor to moderately fertile soil. Position in full sun, in a site that is not prone to excessive winter wet.*

☼ ♢ ♀H4 ↕15cm (6in) ↔to 50cm (20in)

OLEARIA MACRODONTA

An summer-flowering, evergreen shrub or small tree that forms an upright, broadly columnar habit. Large clusters of fragrant, daisy-like white flowers with reddish-brown centres are borne amid the sharply toothed, glossy dark green leaves. A good hedging plant or windbreak in coastal areas with mild climates; in colder regions it is better grown in a warm, sheltered shrub border.

CULTIVATION *Grow in well-drained, fertile soil, in full sun with shelter from cold, drying winds. Prune unwanted or frost-damaged growth in late spring.*

☼ ◊ ♀H3 ↕6m (20ft) ↔5m (15ft)

OMPHALODES CAPPADOCICA

This clump-forming, shade-loving, evergreen perennial bears sprays of small, azure-blue, forget-me-not-like flowers with white centres. These appear in early spring above the pointed, mid-green leaves. Effective planted in groups through a woodland garden.

CULTIVATION *Grow in moist, humus-rich, moderately fertile soil. Choose a site in partial shade.*

☀ ♦ ♀H4 ↕to 25cm (10in) ↔to 40cm (16in)

OMPHALODES CAPPADOCICA 'CHERRY INGRAM'

This clump-forming, evergreen perennial is very similar to the species (see previous page, bottom), but with larger, deep blue flowers which have white centres. These appear in early spring, above the pointed, finely hairy, mid-green leaves. Suits a woodland garden.

CULTIVATION *Best in reliably moist, moderately fertile, humus-rich soil. Choose a site in partial shade.*

☀ 💧 🏆H4 ↕to 25cm (10in) ↔to 40cm (16in)

ONOCLEA SENSIBILIS

The sensitive fern forms a beautifully textured mass of arching, finely divided, broadly lance-shaped, deciduous fronds. The foliage is pinkish-bronze in spring, maturing to pale green. There are no flowers. Thrives at the edge of water, or in a damp, shady border.

CULTIVATION *Grow in moist, humus-rich, preferably acid soil. Site in partial shade, as fronds will scorch in sun.*

☀ 💧💧 🏆H4 ↕60cm (24in) ↔indefinite

OPHIOPOGON PLANISCAPUS 'NIGRESCENS'

This evergreen, spreading perennial forms clumps of grass-like, curving, almost black leaves. It looks very unusual and effective when planted in gravel-covered soil. Spikes of small, tubular, white to lilac flowers appear in summer, followed by round, blue-black fruits in autumn.

CULTIVATION *Grow in moist but well-drained, fertile, slightly acid soil that is rich in humus, in full sun or partial shade. Top-dress with leaf mould in autumn, where practicable.*

☼☀ 💧💧 🏆H4 ↕20cm (8in) ↔30cm (12in)

ORIGANUM LAEVIGATUM

A woody-based, bushy perennial bearing open clusters of small, tubular, purplish-pink flowers from late spring to autumn. The oval, dark green leaves are aromatic, powerfully so when crushed. Suits a rock garden or scree bed; the flowers are attractive to bees. May not survive in cold climates.

CULTIVATION *Grow in well-drained, poor to moderately fertile, preferably alkaline soil. Position in full sun. Trim back flowered stems in early spring.*

☼ 💧 🏆H3 ↕to 60cm (24in) ↔45cm (18in)

ORIGANUM LAEVIGATUM 'HERRENHAUSEN'

A low-growing perennial that is more hardy than the species (see previous page, bottom), with purple-flushed young leaves and denser whorls of pink flowers in summer. The mature foliage is dark green and aromatic. Suits a Mediterranean-style planting or rock garden.

CULTIVATION *Grow in very well-drained, poor to fairly fertile, preferably alkaline soil, in full sun. Trim back flowered stems in early spring.*

☼ ◊ ♈H4 ↕↔ 45cm (18in)

ORIGANUM VULGARE 'AUREUM'

Golden marjoram is a colourful, bushy perennial with tiny, golden-yellow leaves that age to greenish-yellow. Short spikes of tiny, pretty pink flowers are occasionally produced in summer. The highly aromatic leaves can be used in cooking. Good for ground cover on a sunny bank, or in a herb garden, although it tends to spread.

CULTIVATION *Grow in well-drained, poor to moderately fertile, alkaline soil. Position in full sun. Trim back after flowering to maintain a compact form.*

☼ ◊ ♈H4 ↕↔ 30cm (12in)

OSMANTHUS x *BURKWOODII*

A dense and rounded, evergreen shrub, sometimes known as x *Osmarea burkwoodii*, carrying oval, slightly toothed, leathery, dark green leaves. Profuse clusters of small, very fragrant white flowers, with long throats and flat faces, are borne in spring. Ideal for a shrub border, or as a hedge.

CULTIVATION *Grow in well-drained, fertile soil, in sun or partial shade with shelter from cold, drying winds. Prune to shape after flowering, giving hedges a trim in summer.*

☼◑ ◊ ♀H4 ↕↔ 3m (10ft)

OSMANTHUS DELAVAYI

A rounded, fragrant, evergreen shrub with arching branches bearing small, glossy dark leaves. Profuse clusters of sweetly-scented, pure white flowers appear in mid- to late spring. Good for a shrub border or woodland garden, and excellent for hedging. May also be wall-trained.

CULTIVATION *Grow in well-drained, fertile soil, in full sun or partial shade with shelter from cold winds. Tolerates alkaline conditions. Prune to shape after flowering; trim hedges in summer.*

☼◑ ◊ ♀H4 ↕2–6m (6–20ft) ↔4m (12ft) or more

OSMUNDA REGALIS

The royal fern is a stately, clump-forming perennial with bright green, finely divided foliage. Distinctive, rust-coloured fronds are produced at the centre of each clump in summer. There are no flowers. Excellent in a damp border, or at the margins of a pond or stream. There is an attractive version of this fern with crested fronds, 'Cristata', growing slightly less tall, to 1.2m (4ft).

CULTIVATION *Grow in very moist, fertile, humus-rich soil, in semi-shade. Tolerates full sun if conditions are reliably damp.*

☼☀ 💧💧 ♀H4 ↕2m (6ft) ↔4m (12ft)

OSTEOSPERMUM JUCUNDUM

A neat, clump-forming, woody-based perennial bearing large, daisy-like, mauve-pink flowers which are flushed bronze-purple on the undersides. The blooms open in succession from late spring until autumn. Ideal for wall crevices or at the front of a border. Also known as *O. barberae.* 'Blackthorn Seedling', with dark purple flowers, is a striking cultivar.

CULTIVATION *Best in light, well-drained, fairly fertile soil. Choose a site in full sun. Deadhead to prolong flowering.*

☼ 💧 ♀H3–4 ↕10–50cm (4–20in) ↔50–100cm (20–39in)

OSTEOSPERMUM HYBRIDS

These evergreen subshrubs are grown primarily for their daisy-like, bright and cheerful flowerheads, sometimes with pinched petals or centres in contrasting colours. They are borne singly or in open clusters over a long season, which begins in late spring and ends in autumn. Numerous cultivars have been named, varying from deep magenta through to white, pink, or yellow. Osteospermums are ideal for a sunny border – the flowers close in dull conditions. The half-hardy types are best grown as annuals in frost-prone areas, or in containers so that they can easily be moved into a greenhouse or conservatory during winter.

CULTIVATION *Grow in light, moderately fertile, well-drained soil in a warm, sheltered site in full sun. In frost-prone areas, overwinter half-hardy types in frost-free conditions. Regular deadheading will encourage more flowers.*

☼ ◊ ♀H1+3, 3–4

1 ↕↔ 60cm (24in)
2 ↕30cm (12in) ↔ 45cm (18in)
3 ↕↔ 60cm (24in)
4 ↕↔ 45cm (18in)
5 ↕35cm (14in) ↔ 45cm (18in)
6 ↕↔ 60cm (24in)

1 *Osteospermum* 'Buttermilk' (H1+3) **2** *O.* 'Hopleys' (H3–4) **3** *O.* 'Pink Whirls' (H1+3) **4** *O.* 'Stardust' (H3–4) **5** *O.* 'Weetwood' (H3–4) **6** *O.* 'Whirligig' (H1+3)

OXALIS ADENOPHYLLA

A bulbous perennial forming clumps of pretty, grey-green leaves which are divided into many heart-shaped leaflets. In late spring, widely funnel-shaped, purple-pink flowers contrast beautifully with the foliage. Native to the Andes, it suits a well-drained rock garden, trough or raised bed. Pink-flowered *O. enneaphylla* and the blue-flowered *O.* 'Ione Hecker' are very similar, though they grow from rhizomatous roots rather than bulbs.

CULTIVATION *Grow in any moderately fertile soil with good drainage. Choose a position in full sun.*

☼ ◊ ♈H4 ↕10cm (4in) ↔to 15cm (6in)

PACHYSANDRA TERMINALIS

This freely spreading, bushy, evergreen foliage perennial makes a very useful ground cover plant for a shrub border or woodland garden. The oval, glossy dark green leaves are clustered at the tips of the stems. Spikes of small white flowers are produced in early summer. There is a less vigorous version with white-edged leaves, 'Variegata'.

CULTIVATION *Grow in any but very dry soil that is rich in humus. Choose a position in partial or full shade.*

☀☀ ♦ ♈H4 ↕20cm (8in) ↔indefinite

PAEONIA DELAVAYI

An upright, sparsely branched, deciduous shrub bearing nodding, bowl-shaped, rich dark red flowers in early summer. The dark green leaves are deeply cut into pointed lobes and have blue-green undersides. A tall bush peony, good in a shrub border.

CULTIVATION *Grow in deep, moist but well-drained, fertile soil that is rich in humus. Position in full sun or partial shade with shelter from cold, drying winds. Occasionally cut an old, leggy stem back to ground level in autumn, but avoid regular or hard pruning.*

☼☀ 💧💧 🏆H4 ↕2m (6ft) ↔1.2m (4ft)

PAEONIA LACTIFLORA 'BOWL OF BEAUTY'

A herbaceous, clump-forming perennial bearing very large, bowl-shaped flowers in early summer. These have carmine-pink, red-tinted petals arranged around a dense cluster of creamy-white stamens. The leaves are mid-green and divided into many leaflets. Ideal for a mixed or herbaceous border.

CULTIVATION *Grow in deep, moist but well-drained, fertile, humus-rich soil, in full sun or partial shade. Provide support. Resents being disturbed.*

☼☀ 💧💧 🏆H4 ↕↔ 80–100cm (32–39in)

PAEONIA LACTIFLORA 'DUCHESSE DE NEMOURS'

This clump-forming, free-flowering, herbaceous perennial produces large, fully double flowers which open from green-flushed buds in early summer. They are fragrant, pure white, with ruffled, yellow-based inner petals. The leaves are deep green and divided. Ideal for a mixed or herbaceous border. 'Laura Dessert' has similar, creamy-pink flowers.

CULTIVATION *Grow in moist but well-drained, deep, fertile, humus-rich soil., in full sun or semi-shade. Flowerheads may need support. Does not like to be disturbed once established.*

☼☀ ◊◊ ♆H4 ↕↔ 70–80cm (28–32in)

PAEONIA LACTIFLORA 'SARAH BERNHARDT'

This herbaceous peony is similar to 'Duchesse de Nemours' (above), but with very large, rose-pink flowers. The blooms appear during early summer, above the clumps of mid-green, deeply divided leaves. A lovely addition to a summer border.

CULTIVATION *Grow in moist but well-drained, deep, humus-rich, fertile soil. Position in full sun or partial shade and provide support for the flowering stems. Does not respond well to root disturbance.*

☼☀ ◊◊ ♆H4 ↕↔ to 1m (3ft)

PAEONIA LUTEA VAR. *LUDLOWII*

This vigorous, deciduous shrub has an upright and open form. In late spring, large, bright yellow, nodding flowers open amid the bright green foliage. The leaves are deeply divided into several pointed leaflets. Good in a shrub border or planted on its own.

CULTIVATION *Best in deep, well-drained but moist, fertile soil that is rich in humus. Site in sun or semi-shade with shelter from cold, drying winds. Avoid hard pruning, but occasionally cut old, leggy stems to ground level in autumn.*

☼ ◑ ◊◊ 🏆H4 ↕↔ 1.5m (5ft)

PAEONIA OFFICINALIS 'RUBRA PLENA'

This long-lived, clump-forming, herbaceous peony makes a fine early-summer-flowering addition to any border display. The large, fully double, vivid crimson flowers, with ruffled, satiny petals, contrast well with the deep green, divided leaves. 'Rosea Plena', very similar, is also recommended.

CULTIVATION *Grow in well-drained but moist, deep, fertile, humus-rich soil. Choose a site in full sun or partial shade. Support the flowering stems.*

☼ ◑ ◊◊ 🏆H4 ↕↔ 70–75cm (28–30in)

PAPAVER ORIENTALE 'BEAUTY OF LIVERMERE'

This tall oriental poppy with crimson scarlet flowers is an upright, clump-forming perennial. The flowers open during late spring to mid-summer, and develop into large seed pods. Each petal has a bold, black mark at the base. The mid-green, divided leaves are borne on upright, bristly stems. Looks spectacular in a border.

CULTIVATION *Grow in well-drained, poor to moderately fertile soil. Choose a position in full sun.*

☼ ◊ ♀H4 ↕1–1.2m (3–4ft) ↔90cm (36in)

PAPAVER ORIENTALE 'BLACK AND WHITE'

This oriental poppy, white-flowered with crimson-black markings at the petal bases, is a clump-forming perennial. The flowers are borne above the mid-green foliage at the tips of white-bristly, upright stems during early summer; they are followed by distinctive seed pods. Makes a good border perennial.

CULTIVATION *Best in deep, moderately fertile soil with good drainage. Choose a position in full sun.*

☼ ◊ ♀H4 ↕45–90cm (18–36in) ↔60–90cm (24–36in)

PAPAVER ORIENTALE
'CEDRIC MORRIS'

This oriental poppy has very large, soft pink flowers with black-marked bases that are set off well against the grey-hairy foliage. It forms upright clumps which are well-suited to a herbaceous or mixed border. Distinctive seed pods develop after the flowers have faded.

CULTIVATION *Grow in deep, moderately fertile soil with good drainage. Choose a position in full sun.*

☼ 💧 H4 ↕45–90cm (18–36in) ↔60–90cm (24–36in)

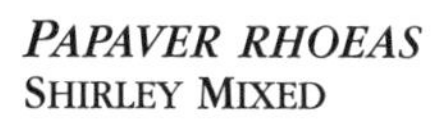

PAPAVER RHOEAS
SHIRLEY MIXED

These field poppies are summer-flowering annuals with single, semi-double or double, bowl-shaped flowers in shades of yellow, orange, pink and red. These appear on upright stems above the finely divided, bright green leaves. Can be naturalized in a wildflower meadow.

CULTIVATION *Best in well-drained, poor to moderately fertile soil, in full sun. Divide and replant clumps in spring.*

☼ 💧 H4 ↕to 1m (3ft) ↔to 30cm (12in)

PARAHEBE CATARRACTAE

A small evergreen subshrub bearing loose clusters of white summer flowers with purple veins and red eyes. The leaves are dark green and oval, purple-tinged when young. Looks very effective tumbling over walls or rocks. May not survive winter in frosty climates. Its cultivar 'Delight' is also recommended.

CULTIVATION *Grow in well-drained, poor to moderately fertile soil, in full sun. In frost-prone climates, provide shelter from cold, drying winds.*

☼ ◊ ♀H3 ↕↔ 30cm (12in)

PARAHEBE PERFOLIATA

Digger's speedwell is a spreading, evergreen perennial bearing short spikes of blue, saucer-shaped flowers in late summer. The blue- or grey-green, overlapping leaves are oval and slightly leathery. Suitable for gaps in old walls or a rock garden.

CULTIVATION *Grow in poor to fairly fertile soil with good drainage, in full sun. In frost-prone climates, provide shelter from cold, drying winds.*

☼ ◊ ♀H3–4 ↕60–75cm (24–30in) ↔45cm (18in)

PARTHENOCISSUS HENRYANA

Chinese Virginia creeper is a woody, twining, deciduous climber with colourful foliage in autumn. The insignificant summer flowers are usually followed by blue-black berries. The conspicuously white-veined leaves, made up of three to five leaflets, turn bright red late in the season. Train over a wall or a strongly built fence.

CULTIVATION *Grow in well-drained but moist, fertile soil. Tolerates sun, but leaf colour is best in deep or partial shade. Young plants may need some support. Prune back unwanted growth in autumn.*

H4 ↕10m (30ft)

PARTHENOCISSUS TRICUSPIDATA

Even more vigorous than Virginia creeper (*P. quinquefolia*), Boston ivy is a woody, deciduous climber with foliage that turns a beautiful colour in autumn. The variably lobed, bright green leaves flush a brilliant red, fading to purple before they fall. Creates a strong textural effect on cold, featureless walls.

CULTIVATION *Best in moist but well-drained, fertile soil that is rich in humus. Position in partial or full shade. Young plants may need some support before they are established. Remove any unwanted growth in autumn.*

H4 ↕20m (70ft)

PASSIFLORA CAERULEA

The blue passion flower is a fast-growing, evergreen climber valued for its large, exotic flowers, crowned with prominent blue- and purple-banded filaments. These are borne from summer to autumn amid the dark green, divided leaves. Of borderline hardiness; in cold climates grow in the protection of a warm wall.

CULTIVATION *Best in moist but well-drained, moderately fertile soil, in a sunny, sheltered site. Remove crowded growth in spring, cutting back flowered shoots at the end of the season.*

☼ ◊◊ ♀H3 ↕10m (30ft) or more

PASSIFLORA CAERULEA 'CONSTANCE ELLIOTT'

A fast-growing, evergreen climber, resembling the species (above), but with white flowers, borne from summer to autumn. The leaves are dark green and deeply divided into three to nine lobes. Good for a pergola in warm climates, but needs the shelter of a warm wall in areas with cold winters.

CULTIVATION *Grow in moist but well-drained, moderately fertile soil. Choose a sheltered site, in full sun. Remove weak growth in spring, cutting back flowered shoots at the end of the season.*

☼ ◊◊ ♀H3 ↕10m (30ft) or more

PASSIFLORA RACEMOSA

The red passion flower is a vigorous climber noted for its large, bright red flowers borne in hanging clusters in summer and autumn. They are followed by deep green fruits. The leathery leaves are glossy and mid-green. It cannot be grown outdoors in cool climates, but suits a greenhouse that is heated over winter, or makes a striking conservatory plant.

CULTIVATION *Grow in a greenhouse border or large tub, in loam-based potting compost. Provide full light, with shade from hot sun. Water sparingly in winter. Prune in early spring. Minimum temperature 16°C (61°F)*

☼◑ ◊◆ ♀H2 ↕5m (15ft)

PAULOWNIA TOMENTOSA

The foxglove tree is named for its fragrant, bell- to-trumpet-shaped, pinkish-lilac flowers, which appear in upright spikes in late spring. The large, heart-shaped leaves are bright green. This is a fast-growing deciduous tree, very tolerant of atmospheric pollution, and flowering best in areas with long, hot summers.

CULTIVATION *Grow in fertile, well-drained soil in full sun. In frost-prone areas, shelter from cold, drying winds; new growth may be damaged by late frost. Prune in late winter.*

☼ ◊ ♀H3 ↕12m (40ft) ↔10m (30ft)

SCENTED-LEAVED PELARGONIUMS

Pelargoniums, often incorrectly called geraniums (see pages 194–197), are tender, evergreen perennials. Many are grown specifically for their pretty, scented foliage. Their flowers, generally in pinks and mauves, are less showy and more delicate in form than those of the zonal and regal pelargoniums, bred more specifically for floral display. Leaf fragrance varies from sweet or spicy through to the citrus of *P. crispum* 'Variegatum' or the peppermint-like *P. tomentosum*. They will perfume a greenhouse or conservatory, or use as summer edging along a path or in pots on a patio, where they will be brushed against and release their scent.

CULTIVATION *Grow in well-drained, fertile, neutral to alkaline soil or compost, in sun. Deadhead regularly. In cold areas, over-winter in frost-free conditions, cutting back top growth by one-third. Repot as growth resumes. Minimum temperature 2°C (35°F).*

☼ ◊ ♀H1+3

1 ↕50cm (20in) ↔ 25–30cm (10–12in)
2 ↕ ↔ 60cm (24in)
3 ↕60–90cm (24–36in) ↔ 60cm (24in)
4 ↕to 1.2m (4ft) ↔ 30cm (12in)
5 ↕45cm (18in) ↔ 20–25cm (8–10in)
6 ↕35–45cm (30–36in) ↔ 15m (6in)

1 *Pelargonium* 'Attar of Roses' **2** *P.* 'Bolero' **3** *P.* 'Charity' **4** *P.* 'Citriodorum'
5 *P.* 'Copthorne' **6** *P. crispum* 'Variegatum'

7 ↕60cm (24in) ↔ 30cm (12in)
8 ↕ ↔ 60cm (24in)
9 ↕30–40cm (12–16in) ↔ 20cm (8in)
10 ↕ ↔ 60cm (24in)
11 ↕30–35cm (12–14in) ↔ 15cm (6in)
12 ↕90cm (36in) ↔ 30cm (12in)
13 ↕30–35cm (12–14in) ↔ 25cm (10in)
14 ↕60cm (24in) ↔ 30cm (12in)
15 ↕30–40cm (12–16in) ↔ 30cm (12in)
16 ↕45–50cm (18–20in) ↔ 25cm (10in)
17 ↕75–90cm (30–36in) ↔ to 75cm (30in)

7 *P.* 'Gemstone' **8** *P.* 'Grace Thomas' **9** *P.* 'Lady Plymouth' **10** *P.* 'Lara Starshine'
11 *P.* 'Mabel Grey' **12** *P.* 'Nervous Mabel' **13** *P.* 'Orsett' **14** *P.* 'Peter's Luck'
15 *P.* 'Royal Oak' **16** *P.* 'Sweet Mimosa' **17** *P. tomentosum*

FLOWERING PELARGONIUMS

Pelargoniums grown for their bold flowers are tender, evergreen perennials: most cultivars are bushy, but there are also trailing types for window boxes and hanging baskets. Flower forms vary from the tightly frilled 'Apple Blossom Rosebud' to the delicate, narrow-petalled 'Bird Dancer'; colours range from shades of orange through pink and red to rich purple, and some have coloured foliage. A popular choice as bedding plants, flowering from spring into summer, although many will flower throughout the year if kept above 7°C (45°F), making them excellent for a conservatory and as houseplants.

CULTIVATION *Grow in well-drained, fertile, neutral to alkaline soil or compost, in sun or partial shade. Deadhead regularly to prolong flowering. In cold areas, overwinter in frost-free conditions, cutting back by one-third. Repot in late winter as new growth resumes. Minimum temperature 2°C (35°F).*

☼ ◑ ◊ ♈H1+3

1 ↕25–30cm (10–12in) ↔ to 25cm (10in)

2 ↕to 30cm (12in) ↔ 25cm (10in)

3 ↕30–40cm (12–16in) ↔ 20–25cm (8–10in)

1 *Pelargonium* 'Alice Crousse' 2 *P.* 'Amethyst' 3 *P.* 'Apple Blossom Rosebud'

4 ↕15–20cm (6–8in) ↔ 15cm (6in)

5 ↕25–30cm (10–12in) ↔ 15cm (6in)

6 ↕45–60cm (18–24in) ↔ to 25cm (10in)

7 ↕10–12cm (4–5in) ↔ 7–10cm (3–4in)

8 ↕40–45cm (16–18in) ↔ 25cm (10in)

4 *P.* 'Bird Dancer' **5** *P.* 'Dolly Varden' **6** *P.* 'Flower of Spring' **7** *P.* 'Francis Parrett'
8 *P.* 'Happy Thought'

9 ↕40–45cm 16–18in) ↔ to 30cm (12in)

10 ↕25–30cm (10–12in) ↔ 12cm (5in)

MORE CHOICES

'Ashfield Serenade' Mauve with white-based pink upper petals.

'Belinda Adams' Miniature, with pink-flushed white flowers.

'Ben Franklin' Rose-pink flowers, silver leaves.

'Dame Anna Neagle' Light pink flowers and gold and bronze leaves.

'Madame Crousse' Pale pink flowers, trailing.

'Vancouver Centennial' Orange-red flowers, bronze/brown leaves.

11 ↕to 30cm (12in) ↔ 30cm (12in)

12 ↕40–45cm (16–18in) ↔ 20cm (8in)

13 ↕to 60cm (24in) ↔ to 25cm (10in)

14 ↕to 60cm (24in) ↔ 25cm (10in)

15 ↕20cm (8in) ↔ 18cm (7in)

9 *Pelargonium* 'Irene' **10** *P.* 'Mr Henry Cox' **11** *P.* Multibloom Series **12** *P.* 'Paton's Unique' **13** *P.* 'The Boar' **14** *P.* 'Voodoo' **15** *P.* Video Series

REGAL PELARGONIUMS

These bushy, half-hardy perennials and shrubs have gloriously showy, blowsy flowers, earning them the sobriquet "the queen of pelargoniums". Their main flowering period is from spring to early summer, with blooms carried in clusters in reds, pinks, and purples, orange, white, and reddish-black; colours are combined in some cultivars. Often grown as container plants in frost-prone areas both for indoor and outdoor display, although they may also be used in summer bedding schemes.

CULTIVATION *Best in good-quality, moist but well-drained potting compost in full light with shade from strong sun. Water moderately during growth, feeding every two weeks with a liquid fertiliser. Water sparingly if overwintering in a frost-free greenhouse. Cut back by one-third and repot in late winter. Outdoors, grow in fertile, neutral to alkaline, well-drained soil in full sun. Deadhead regularly. Minimum temperature 7°C (45°F).*

☼ ◊◊ ♀H1+3

1 ↕45cm (18in) ↔ to 25cm (10in)

2 ↕45cm (18in) ↔ to 25cm (10in)

3 ↕45cm (18in) ↔ to 25cm (10in)

MORE CHOICES

'Chew Magna' Pink petals with a red blaze.

'Leslie Judd' Salmon-pink and wine-red.

'Lord Bute' Dark reddish-black flowers.

'Spellbound' Pink with wine-red petal markings.

4 ↕30–40cm (12–16in) ↔ to 20cm (8in)

5 ↕30–40cm (12–16in) ↔ to 20cm (8in)

1 *Pelargonium* 'Ann Hoystead' **2** *P.* 'Bredon' **3** *P.* 'Carisbrooke' **4** *P.* 'Lavender Grand Slam' **5** *P.* 'Sefton'

PENSTEMONS

Penstemons are elegant, semi-evergreen perennials valued for their spires of tubular, foxglove-like flowers, in white and shades of pink, red, and purple, held above lance-shaped leaves. Smaller penstemons, such as *P. newberryi*, are at home in a rock garden or as edging plants, while larger cultivars make reliable border perennials which flower throughout summer and into autumn. 'Andenken an Friedrich Hahn' and 'Schoenholzeri' are among the hardiest cultivars, although most types benefit from a dry winter mulch in areas that are prone to heavy frosts. Penstemons tend to be short-lived, and are best replaced after a few seasons.

CULTIVATION *Grow border plants in well-drained, humus-rich soil, and dwarf cultivars in sharply drained, gritty, poor to moderately fertile soil. Choose a site in full sun or partial shade. Deadhead regularly to prolong the flowering season.*

☼ ◑ ◊ 🏆H3, H3–4, H4

1 ↕1.2m (4ft) ↔ 45cm (18in)

2 ↕↔ 45–60cm (18–24in)

3 ↕75cm (30in) ↔ 60cm (24in)

1 *P.* 'Alice Hindley' (H3) **2** *P.* 'Apple Blossom' (H3)
3 *P.* 'Andenken an Friedrich Hahn' (syn. *P.* 'Garnet') (H4)

4 ↕90cm (36in) ↔ 75cm (30in)

5 ↕45–60cm (18–24in) ↔ 30cm (12in)

6 ↕25cm (10in) ↔ 30cm (12in)

7 ↕90cm (36in) ↔ 60cm (24in)

8 ↕60cm (24in) ↔ 45cm (18in)

4 *P.* 'Chester Scarlet' (H3) **5** *P.* 'Evelyn' (H3–4) **6** *P. newberryi* (H4) **7** *P.* 'Schoenholzeri' (H4) (syn. *P.* 'Firebird', *P.* 'Ruby') **8** *P.* 'White Bedder' (H3) (syn. *P.* 'Snowstorm')

PERILLA FRUTESCENS VAR. *CRISPA*

An upright, bushy annual grown for its frilly, pointed, mid green leaves which are flecked with deep purple. Spikes of tiny white flowers appear in summer. The dark foliage makes a good contrasting background to the bright flowers of most summer bedding plants.

CULTIVATION *Grow in moist but well-drained, fertile soil. Position in sun or partial shade. Plant in spring after the danger of frost has passed.*

☼ ◑ 💧 ♀H3 ↕to 1m (3ft) ↔to 30cm (12in)

PEROVSKIA 'BLUE SPIRE'

This upright, deciduous subshrub, grown for its foliage and flowers, suits a mixed or herbaceous border. Branching, airy spikes of tubular, violet-blue flowers are borne in profusion during late summer and early autumn, above the silvery-grey, divided leaves. Tolerates coastal conditions.

CULTIVATION *Best in poor to moderately fertile soil that is well-drained. Tolerates chalky soil. For vigorous, bushy growth, prune back hard each spring to a low framework. Position in full sun.*

☼ 💧 ♀H4 ↕1.2m (4ft) ↔1m (3ft)

PERSICARIA AFFINIS 'SUPERBA'

A vigorous, evergreen perennial, formerly in the genus *Polygonum*, that forms mats of lance-shaped, deep green leaves which turn rich brown in autumn. Dense spikes of long-lasting, pale pink flowers, ageing to dark pink, are borne from mid-summer to mid-autumn. Plant in groups at the front of a border or use as ground cover. 'Darjeeling Red' is similar and equally good.

CULTIVATION *Grow in any moist soil, in full sun or partial shade. Dig out invasive roots in spring or autumn.*

☼☀ 💧 🏆H3–4 ↕to 25cm (10in) ↔60cm (24in)

PERSICARIA BISTORTA 'SUPERBA'

A fast-growing, semi-evergreen perennial, formerly in the genus *Polygonum*, that makes good ground cover. Dense, cylindrical, soft pink flowerheads are produced over long periods from early summer to mid-autumn, above the clumps of mid-green foliage.

CULTIVATION *Best in any well-drained, reliably moist soil, in full sun or partial shade. Tolerates dry soil.*

☼☀ 💧💧 🏆H4 ↕75cm (30in) ↔90cm (36in)

PERSICARIA VACCINIIFOLIA

This creeping, evergreen perennial, formerly in the genus *Polygonum*, bears glossy mid-green leaves which flush red in autumn. In late summer and autumn, spikes of deep pink flowers appear on branching, red-tinted stems. Suits a rock garden by water, or at the front of a border. Good for ground cover.

CULTIVATION *Grow in any moist soil, in full sun or semi-shade. Control spread by digging up invasive roots in spring or autumn.*

☼☀ ◑ ♀H4 ↕20cm (8in) ↔50cm (20in) or more

PETUNIA CARPET SERIES

These very compact, spreading petunias are usually grown as annuals for their abundant, trumpet-shaped flowers. These open over a long period from late spring to early autumn, in shades of pink through strong reds and oranges to yellow and white. The leaves are oval and dark green. Ideal as dense, colourful summer bedding, particularly in poor soil.

CULTIVATION *Grow in well-drained soil, in a sunny, sheltered site. Deadhead to prolong flowering. Plant out only once any danger of frost has passed.*

☼ ◊ ♀H3 ↕15cm (6in) ↔to 1m (3ft)

PETUNIA
MIRAGE SERIES

These dense, spreading annuals are very similar to the Carpet Series (facing page, below), but the trumpet-shaped flowers come in shades of not only red and pink, but also blues and purples. These are borne from late spring until early autumn above the dark green, oval leaves. Particularly useful as a bedding plant on poor soils.

CULTIVATION *Grow in any light, well-drained soil, in full sun with shelter from wind. Deadhead to prolong flowering. Plant out after the danger of frost has passed.*

☼ ◊ 🏆H3 ↕to 30cm (12in) ↔to 1m (3ft)

PHALARIS ARUNDINACEA
'PICTA'

Gardeners' garters is an evergreen, clump-forming perennial grass with narrow, white-striped leaves. Tall plumes of pale green flowers, fading to buff as they mature, are borne on upright stems during early to mid-summer. Good for ground cover, but can be invasive.

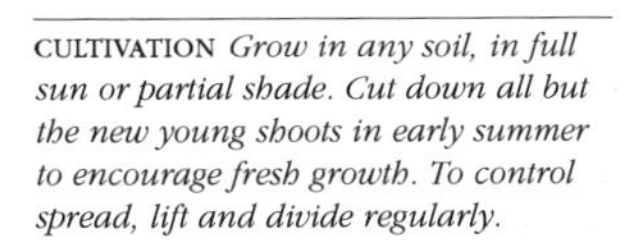

CULTIVATION *Grow in any soil, in full sun or partial shade. Cut down all but the new young shoots in early summer to encourage fresh growth. To control spread, lift and divide regularly.*

☼ ◐ ♦ 🏆H4 ↕to 1m (3ft) ↔indefinite

PHILADELPHUS 'BEAUCLERK'

A slightly arching, deciduous shrub valued for its clusters of fragrant, large white flowers with slightly pink-flushed centres. These are borne in early and mid-summer, amid the broadly oval, dark green leaves. Grow in a shrub border, on its own or as a screen.

CULTIVATION *Grow in any well-drained, moderately fertile soil. Tolerates shallow, chalky soil and light shade, but flowers are best in full sun. Cut back 1 in 4 stems to the ground after flowering, to stimulate strong growth.*

☼ ◐ ◊ ♀H4 ↕↔ 2.5m (8ft)

PHILADELPHUS 'BELLE ETOILE'

This arching, deciduous shrub is similar to, but more compact than 'Beauclerk' (above). An abundance of very fragrant, large white flowers with bright yellow centres appear during late spring to early summer. The leaves are tapered and dark green. Good in a mixed border.

CULTIVATION *Grow in any moderately fertile soil with good drainage. Flowers are best in full sun, but tolerates partial shade. After flowering, cut 1 in 4 stems back to the ground, to stimulate strong new growth.*

☼ ◐ ◊ ♀H4 ↕1.2m (4ft) ↔2.5m (8ft)

PHILADELPHUS CORONARIUS 'VARIEGATUS'

This upright, deciduous shrub has attractive, mid-green leaves that are heavily marked with white around the edges. Short clusters of very fragrant white flowers open in early summer. Use to brighten up the back of a mixed border or woodland garden, or on its own.

CULTIVATION *Grow in any fairly fertile soil with good drainage, in sun or semi-shade. For the best foliage, grow in light shade and prune in late spring. For the best flowers, grow in sun, cutting some stems to the ground after flowering.*

☼ ◑ ◊ 🏆H4 ↕2.5m (8ft) ↔2m (6ft)

PHILADELPHUS 'MANTEAU D'HERMINE'

This flowering, deciduous shrub is low and spreading in habit, with long-lasting, double, very fragrant, creamy-white flowers. These appear from early to mid-summer amid the pale to mid-green, elliptic leaves. Suits a mixed border.

CULTIVATION *Grow in any well-drained, fairly fertile soil. Tolerates partial shade, but flowers are best in full sun. Cut back 1 in 4 stems to the ground after flowering, for strong new growth.*

☼ ◑ ◊ 🏆H4 ↕1m (3ft) ↔1.5m (5ft)

PHLOMIS FRUTICOSA

Jerusalem sage is a mound-forming, spreading, evergreen shrub, carrying sage-like and aromatic, grey-green leaves with woolly undersides. Short spikes of hooded, dark golden-yellow flowers appear from early to mid-summer. Effective when massed in a border.

CULTIVATION *Best in light, well-drained, poor to fairly fertile soil, in sun. Prune out any weak or leggy stems in spring.*

☼ ◊ H4 ↕1m (3ft) ↔1.5m (5ft)

PHLOMIS RUSSELIANA

An upright, evergreen border perennial, sometimes known as either *P. samia* or *P. viscosa*, bearing pointed, hairy, mid-green leaves. Spherical clusters of hooded, pale yellow flowers appear along the stems from late spring to autumn. The flowers last well into winter.

CULTIVATION *Grow in any well-drained, moderately fertile soil, in full sun or light shade. May self-seed.*

☼ ◐ ◊ H4 ↕to 1m (3ft) ↔75cm (30in)

PHLOX DIVARICATA 'CHATTAHOOCHEE'

A short-lived, semi-evergreen border perennial bearing many flat-faced, long-throated, lavender-blue flowers with red eyes. These are produced over a long period from summer to early autumn, amid the lance-shaped leaves which are carried on purple-tinted stems.

CULTIVATION *Grow in moist but well-drained, humus-rich, fertile soil. Chose a site in partial shade.*

◑ ◊◆ 🏆H4 ↕15cm (6in) ↔30cm (12in)

PHLOX DOUGLASII 'BOOTHMAN'S VARIETY'

A low and creeping, evergreen perennial that forms mounds of narrow, dark green leaves. Dark-eyed, violet-pink flowers with long throats and flat faces appear in late spring or early summer. Good in a rock garden or wall, or as edging in raised beds.

CULTIVATION *Grow in well-drained, fertile soil, in full sun. In areas with low rainfall, position in dappled shade.*

☼◑ ◊ 🏆H4 ↕20cm (8in) ↔30cm (12in)

PHLOX DOUGLASII 'CRACKERJACK'

This low-growing perennial is covered in reddish-magenta flowers in late spring or early summer. The evergreen foliage is dense and dark green. An attractive, compact plant, ideal for a rock garden, in a dry wall, as edging, or as ground cover.

CULTIVATION *Grow in well-drained, fertile soil in full sun or, in areas with low rainfall, in dappled shade.*

☼ ◊ ♈H4 ↕to 12cm (5in) ↔to 20cm (8in)

PHLOX DOUGLASII 'RED ADMIRAL'

A low-growing, spreading, evergreen perennial that looks good trailing over rocks or spilling over the edge of a border. The mounds of dark green foliage are carpeted with deep crimson, flat-faced flowers during late spring and early summer.

CULTIVATION *Grow in well-drained, fertile soil, in full sun. In areas with low rainfall, position in partial shade.*

☼ ◐ ◊ ♈H4 ↕20cm (8in) ↔30cm (12in)

PHLOX DRUMMONDII CULTIVARS

The Greek name *Phlox* means "a flame", referring to the very bright colours of the blooms. *P. drummondii*, the annual phlox, is the parent of many named cultivars used primarily for bold summer bedding. They are upright to spreading, bushy annuals, which bear bunched clusters of hairy purple, pink, red, lavender-blue, or white flowers in late spring. The flowers are often paler at the centres with contrasting marks at the bases of the petal lobes. The stem-clasping leaves are mid-green. Useful in rock gardens, herbaceous borders, flower beds, or in containers; if raised in a greenhouse, they may come into bloom earlier.

CULTIVATION *Grow in reliably moist but well-drained, sandy soil in full sun. Enrich the soil with plenty of organic matter before planting and feed once a week with a dilute liquid fertilizer. Slugs and snails are attracted by young plants.*

☼ ◊♦ ♀H3

1 ↕10–45cm (4–18in) ↔ 25cm (10in) 2 ↕10–45cm (4–18in) ↔ 25cm (10in) 3 ↕10–45cm (4–18in) ↔ 25cm (10in)

4 ↕10–45cm (4–18in) ↔ 25cm (10in) 5 ↕10–45cm (4–18in) ↔ 25cm (10in) 6 ↕10–45cm (4–18in) ↔ 25cm (10in)

1 *Phlox drummondii* 'Beauty Mauve' **2** 'Beauty Pink' **3** 'Brilliancy Mixed' **4** 'Brilliant' **5** 'Buttons Salmon with Eye' **6** 'Phlox of Sheep'

PHLOX 'KELLY'S EYE'

A vigorous, evergreen, mound-forming perennial producing a colourful display of long-throated, flat-faced, pale pink flowers with red-purple centres in late spring and early summer. The leaves are dark green and narrow. Suitable for a rock garden or wall crevices.

CULTIVATION *Grow in fertile soil that has good drainage, in full sun. In low-rainfall areas, site in partial shade.*

☼ ◑ ◇ ♀H4 ↕15cm (6in) ↔20cm (8in)

PHLOX MACULATA 'ALPHA'

A cultivar of meadow phlox with tall, fat clusters of lilac-pink flowers, which appear above the foliage in the first half of summer. It is an upright herbaceous perennial with wiry stems. Ideal in a moist border, and a good source of fragrant flowers for cutting.

CULTIVATION *Grow in fertile, moist soil in full sun or partial shade. Remove spent flowers to prolong flowering, and cut back to the ground in autumn. May need staking.*

☼ ◑ ♦ ♀H4 ↕to 90cm (36in) ↔45cm (18in)

PHLOX PANICULATA CULTIVARS

Cultivars of the perennial phlox, *P. paniculata*, are herbaceous plants bearing dome-shaped or conical clusters of flowers above lance-shaped, toothed, mid-green leaves. Appearing throughout summer and into autumn, the flat-faced, long-throated, delicately fragranced flowers are white, pink, red, purple or blue, often with contrasting centres; they are long-lasting when cut. Larger flowers can be encouraged by removing the weakest shoots in spring when the plant is still quite young. All types are well-suited to a herbaceous border; most cultivars need staking, but others, such as 'Fujiyama', have particularly sturdy stems.

CULTIVATION *Grow in any reliably moist, fertile soil. Choose a position in full sun or partial shade. Feed with a balanced liquid fertiliser in spring and deadhead regularly to prolong flowering. After flowering, cut back all foliage to ground level.*

☼ ◑ ♦ ♈H4

1 ↕1.2m (4ft) ↔ 60cm (24in)

2 ↕90cm (36in) ↔ 45cm (18in)

3 ↕1.2m (4ft) ↔ 60cm (24in)

4 ↕1.1m (3½ft) ↔ to 1m (3ft)

5 ↕1m (3ft) ↔ 45cm (18in)

6 ↕1.2m (4ft) ↔ 60–100cm (24–39in)

1 *Phlox paniculata* 'Brigadier' **2** 'Eventide' **3** 'Fujiyama' **4** 'Le Mahdi' **5** 'Mother of Pearl' **6** 'Windsor'

PHORMIUM COOKIANUM SUBSP. *HOOKERI* 'CREAM DELIGHT'

This mountain flax forms a clump of broad, arching leaves, to 1.5m (5ft) long, each with broad vertical bands of creamy yellow. Tall, upright clusters of tubular, yellow-green flowers appear in summer. An unusual plant, often used as a focal point. Good in a large container, and tolerant of coastal exposure.

CULTIVATION *Grow in fertile, moist but well-drained soil in full sun. In frost-prone areas, provide a deep, dry winter mulch. Divide crowded clumps in spring.*

☼ ◊♦ ♀H3–4 ↕to 2m (6ft) ↔3m (10ft)

PHORMIUM COOKIANUM SUBSP. *HOOKERI* 'TRICOLOR'

This striking perennial, very useful as a focal point in a border, forms arching clumps of broad, light green leaves, to 1.5m (5ft) long; they are boldly margined with creamy-yellow and red stripes. Tall spikes of tubular, yellow-green flowers are borne in summer. In cold climates, grow in a container during the summer, and overwinter under glass.

CULTIVATION *Grow in moist but well-drained soil, in sun. Provide a deep, dry winter mulch in frost-prone areas.*

☼ ◊♦ ♀H3–4 ↕0.6–2m (2–6ft) ↔0.3–3m (1–10ft)

PHORMIUM 'SUNDOWNER'

A clump-forming, evergreen perennial valued for its form and brilliant colouring. It has broad, bronze-green leaves, to 1.5m (5ft) long, with creamy rose-pink margins. In summer, tall, upright clusters of tubular, yellow-green flowers are borne, to be followed by decorative seedheads that persist through winter. Suits coastal gardens.

CULTIVATION *Grow in any deep soil that does not dry out, but best in a warm, wet, sheltered site in full sun. Provide a deep mulch in winter, and divide overcrowded clumps in spring.*

☼ ◊◆ 🏆H3 ↕↔to 2m (6ft)

PHORMIUM TENAX

The New Zealand flax is an evergreen perennial that forms clumps of very long, tough leaves, to 3m (10ft) in length in ideal conditions. They are dark green above, and blue-green beneath. Dusky red flowers appear in summer on very tall, stout and upright spikes. One of the largest phormiums – a striking plant with decorative seedheads in winter.

CULTIVATION *Best in deep, reliably moist but well-drained soil in a warm, sheltered site in full sun. Mulch deeply for winter and divide overcrowded clumps in spring.*

☼ ◊◆ 🏆H4 ↕4m (12ft) ↔2m (6ft)

PHORMIUM TENAX PURPUREUM GROUP

An evergreen perennial that forms clumps of long, stiff, sword-shaped, deep copper to purple-red leaves. Large spikes of dark red, tubular flowers on blue-purple stems appear in summer. Ideal for coastal gardens, but will suffer in areas with cold winters. Can be container-grown, or choose the similar but much smaller 'Nanum Purpureum'.

CULTIVATION *Grow in deep, fertile, humus-rich soil that is reliably moist. Position in full sun with shelter from cold winds. In frost-prone areas, provide a deep, dry mulch in winter.*

☼ ◊ ♈H3–4 ↕2–2.5m (6–8ft) ↔1m (3ft)

PHORMIUM 'YELLOW WAVE'

An evergreen perennial, forming clumps of broad, arching, yellow-green leaves with mid-green vertical stripes; they take on chartreuse tones in autumn. Spikes of tubular red flowers emerge from the centre of each leaf clump in summer. Especially good for seaside gardens, it will add a point of interest to any border, especially in winter when it remains bold and attractive.

CULTIVATION *Best in fertile, moist but well-drained soil in full sun. Provide a deep, dry mulch for winter in frost-prone areas. Divide crowded clumps in spring.*

☼ ◊♦ ♈H3 ↕3m (10ft) ↔2m (6ft)

PHOTINIA X *FRASERI* 'RED ROBIN'

An upright, compact, evergreen shrub often grown as a formal or semi-formal hedge for its bright red young foliage, the effect of which is prolonged by clipping. The mature leaves are leathery, lance-shaped and dark green. Clusters of small white flowers appear in mid-spring. 'Robusta' is another recommended cultivar.

CULTIVATION *Grow in moist but well-drained, fertile soil, in full sun or semi-shade. Clip hedges 2 or 3 times a year to perpetuate the colourful foliage.*

☼☀ 💧💧 🏆H4 ↕↔ 5m (15ft)

PHOTINIA VILLOSA

A spreading, shrubby tree grown for its foliage, flowers and fruits. Flat clusters of small white flowers appear in late spring, and develop into attractive red fruits. The dark green leaves are bronze when young, turning orange and red in autumn before they fall. Attractive all year round, and tolerant of permanently damp soil.

CULTIVATION *Grow in fertile, moist but well-drained, neutral to acid soil in full sun or partial shade. Remove any congested or diseased growth in late winter.*

☼☀ 💧💧 🏆H4 ↕↔5m (15ft)

PHYGELIUS AEQUALIS 'YELLOW TRUMPET'

An upright, evergreen shrub forming loose spikes of hanging, tubular, pale cream-yellow flowers during summer. The leaves are oval and pale green. Suits a herbaceous or mixed border; in cold areas, grow against a warm, sunny wall.

CULTIVATION *Best in moist but well-drained soil, in sun. In cold climates, shelter from wind and cut frost-damaged stems back to the base in spring. Dead-head to prolong flowering. Dig up unwanted shoots to contain spread.*

☼ 💧 H3–4 ↕↔ 1m (3ft)

PHYGELIUS CAPENSIS

The Cape figwort is an evergreen shrub valued for its summer display of upright spikes of orange flowers. The foliage is dark green, and the plant is often mistaken for a novel-coloured fuchsia. Grow near the back of a herbaceous border, against a warm, sunny wall in cold climates. Birds may find the flowers attractive.

CULTIVATION *Grow in fertile, moist but well-drained soil in full sun with shelter from cold, drying winds. Remove spent flower clusters to encourage more blooms, and provide a dry winter mulch in frost-prone areas. Cut back to the ground in spring.*

☼ 💧 H3 ↕1.2m (4ft) ↔1.5m (5ft)

PHYGELIUS X *RECTUS* 'AFRICAN QUEEN'

An upright, evergreen border shrub that bears long spikes of hanging, tubular, pale red flowers with orange to yellow mouths. These appear in summer above the oval, dark green leaves. Best against a warm wall in cold areas.

CULTIVATION *Grow in moist but well-drained, fertile soil, in sun with shelter from cold, drying winds. Cut frost-damaged stems back to the base in spring. Deadhead to prolong flowering.*

☼ ◊◊ ♀H3–4 ↕1m (3ft) ↔1.2m (4ft)

PHYGELIUS X *RECTUS* 'DEVIL'S TEARS'

An upright, evergreen shrub with dark green foliage. In summer, it carries spikes of hanging, red-pink flowers with yellow throats. A reasonably compact shrub with abundant flowers, ideal for a herbaceous or mixed border.

CULTIVATION *Best in moist but well-drained, reasonably fertile soil in full sun. Remove spent flower clusters to encourage further blooming, and cut the plant back to ground level in spring, if damaged over winter; otherwise trim to shape.*

 ♀H4 ↕↔1.5m (5ft)

PHYGELIUS X *RECTUS* 'SALMON LEAP'

This upright shrub is very similar to 'Devil's Tears' (see previous page), but with orange flowers, which turn slightly back toward the stems. They appear in large sprays above the dark green foliage. Good in a mixed or herbaceous border.

CULTIVATION *Grow in moist but well-drained, fertile soil in full sun. Remove spent flower clusters to encourage more blooms. Cut the plant back to ground level in spring if damaged by winter weather; otherwise trim to shape.*

☼ ◊♦ 🏆H4 ↕1.2m (4ft) ↔1.5m (5ft)

PHYLLOSTACHYS NIGRA

Black bamboo is an arching, clump-forming, evergreen shrub. The gentle, lance-shaped, dark green leaves are produced on slender green canes which turn black in their second or third year. Use as a screen or as a large feature plant.

CULTIVATION *Grow in moist but well-drained soil, in sun or partial shade with shelter from cold winds. Mulch over winter. Cut out damaged and overcrowded canes in spring or early summer. Confine spread by burying a barrier around the roots.*

☼ ◐ ◊♦ 🏆H4 ↕3–5m (10–15ft) ↔2–3m (6–10ft)

PHYLLOSTACHYS NIGRA VAR. *HENONIS*

This clump-forming, evergreen bamboo is similar to the black bamboo in habit (see facing page, below), but has bright green canes that mature to yellow-green in the second or third year. The lance-shaped, dark green leaves are downy and rough when young.

CULTIVATION *Grow in well-drained but moist soil, in sun or semi-shade. Shelter from cold winds and mulch over winter. Thin crowded clumps in late spring. Bury a barrier around the roots to confine spread.*

☼☀ ◊♦ ♀H4 ↕3–5m (10–15ft) ↔2–3m (6–10ft)

PHYSOCARPUS OPULIFOLIUS 'DART'S GOLD'

A rounded and thicket-forming, deciduous shrub valued for its three-lobed leaves which are a spectacular golden-yellow when young. Dense clusters of small, white or pale pink flowers appear in spring. Lightens up a shrub border.

CULTIVATION *Best in moist but well-drained, acid soil, but tolerates most conditions except for shallow chalk. Site in sun or partial shade. Cut old stems back to the base after flowering; dig out spreading shoots to confine spread.*

☼☀ ◊♦ ♀H4 ↕2m (6ft) ↔2.5m (8ft)

PHYSOSTEGIA VIRGINIANA 'VIVID'

An upright, densely clump-forming border perennial bearing spikes of bright purple-pink, hooded flowers from mid-summer to early autumn above the narrow, mid-green leaves. The flowers are good for cutting and will remain in a new position if they are moved on the stalks; because of this, it is known as the obedient plant. Looks good mixed with the white cultivar, 'Summer Snow'.

CULTIVATION *Grow in reliably moist, fertile soil that is rich in humus. Position in full sun or partial shade.*

H4 ↕30–60cm (12–24in) ↔30cm (12in)

PICEA GLAUCA VAR. *ALBERTIANA* 'CONICA'

A conical, slow-growing, evergreen conifer carrying dense, blue-green foliage. The short, slender needles are borne on buff-white to ash-grey stems. Oval cones appear during summer, green at first, then maturing to brown. Makes an excellent neat specimen tree for a small garden.

CULTIVATION *Grow in deep, moist but well-drained, preferably neutral to acid soil, in full sun. Prune in winter if necessary, but keep to a minimum.*

H4 ↕2–6m (6–20ft) ↔1–2.5m (3–8ft)

PICEA MARIANA 'NANA'

This dwarf, low-growing form of black spruce is a mound-forming conifer with scaly, grey-brown bark. The evergreen, bluish-grey needles are short, soft and slender. Useful in a rock garden or conifer bed, or as an edging plant.

CULTIVATION *Best in deep, moist but well-drained, fertile, humus-rich soil, in partial shade. Completely remove any shoots that show vigorous upright growth as soon as they are seen.*

◐ ● 🏆H4 ↕↔ 50cm (20in)

PICEA PUNGENS 'KOSTER'

This conical evergreen conifer, with attractive horizontal branches, becomes more columnar with age, and the young growth is clothed in silvery-blue foliage. The long, sharp-pointed needles turn greener as they mature. Cylindrical green cones are borne during the summer, aging to pale brown. Good in large gardens as a prominent specimen tree. 'Hoopsii' is very similar indeed, with blue-white foliage.

CULTIVATION *Grow in well-drained, fertile, neutral to acid soil, in full sun. Prune in late autumn or winter if necessary, but keep to a minimum.*

☼ ○ 🏆H4 ↕15m (50ft) ↔5m (15ft)

PIERIS 'FOREST FLAME'

An upright, evergreen shrub valued for its slender, glossy, lance-shaped leaves that are bright red when young; they mature through pink and creamy-white to dark green. Upright clusters of white flowers enhance the effect in early to mid-spring. Ideal for a shrub border or a peaty, lime-free soil; it will not thrive in alkaline soil.

CULTIVATION *Grow in moist but well-drained, fertile, humus-rich, acid soil, in full sun or partial shade. In frost-prone areas, shelter from cold, drying winds. Trim lightly after flowering.*

☼☀ ♁♁ ♈H4 ↕4m (12ft) ↔2m (6ft)

PIERIS FORMOSA VAR. *FORRESTII* 'WAKEHURST'

This upright, evergreen, acid-soil-loving shrub has brilliant red young foliage that matures to dark green. The large, slightly drooping clusters of small, fragrant white flowers from mid-spring are also attractive. Grow in a woodland garden or shrub border; may not survive in areas with cold winters. 'Jermyns' is very similar, with darker red young leaves.

CULTIVATION *Best in well-drained but moist, fertile, humus-rich, acid soil. Site in full sun or partial shade with shelter from cold winds in frost-prone climates. Trim lightly after flowering.*

☼☀ ♁♁ ♈H3 ↕5m (15ft) ↔4m (12ft)

PIERIS JAPONICA 'BLUSH'

A rounded, evergreen shrub that bears small, pink-flushed white flowers in late winter and early spring. These are carried in long, drooping clusters amid the glossy dark green foliage. A good early-flowering border shrub, although it will not thrive in alkaline soil.

CULTIVATION *Grow in well-drained but moist, fertile, humus-rich, acid soil. Site in full sun or partial shade. Trim lightly after flowering, removing any dead, damaged or diseased shoots.*

H4 ↕4m (12ft) ↔3m (10ft)

PILEOSTEGIA VIBURNOIDES

A slow-growing, woody, evergreen climber that is dusted with feathery clusters of tiny, creamy-white flowers in late summer and autumn. The glossy dark green, leathery leaves look very attractive against a large tree trunk or shady wall.

CULTIVATION *Grow in well-drained, fertile soil, in full sun or shade. Shorten stems after flowering as the plant begins to outgrow the allotted space.*

 H4 ↕6m (20ft)

PINUS MUGO 'MOPS'

This dwarf pine is an almost spherical conifer with scaly, grey bark and thick, upright branches. The shoots are covered with long, well-spaced, dark to bright green needles. The dark brown, oval cones take a few years to ripen. Effective in a large rock garden or, where space allows, planted in groups.

CULTIVATION *Grow in any well-drained soil, in full sun. Very little pruning is required since growth is slow.*

☼ ♢ ♈H4 ↕to 1m (3ft) ↔to 2m (6ft)

PITTOSPORUM TENUIFOLIUM

A columnar, evergreen shrub, much valued for its glossy green leaves with wavy edges. Fast growing at first, it then broadens out into a tree. Tiny, honey-scented, purple-black, bell-shaped flowers open from late spring. Makes a good hedge, but may not survive cold winters. There is a gold-leaved version, 'Warnham Gold'.

CULTIVATION *Grow in well-drained but moist, fertile soil, in full sun or partial shade. In frost-prone areas, shelter from cold winds. Trim to shape in spring; avoid pruning after mid-summer.*

☼ ◑ ♢♦ ♈H3 ↕4–10m (12–30ft) ↔2–5m (6–15ft)

PITTOSPORUM TENUIFOLIUM 'TOM THUMB'

A compact, rounded, evergreen foliage shrub that would suit a mixed border designed for year-round interest. The glossy bronze-purple leaves are elliptic and wavy-edged. Tiny, honey-scented, purple flowers are borne in late spring and early summer. Shelter against a warm wall in cold areas.

CULTIVATION *Grow in well-drained but moist, fertile soil, in full sun for best colour. Shelter from cold winds in frost-prone areas. Trim to shape in spring; established plants need little pruning.*

H3 ↕to 1m (3ft) ↔60cm (24in)

PITTOSPORUM TOBIRA

Japanese mock orange is a rounded, evergreen shrub or small tree with leathery, dark green, oval leaves. Clusters of sweet-scented, creamy-white flowers are borne in late spring and early summer, followed by yellow-brown seed capsules. A fine specimen plant that can be pot-grown and overwintered under glass in cold climates. 'Variegatum' has smaller leaves edged in white.

CULTIVATION *Best in moist but well-drained, fertile soil or compost, in full sun or partial shade. Prune back to restrict growth in winter or early spring.*

H3 ↕2–10m (6–30ft) ↔1.5–3m (5–10ft)

PLATYCODON GRANDIFLORUS

The balloon flower is a clump-forming perennial producing clusters of large purple to violet-blue flowers that open from balloon-shaped buds in late summer, above bluish-green, oval leaves. For a rock garden or herbaceous border. 'Apoyama', with deep-coloured flowers, and 'Mariesii' are recommended cultivars.

CULTIVATION *Best in deep, well-drained, loamy, fertile soil that does not dry out, in full sun or partial shade. Flower stems may need staking. Established plants dislike root disturbance.*

☼ ◐ ◊ 🏆H4 ↕to 60cm (24in) ↔30cm (12in)

PLEIOBLASTUS AURICOMUS

This upright bamboo is an evergreen shrub grown for its brilliant green, yellow-striped foliage. The bristly-edged, lance-shaped leaves are carried on purple-green canes. Effective in an open glade in a woodland garden. Good in sun, backed by trees or tall shrubs.

CULTIVATION *Grow in moist but well-drained, fertile, humus-rich soil, in full sun for best leaf colour. Provide shelter from cold, drying winds. Thin over-crowded clumps in late spring or early summer. Confine spread by burying a barrier around the roots.*

☼ ◊◆ 🏆H4 ↕↔ to 1.5m (5ft)

PLEIOBLASTUS VARIEGATUS

This upright, evergreen bamboo is much shorter than *P. auricomis* (see facing page, below), with cream- and green-striped foliage. The lance-shaped leaves, borne on hollow, pale green canes, are covered in fine white hairs. Suits a sunny border backed by shrubs, and where it has space to spread; it will swamp less vigorous neighbours unless confined.

CULTIVATION *Grow in moist but well-drained, fertile, humus-rich soil, in sun with shelter from cold winds. Thin out clumps in late spring. Bury a barrier around the roots to confine spread.*

☼ ◊◆ ♕H4 ↕75cm (30in) ↔1.2m (4ft)

PLUMBAGO AURICULATA

Cape leadwort is a scrambling, semi-evergreen, frost-tender shrub often trained as a climber. It bears dense trusses of long-throated, sky-blue flowers from summer to late autumn, amid the oval leaves. In cold climates, plants over-wintered under glass can be moved outside in summer. Often sold as *P. capensis.*

CULTIVATION *Grow in well-drained, fertile soil or compost, in full sun or light shade. Pinch out the tips of young plants to promote bushiness, and tie climbing stems to a support. Cut back to a permanent framework in early spring.*

☼ ◐ ◊ ♕H1–2 ↕3–6m (10–20ft) ↔1–3m (3–10ft)

POLEMONIUM 'LAMBROOK MAUVE'

This clump-forming perennial, a garden variety of Jacob's ladder, forms rounded mounds of neat, divided, mid-green leaves. An abundance of funnel-shaped, sky-blue flowers cover the foliage from late spring to early summer. Good in any border, or in a wild garden.

CULTIVATION *Grow in well-drained but moist, moderately fertile soil, in full sun or partial shade. Deadhead regularly.*

☼◐ 💧💧 ♈H4 ↕↔ to 45cm (18in)

POLYGONATUM x *HYBRIDUM*

Solomon's seal is a perennial for a shady border, bearing hanging clusters of tubular, small white flowers with green mouths, along slightly arching stems. These appear in late spring among elliptic, bright green leaves. Round blue-black fruits develop after the flowers. For a woodland garden. *P.* x *odoratum* 'Flore Pleno' is a very similar but smaller plant.

CULTIVATION *Grow in moist but well-drained, fertile soil that is rich in humus. Position in partial or full shade.*

◐● 💧💧 ♈H4 ↕to 1.5m (5ft) ↔30cm (12in)

POLYSTICHUM ACULEATUM

The prickly shield fern is an elegant, evergreen perennial producing a shuttlecock of finely divided, dark green fronds. There are no flowers. An excellent foliage plant for shady areas in a rock garden or well-drained border.

CULTIVATION *Grow in fertile soil that has good drainage, in partial or deep shade. Choose a site sheltered from excessive winter wet. Remove the previous year's dead fronds before the new growth unfurls in spring.*

H4 ↕60cm (24in) ↔1m (3ft)

POLYSTICHUM SETIFERUM

The soft shield fern is a tall, evergreen perennial, with finely divided, dark green fronds which are soft to the touch. The foliage forms splayed, shuttlecock-like clumps. Suits a shady border or rock garden. 'Divisilobum Densum' is a handsome cultivar with densely feathery fronds; 'Pulcherrimum Bevis' has tall, sweeping fronds.

CULTIVATION *Grow in well-drained, fertile soil, in partial or deep shade. Choose a site that is protected from excessive winter wet. Remove any dead or damaged fronds in spring.*

 ↕1.2m (4ft) ↔1m (3ft)

POTENTILLA FRUTICOSA

Cultivars of *P. fruticosa* are compact and rounded, deciduous shrubs that produce an abundance of flowers over a long period from late spring to mid-autumn. The leaves are dark green and composed of several oblong leaflets. Flowers are saucer-shaped and wild-rose-like, sometimes borne singly, but often in clusters of three. Most cultivars are yellow-flowered, but blooms may also be white, as with 'Abbotswood', or flushed with pink, as in 'Daydawn'. These are undemanding shrubs that make invaluable additions to mixed or shrub borders; they can also be grown as attractive low hedges.

CULTIVATION *Grow in well-drained, poor to moderately fertile soil. Best in full sun, but many tolerate partial shade. Trim lightly after flowering, cutting older wood to the base and removing weak, twiggy growth. Old shrubs sometimes respond well to renovation, but may be better replaced.*

☼ 💧 ♈H4

1 ↕75cm (30in) ↔ 1.2m (4ft)

2 ↕1m (3ft) ↔ 1.5m (5ft)

3 ↕1m (3ft) ↔ 1.2m (4ft)

4 ↕1m (3ft) ↔ 1.5m (5ft)

1 *P. fruticosa* 'Abbotswood' **2** *P. fruticosa* 'Elizabeth' **3** *P. fruticosa* 'Daydawn'
4 *P. fruticosa* 'Primrose Beauty'

POTENTILLA 'GIBSON'S SCARLET'

A dense, clump-forming herbaceous perennial grown for its very bright scarlet flowers, borne in succession throughout summer. The soft green leaves are divided into five leaflets. Good for a rock garden or for a bold summer colour in a mixed or herbaceous border. 'William Rollisson' is a similar plant, with more orange-red, semi-double flowers.

CULTIVATION *Grow in well-drained, poor to moderately fertile soil. Choose a position in full sun.*

☼ ◊ 🏆H4 ↕to 45cm (18in) ↔60cm (24in)

POTENTILLA MEGALANTHA

A compact, clump-forming perennial bearing a profusion of upright, cup-shaped, rich yellow flowers during mid- to late summer. The slightly hairy, mid-green leaves are divided into three coarsely scalloped leaflets. Suits the front of a border.

CULTIVATION *Grow in poor to fairly fertile soil that has good drainage. Choose a position in full sun.*

☼ ◊ 🏆H4 ↕15–30cm (6–12in) ↔15cm (6in)

POTENTILLA NEPALENSIS 'MISS WILMOTT'

A summer-flowering perennial that forms clumps of mid-green, divided leaves on wiry, red-tinged stems. The small pink flowers, with cherry-red centres, are borne in loose clusters. Good for the front of a border or in cottage-style plantings.

CULTIVATION *Grow in any well-drained, poor to moderately fertile soil. Position in full sun or light dappled shade.*

☼☀ ◊ ♥H4 ↕30–45cm (12–18in) ↔60cm (24in)

PRIMULA DENTICULATA

The drumstick primula is a robust, clump-forming perennial bearing spherical clusters of small purple flowers with yellow centres. They are carried on stout, upright stalks from spring to summer, above the basal rosettes of oblong to spoon-shaped, mid-green leaves. Thrives in damp, but not water-logged, soil; ideal for a waterside planting.

CULTIVATION *Best in moist, humus-rich, neutral to acid or peaty soil. Choose a site in partial shade, but tolerates full sun where soil is reliably damp.*

☼☀ ♦ ♥H4 ↕↔ 45cm (18in)

PRIMULA ELATIOR

The oxlip is a semi-evergreen perennial wildflower that can vary in appearance. Clusters of tubular yellow flowers emerge from the basal rosettes of scalloped, mid-green leaves on stiff, upright stems in spring and summer. Plant in groups to naturalize in a moist meadow.

CULTIVATION *Grow in moderately fertile, deep, moist but well-drained soil. Site in partial shade, but full sun is tolerated as long as the soil remains moist at all times.*

☼ ◑ ◊♦ ♈H4 ↕30cm (12in) ↔25cm (10in)

PRIMULA FLACCIDA

A rosette-forming, deciduous perennial for an open woodland garden or alpine house. In summer, tall flowering stems bear conical clusters of funnel-shaped and downward-pointing, floury, lavender-blue flowers, carried above the pale to mid-green leaves.

CULTIVATION *Grow in deep or partial shade in peaty, gritty, moist but sharply drained, acid soil. Protect from excessive winter wet.*

◑ ● ◊♦ ♈H4 ↕50cm (20in) ↔30cm (12in)

PRIMULA FLORINDAE

The giant cowslip is a deciduous, summer-flowering perennial wildflower that grows naturally by pools and streams. It forms clumps of oval, toothed, mid-green leaves which are arranged in rosettes at the base of the plant. Drooping clusters of up to 40 funnel-shaped, sweetly-scented yellow flowers are borne well above the foliage on upright stems. Good in a bog garden or waterside setting.

CULTIVATION *Grow in deep, reliably moist, humus-rich soil, in partial shade. Tolerates full sun if soil remains moist.*

☼◑ ◊ ♀H4 ↕to 1.2m (4ft) ↔ 1m (3ft)

PRIMULA FRONDOSA

This deciduous border perennial, with its rosettes of spoon-shaped, mid-green leaves, carries yellow-eyed, pink to purple flowers in late spring or early summer. They are carried in loose clusters of up to 30 at the top of upright stems, and each has a pale yellow eye at its centre.

CULTIVATION *Best in deep, moist, neutral to acid loam or peaty soil enriched with organic matter. Site ideally in partial shade, but it tolerates full sun if the soil is reliably moist.*

☼◑ ♦ ♀H4 ↕15cm (6in) ↔25cm (10in)

PRIMULA GOLD-LACED GROUP

A group of semi-evergreen primroses grown for their showy spring flowers in the border or in containers. Clusters of golden-eyed, very dark mahogany-red or black flowers, with a thin gold margin around each petal, are carried atop upright flowering stems above the sometimes reddish foliage.

CULTIVATION *Grow in moist, deep, neutral to acid loam or peaty soil that is rich in organic matter. Choose a site in partial shade, or in full sun if the soil is reliably moist.*

☼ ◐ ♦ ♀H4 ↕25cm (10in) ↔30cm (12in)

PRIMULA 'GUINEVERE'

A fast-growing, evergreen, clump-forming perennial, sometimes called 'Garryarde Guinevere', that bears clusters of pale purplish-pink flowers. They have yellow centres, flat faces and long throats, and are carried above the deep bronze, oval leaves in spring. Suits damp, shady places.

CULTIVATION *Grow in moist, neutral to acid soil that is well-drained, in partial shade. Tolerates full sun, but only if the soil remains damp at all times.*

☼ ◐ ♦ ♀H4 ↕12cm (5in) ↔25cm (10in)

POLYANTHUS PRIMROSES (*PRIMULA*)

Garden polyanthus are rosette-forming, evergreen perennials with a complex parentage, which probably includes *P. veris* (see p.419) and *P. elatior* (see p.413). The plants form sturdy, basal rosettes of oval, heavily veined leaves, overshadowed by the colourful clusters of flat-faced flowers from late winter to early spring. A splendid array of primary and pastel colours is available, mostly red, blue-violet, orange, yellow, white or pink, with yellow centres. Ranges like the Crescendo and Rainbow series are among the most popular and easily grown bedding and container plants, brightening up gardens, patios and windowsills in winter.

CULTIVATION *Grow in moderately fertile, moist but well-drained, humus-rich soil or loam-based compost in a cool site in full sun or partial shade. Sow seed in summer, and plant out in autumn in well-prepared soil. Divide large clumps in autumn.*

☼ ◑ ◊ ♦ ♈H4

1 ↕15cm (6in) ↔ 30cm (12in)

2 ↕15cm (6in) ↔ 30cm (12in)

3 ↕15cm (6in) ↔ 30cm (12in)

4 ↕15cm (6in) ↔ 30cm (12in)

1 *Primula* 'Crescendo Bright Red' **2** *P.* 'Crescendo Pink and Rose Shades'
3 *P.* 'Rainbow Blue Shades' **4** *P.* 'Rainbow Cream Shades'

PRIMULA KEWENSIS

A primrose for a windowsill, greenhouse or conservatory, this is an evergreen perennial which forms rosettes of noticeably toothed, slightly floury, mid-green leaves. It gives an early spring display of fragrant yellow flowers in whorls along upright flowering stems.

CULTIVATION *In a container, grow in loam-based potting compost with added grit or peat. As a houseplant, choose a site with bright filtered light.*

☼☀ ◊◊ ♈H2 ↕to 45cm (18in) ↔20cm (8in)

PRIMULA OBCONICA 'LIBRE MAGENTA'

In cold climates, the many obconicas make colourful display plants for a cool greenhouse. This one has branched heads of magenta flowers that darken with age, on dark, hairy stems emerging from among coarse, mid-green leaves. Also makes a good house- or cool conservatory plant; min. temp. 2°C (36°F).

CULTIVATION *In a container, grow in a loam-based potting compost with added grit or peat. Choose a spot with bright filtered light. Water freely when in growth and feed weekly with a half-strength liquid fertilizer.*

☼☀ ◊ ♈H1 ↕↔ 30cm (12in)

CANDELABRA PRIMROSES (*PRIMULA*)

Candelabra primroses are robust, herbaceous perennials, so-called because their flowers are borne in tiered clusters, which rise above the basal rosettes of broadly oval leaves in late spring or summer. Depending on the species, the foliage may be semi-evergreen, evergreen or deciduous. The flowers have flat faces and long throats; as with most cultivated primroses, there is a wide choice of colours, from the brilliant red 'Inverewe' to the golden-yellow *P. prolifera.* Some flowers change colour as they mature; those of *P. bulleyana* fade from crimson to orange. Candelabra primroses look most effective when grouped together in a bog garden or waterside setting.

CULTIVATION *Grow in deep, moist, neutral to acid soil that is rich in humus. Site in partial shade, although full sun is tolerated if the soil remains moist at all times. Divide and replant clumps in early spring.*

☀ 💧 ♀H4

1 ↕↔ 60cm (24in)

2 ↕↔ 60cm (24in)

3 ↕ 75cm (30in) ↔ 60cm (24in)

4 ↕ to 1m (3ft) ↔ 60cm (24in)

5 ↕ to 1m (3ft) ↔ 60cm (24in)

1 *P. bulleyana* **2** *P. prolifera* (syn. *P. helodoxa*) **3** *P.* 'Inverewe' **4** *P. pulverulenta* **5** *P. pulverulenta* Bartley Hybrids

PRIMULA ROSEA

A deciduous perennial that bears rounded clusters of glowing pink, long-throated flowers on upright stalks in spring. Clumps of oval, toothed, mid-green leaves emerge after the flowers; these are tinted red-bronze when young. Good for a bog garden or waterside planting.

CULTIVATION *Grow in deep, reliably moist, neutral to acid or peaty soil that is rich in humus. Prefers partial shade, but tolerates full sun if the soil is moist at all times.*

◐ 💧 🏆H4 ↕↔ 20cm (8in)

PRIMULA VERIS

The cowslip is a semi-evergreen, spring-flowering perennial wildflower with a variable appearance. Stout flower stems carry dense clusters of small, funnel-shaped, sweetly scented, nodding yellow flowers, above the clumps of lance-shaped, crinkled leaves. Lovely naturalized in damp grassed areas.

CULTIVATION *Best in deep, moist but well-drained, fertile, peaty soil that is rich in humus, in semi-shade or full sun, if soil remains reliably damp.*

☼ ◐ 💧 🏆H4 ↕↔ to 25cm (10in)

PRIMULA 'WANDA'

A very vigorous, semi-evergreen perennial that bears clusters of flat-faced, claret-red flowers with yellow centres over a long period in spring. The oval, toothed, purplish-green leaves are arranged in clumps at the base of the plant. Good in a waterside setting.

CULTIVATION *Best in deep, moist but well-drained, fertile soil that is enriched with humus. Prefers partial shade, but tolerates full sun if soil remains damp.*

☼☀ ◊♦ ♀H4 ↕10–15cm (4–6in) ↔30–40cm (12–16in)

PRUNELLA GRANDIFLORA 'LOVELINESS'

This vigorous, spreading perennial bears dense, upright spikes of light purple, tubular flowers in summer. The lance-shaped, deep-green leaves are arranged in clumps at ground level. Versatile ground cover when planted in groups; the flowers are attractive to beneficial insects.

CULTIVATION *Grow in any soil, in sun or partial shade. May swamp smaller plants, so allow room to expand. Divide clumps in spring or autumn to maintain vigour. Deadhead to prevent self-seeding.*

☼☀ ◊♦ ♀H4 ↕15cm (6in) ↔to 1m (3ft) or more

PRUNUS X *CISTENA*

An upright, slow-growing, deciduous shrub valued in particular for its foliage, which is red when young, maturing to red-purple. Bowl-shaped, pinkish-white flowers open from mid- to late spring, sometimes followed by small, cherry-like, purple-black fruits. Good as a windbreak hedge.

CULTIVATION *Grow in any but water-logged soil, in full sun. Prune back overcrowded shoots after flowering. To grow as a hedge, prune the shoot tips of young plants then trim in mid-summer to encourage branching.*

☼ ◊♦ ♈H4 ↕↔ 1.5m (5ft)

PRUNUS GLANDULOSA 'ALBA PLENA'

This small cherry is a neat, rounded, deciduous shrub producing dense clusters of pure white, bowl-shaped, double flowers during late spring. The narrowly oval leaves are pale to mid-green. Brings beautiful spring blossom to a mixed or shrub border. The cultivar 'Sinensis' has double pink flowers.

CULTIVATION *Grow in any moist but well-drained, moderately fertile soil, in sun. Can be pruned to a low framework each year after flowering to enhance the flowering performance.*

☼ ◊♦ ♈H4 ↕↔ 1.5m (5ft)

PRUNUS 'KIKU-SHIDARE-ZAKURA'

Also known as 'Cheal's Weeping', this small deciduous cherry tree is grown for its weeping branches and clear pink blossom. Dense clusters of large, double flowers are borne in mid- to late spring, with or before the lance-shaped, mid-green leaves, which are flushed bronze when young. Excellent in a small garden.

CULTIVATION *Best in any moist but well-drained, moderately fertile soil, in full sun. Tolerates chalk. After flowering, prune out only dead, diseased or damaged wood; remove any shoots growing from the trunk as they appear.*

☼ ◊◊ H4 ↕↔ 3m (10ft)

PRUNUS LAUROCERASUS 'OTTO LUYKEN'

This compact cherry laurel is an evergreen shrub with dense, glossy dark green foliage. Abundant spikes of white flowers are borne in mid- to late spring and often again in autumn, followed by conical red fruits that ripen to black. Plant in groups as a low hedge, or to cover bare ground.

CULTIVATION *Grow in any moist but well-drained, moderately fertile soil, in full sun. Prune in late spring or early summer to restrict size.*

☼ ◊◊ H4 ↕1m (3ft) ↔1.5m (5ft)

PRUNUS LUSITANICA SUBSP. *AZORICA*

This Portugal laurel is a slow-growing, evergreen shrub bearing slender spikes of small, fragrant white flowers in early summer. The oval, glossy dark green leaves have red stalks. Purple berries appear later in the season. Attractive year-round as a dense screen or hedge, except where winds are very cold.

CULTIVATION *Grow in any moist but well-drained, fairly fertile soil, in sun with shelter from cold, drying winds. In late spring, prune to restrict size, or to remove old or overcrowded shoots.*

☼ ◊♦ 🏆H3–4 ↕↔ to 20m (70ft)

PRUNUS SERRULA

A rounded, deciduous tree that is valued for its striking, glossy, copper-brown to mahogany-red bark, which peels with age. Small, bowl-shaped white flowers in spring are followed by cherry-like fruits in autumn. The leaves are lance-shaped and dark green, turning yellow in autumn. Best seen as a specimen tree.

CULTIVATION *Best in moist but well-drained, moderately fertile soil, in full sun. Remove dead or damaged wood after flowering, and remove any shoots growing from the trunk as they appear.*

☼ ◊♦ 🏆H4 ↕↔ 10m (30ft)

FLOWERING CHERRY TREES (*PRUNUS*)

Ornamental cherries are cultivated primarily for their white, pink or red flowers which create a mass of blossom, usually on bare branches, from late winter to late spring; cultivars of *P.* x *subhirtella* flower from late autumn. Most popular cultivars not only bear dense clusters of showy, double flowers but have other ornamental characteristics to extend their interest beyond the flowering season: *P. sargentii*, for example, has brilliant autumn foliage colour, and some have shiny, coloured bark. All of these features make flowering cherries superb specimen trees for small gardens.

CULTIVATION *Grow in any moist but well-drained, fairly fertile soil, in sun. Keep all pruning to an absolute minimum; restrict formative pruning to shape to young plants only. Remove any damaged or diseased growth in mid-summer, and keep trunks clear of sprouting shoots.*

☼ ◊◆ ♀H4

1 ↕20m (70ft) ↔ 10m (30ft)
2 ↕↔ 12m (40ft)
3 ↕10m (30ft) ↔ 8m (25ft)
4 ↕10m (30ft) ↔ 8m (25ft)
5 ↕15m (50ft) ↔ 10m (30ft)
6 ↕10m (30ft) ↔ 8m (25ft)

1 *P. avium* **2** *P. avium* 'Plena' **3** *P.* 'Kanzan' **4** *P.* 'Okame' **5** *P. padus* 'Colorata' **6** *P.* 'Pandora'

7 ↕15m (50ft) ↔ 10m (30ft)

8 ↕↔ 8m (25ft)

9 ↕ to 20m (70ft) ↔ 15m (50ft)

10 ↕8m (25ft) ↔ 10m (30ft)

11 ↕5m (15ft) ↔ 8m (25ft)

12 ↕↔ 8m (25ft)

13 ↕10m (30ft) ↔ 6m (20ft)

14 ↕8m (25ft) ↔ 10m (30ft)

15 ↕8m (25ft) ↔ 10m (30ft)

16 ↕ to 15m (50ft) ↔ 10m (30ft)

7 *P. padus* 'Watereri' **8** *P.* 'Pink Perfection' **9** *P. sargentii* **10** *P.* 'Shirofugen' **11** *P.* 'Shôgetsu' **12** *P.* x *subhirtella* 'Autumnalis Rosea' **13** *P.* 'Spire' **14** *P.* 'Taihaku' **15** *P.* 'Ukon' **16** *P.* x *yedoensis*

PSEUDOPANAX LESSONII 'GOLD SPLASH'

This evergreen, upright to spreading shrub or tree bears yellow-splashed, deep green foliage. In summer, less conspicuous clusters of yellow-green flowers are carried amid toothed leaves which are divided into teardrop-shaped leaflets. Purple-black fruits appear later in the season. In frost-prone areas, grow in a container as a foliage plant for a conservatory.

CULTIVATION *Best in well-drained, fertile soil or compost, in sun or partial shade. Prune to restrict spread in early spring. Minimum temperature 2°C (35°F).*

☼☀ ◊ ♀H1 ↕3–6m (10–20ft) ↔2–4m (6–12ft)

PULMONARIA 'LEWIS PALMER'

This lungwort, sometimes called *P.* 'Highdown', is a deciduous perennial that forms clumps of upright, flowering stems. These are topped by open clusters of pink then blue, funnel-shaped flowers in early spring. The coarse, softly hairy leaves, dark green with white spots, are arranged along the stems. Grow in a wild or woodland garden.

CULTIVATION *Best in moist but not waterlogged, fertile, humus-rich soil, in deep or light shade. Divide and replant clumps, after flowering, every few years.*

☀● ♦ ♀H4 ↕35cm (14in) ↔45cm (18in)

PULMONARIA 'MARGERY FISH'

This lungwort is a clump-forming, deciduous perennial with attractive foliage and spring flowers. The leaves, silvered above with spotted edges and midribs, are often at their best in summer, after the coral to red-violet flowers have died away. For woodland, among shrubs, in a wild garden, or at the front of a border.

CULTIVATION *Grow in humus-rich, fertile, moist but not wet soil, in full or partial shade. Remove old leaves after flowering, and divide large clumps at the same time, or in autumn.*

☀☀ 💧 🏆H4 ↕18–28cm (7–11in) ↔60cm (24in)

PULMONARIA OFFICINALIS 'SISSINGHURST WHITE'

A neat, clump-forming, evergreen perennial valued for its pure white spring flowers and white-spotted foliage. The elliptic, hairy, mid- to dark green leaves are carried on upright stems, below funnel-shaped flowers which open from pale pink buds in early spring. Plant in groups as ground cover in a shady position.

CULTIVATION *Grow in moist but not waterlogged, humus-rich soil. Best in deep or light shade, but tolerates full sun. Divide and replant clumps every 2 or 3 years, after flowering.*

☀☀ 💧 🏆H4 ↕to 30cm (12in) ↔45cm (18in)

PULMONARIA RUBRA

This lovely clump-forming, evergreen perennial is a good ground cover plant for a shady position. It has attractive, bright green foliage, and brings early colour with its funnel-shaped, bright brick- to salmon-coloured flowers from late winter to mid-spring. They are attractive to bees and other beneficial insects.

CULTIVATION *Grow in humus-rich, fertile, moist but not waterlogged soil, in full or partial shade. Remove old leaves after flowering. Divide large or crowded clumps after flowering or in autumn.*

◑● 💧 ♈H4 ↕to 40cm(16in) ↔90cm (36in)

PULMONARIA SACCHARATA ARGENTEA GROUP

This group of evergreen perennials has almost completely silver leaves. A striking colour contrast is seen from late winter to early spring, when funnel-shaped red flowers appear; these age to a dark violet. They make good clumps at the front of a shady mixed border.

CULTIVATION *Best in fertile, moist but not waterlogged soil that is rich in humus, sited in full or partial shade. After flowering, remove old leaves and divide congested clumps.*

◑● 💧 ♈H4 ↕30cm (12in) ↔60cm (24in)

PULSATILLA HALLERI

This silky-textured herbaceous perennial, ideal for a rock garden, is densely covered in long silver hairs. It bears upright, bell-shaped, pale violet-purple flowers in late spring above finely divided, light green leaves. The first flowers often open before the new spring leaves have fully unfurled.

CULTIVATION *Grow in fertile, very well-drained gritty soil in full sun. May resent disturbance, so leave established plants undisturbed.*

☼ ◊ 🏆H4 ↕20cm (8in) ↔15cm (6in)

PULSATILLA VULGARIS

The pasque flower is a compact perennial forming tufts of finely divided, light green foliage. Its bell-shaped, nodding, silky-hairy flowers are carried above the leaves in spring; they are deep to pale purple or occasionally white, with golden centres. Good in a rock garden, scree bed or trough, or between paving.

CULTIVATION *Best in fertile soil with very good drainage. Site in full sun, where it will not be prone to excessive winter wet. Do not disturb once planted.*

☼ ◊ 🏆H4 ↕10–20cm (4–8in) ↔20cm (8in)

PULSATILLA VULGARIS 'ALBA'

This clump-forming perennial, a white form of the pasque flower, bears nodding, bell-shaped, silky-hairy white flowers with bold yellow centres in spring. These are carried above the finely divided, light green foliage which is hairy when young. Very pretty in a rock garden or scree bed.

CULTIVATION *Best in fertile soil with very good drainage. Position in full sun with protection from excessive winter wet. Resents disturbance once planted.*

☼ ◊ ♀H4 ↕10–20cm (4–8in) ↔20cm (8in)

PYRACANTHA 'ORANGE GLOW'

An upright to spreading, spiny, evergreen shrub bearing profuse clusters of tiny white flowers in late spring. Orange-red to dark orange berries follow in autumn, and persist well into winter. The leaves are oval and glossy dark green. Excellent as a vandal-resistant barrier hedge, which may also attract nesting birds.

CULTIVATION *Grow in well-drained, fertile soil, in full sun to deep shade. Shelter from cold, drying winds. Prune in mid-spring, and trim new leafy growth again in summer to expose the berries.*

☼ ☀ ● ◊ ♀H4 ↕↔ 3m (10ft)

PYRACANTHA 'WATERERI'

A vigorous, upright, spiny shrub that forms a dense screen of evergreen foliage, ornamented by its abundance of white spring flowers and bright red berries in autumn. The leaves are elliptic and dark green. Good as a barrier hedge or in a shrub border; can also be trained against a shady wall. Attractive to nesting birds. For yellow berries, look for 'Soleil d'Or' or *P. rogersiana* 'Flava'.

CULTIVATION *Grow in well-drained, fertile soil, in sun or shade with shelter from cold winds. Cut back unwanted growth in mid-spring, and trim leafy growth in summer to expose the berries.*

☼ ☀ ◊ ♈H4 ↕↔ 2.5m (8ft)

PYRUS CALLERYANA 'CHANTICLEER'

This very thorny ornamental pear tree has a narrowly conical shape and makes a good specimen tree for a small garden. Attractive sprays of small white flowers in mid-spring are followed by spherical brown fruits in autumn. The oval, finely scalloped leaves are glossy dark green and deciduous; they turn red before they fall. Tolerates urban pollution.

CULTIVATION *Grow in any well-drained, fertile soil, in full sun. Prune in winter to maintain a well-spaced crown.*

☼ ◊ ♈H4 ↕15m (50ft) ↔6m (20ft)

PYRUS SALICIFOLIA 'PENDULA'

This weeping pear is a deciduous tree with silvery-grey, willow like leaves that are downy when young. Dense clusters of small, creamy-white flowers appear during spring, followed by pear-shaped green fruits in autumn. A fine, pollution-tolerant tree for an urban garden.

CULTIVATION *Grow in fertile soil with good drainage, in sun. Prune young trees in winter to create a well-spaced, balanced framework of branches.*

☼ ◊ ♀H4 ↕8m (25ft) ↔6m (20ft)

RAMONDA MYCONI

A tiny, neat, evergreen perennial with basal rosettes of dark green, slightly crinkled, broadly oval leaves. In late spring and early summer,, deep violet-blue flowers are borne above the foliage on short stems. Pink-and white-flowered variants also occur. Grow in a rock garden or on a dry wall. *R. nathaliae* is very similar, with paler leaves.

CULTIVATION *Plant in moist but well-drained, moderately fertile, humus-rich soil, in partial shade. Set plants at an angle to avoid water pooling in the rosettes and causing rot. Leaves wither if too dry, but recover with watering.*

◑ ◊♦ ♀H4 ↕10cm (4in) ↔to 20cm (8in)

RANUNCULUS ACONITIFOLIUS 'FLORE PLENO'

White bachelor's buttons is a clump-forming, herbaceous perennial bearing small, almost spherical, fully double white flowers which last for a long time during late spring and early summer. The toothed leaves are deeply lobed and glossy dark green. Good for a woodland garden.

CULTIVATION *Grow in moist but well-drained soil that is rich in humus. Site in deep or partial shade.*

H4 ↕60cm (24in) ↔45cm (18in)

RANUNCULUS CALANDRINIOIDES

This clump-forming perennial produces clusters of up to three cup-shaped, white or pink-flushed flowers from late winter to early spring. The lance-shaped, blue-green leaves emerge from the base in spring and die down in summer. Grow in a rock garden, scree bed, or alpine house. Will not survive outside in frosty, wet winters.

CULTIVATION *Best in gritty, sharply drained, humus-rich soil, in sun. Water sparingly when dormant in summer.*

H2–3 ↕20cm (8in) ↔15cm (6in)

RANUNCULUS GRAMINEUS

This clump-forming buttercup is a perennial that is equally at home in a herbaceous border or rock garden. Cup-shaped, lemon-yellow flowers are carried above the grass-like, finely hairy leaves in late spring and early summer.

CULTIVATION *Grow in moist but well-drained, fertile soil. Choose a position in full sun or partial shade.*

☼ ◑ ◊◊ ♀H4 ↕to 30cm (12in) ↔to 15cm (6in)

RHAMNUS ALATERNUS 'ARGENTEOVARIEGATA'

This Italian buckthorn is a fast-growing, upright to spreading, evergreen shrub bearing oval, leathery, grey-green leaves with creamy-white margins. Clusters of tiny, yellow-green flowers are borne in spring, followed by spherical red fruits which ripen to black.

CULTIVATION *Grow in any well-drained soil, in full sun. Prune out unwanted growth in early spring; remove any shoots with all-green leaves as seen.*

☼ ◊ ♀H4 ↕5m (15ft) ↔4m (12ft)

RHODANTHEMUM HOSMARIENSE

A spreading subshrub valued for its profusion of daisy-like flowerheads with white petals and yellow eyes. These are borne from early spring to autumn, covering the silver, softly hairy, finely divided leaves. Grow at the base of a warm wall or in a rock garden. In an alpine house, it will flower year-round if deadheaded.

CULTIVATION *Grow in very well-drained soil, in a sunny position. Deadhead regularly to prolong flowering.*

☼ ◊ 🏆H4 ↕10–30cm (4–12in) ↔30cm (12in)

RHODOCHITON ATROSANGUINEUS

An evergreen, slender-stemmed climber, also known as *R. volubilis,* that can be grown as an annual in climates with frosty winters. Hanging, tubular, black to purple flowers with red-purple, bell-shaped "skirts" are borne during summer and autumn, amid the heart-shaped, rich green leaves. Support larger plants with wiring or trellis.

CULTIVATION *Grow in moist but well-drained, fertile, humus-rich soil, in full sun; the roots must be in shade. Pinch out shoot tips to promote a bushy habit; does not tolerate hard pruning.*

☼ ◊◊ 🏆H1–2 ↕3m (10ft)

EVERGREEN AZALEAS (*RHODODENDRON*)

Azaleas can be distinguished from true rhododendrons by their smaller dark green leaves and more tubular flowers. They also tend to make smaller, more spreading, twiggy shrubs. Botanically, rhododendrons have at least ten stamens per flower, and azaleas just five. Evergreen azaleas make beautiful, spring-flowering shrubs, blooming in almost every colour. They suit a variety of uses: dwarf or compact types are excellent in containers or tubs on shaded patios, and larger varieties will brighten up areas in permanent light shade. They do well in sun provided that the soil is not allowed to dry out.

CULTIVATION *Ideal in moist but well-drained, humus-rich, acid soil in part-day shade. Shallow planting is essential. Little formative pruning is necessary. If older plants become congested, thin in early summer. Maintain a mulch of leaf mould, but do not cultivate around the root area.*

☼ ◑ ◊ ♀H4

1 ↕↔ 1.2m (4ft)
2 ↕↔ 1.3m (3½ft)
3 ↕↔ 60cm (24in)
4 ↕↔ 60cm (24in)
5 ↕↔ 60cm (24in)
6 ↕↔ 60cm (24in)

1 *Rhododendron* 'Azuma-kagami' **2** *R.* 'Beethoven' **3** *R.* 'Hatsugiri' **4** *R.* 'Hinode-giri' **5** *R.* 'Hinomayo' **6** *R.* 'Irohayama'

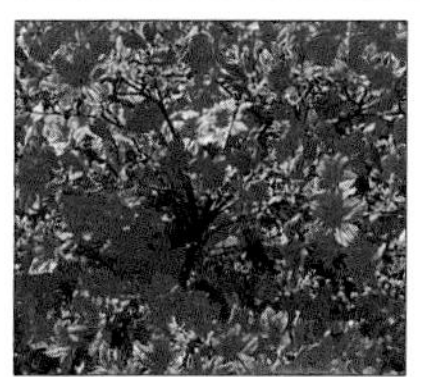

7 ↕↔ 1.5m (5ft)

8 ↕↔ 1.5m (5ft)

9 ↕↔ 1.2m (4ft)

MORE CHOICES

'Addy Wery' Vermilion-red flowers.

'Elsie Lee' Light reddish-mauve flowers.

'Greeting' Red-orange.

'Gumpo' Pinky-white.

'Hexe' Crimson.

'Hino-crimson' Brilliant red flowers.

'Kure-no-yuki' White, also called 'Snowflake'.

'Louise Dowdle' Vivid red-purple flowers.

'Vida Brown' Rose-red.

'Wombat' Pink flowers.

10 ↕↔ 60–90cm (24–36in)

11 ↕↔ 1.2m (4ft)

12 ↕↔ 1.2m (4ft)

7 *R.* 'John Cairns' **8** *R.* 'Kirin' **9** *R.* 'Palestrina' **10** *R.* 'Rosebud'
11 *R.* 'Vuyk's Scarlet' **12** *R.* 'Vuyk's Rosyred'

DECIDUOUS AZALEAS (*RHODODENDRON*)

A group of very hardy flowering shrubs, the only rhododendrons whose dark green leaves are deciduous, often colouring brilliantly before they fall. Deciduous azaleas are perhaps the most beautiful types of rhododendron, with large clusters of sometimes fragrant, white to yellow, orange, pink, or red flowers in spring and early summer. There is quite a variety in size, shape and growth habit, but they suit most garden uses well, especially in light shade. *R. luteum* thrives in sun where the soil is reliably moist. Grow them in tubs if you do not have suitable soil.

CULTIVATION *Grow in moist but well-drained, acid soil enriched with plenty of organic matter, ideally in partial shade. Shallow planting is essential; maintain a thick mulch of leaf mould, which will nourish the plant. Little or no pruning is necessary. Do not cultivate around the base of the plant as this will damage the roots.*

☼ ◑ ◊♦ ♀H4

1 ↕↔ 2.5m (8ft) **2** ↕↔ 3m (10ft) **3** ↕↔ 2.2m (7ft)

4 ↕↔ 1.5–2.5m (5–8ft) **5** ↕↔ 1.5m (5ft)

1 *Rhododendron albrechtii* **2** *R. austrinum* **3** *R.* 'Cecile' **4** *R.* 'Corneille' **5** *R.* 'Homebush'

7 ↕↔ 4m (12ft)

6 ↕↔ 2m (6ft)

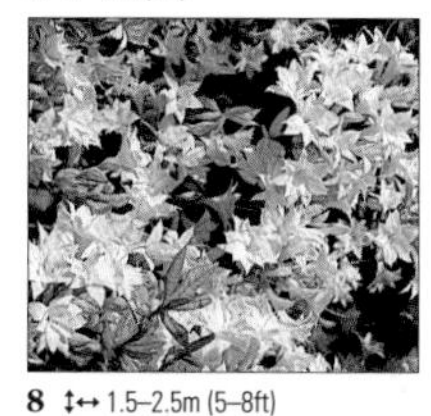

8 ↕↔ 1.5–2.5m (5–8ft)

MORE CHOICES

'Coccineum Speciosum' Orange-red flowers.

'Daviesii' White.

'Gibraltar' Crimson buds opening to orange with a yellow flash.

R. kaempferi Semi-evergreen; red flowers.

'Klondyke' Red buds opening to orange-gold.

'Satan' Bright red.

'Silver Slipper' White flushed pink with an orange flare.

'Spek's Brilliant' Bright orange-scarlet.

9 ↕↔ 2m (6ft)

10 ↕↔ 2.5m (8ft)

11 ↕↔ 2m (6ft)

6 *R.* 'Irene Koster' **7** *R. luteum* **8** *R.* 'Narcissiflorum' **9** *R.* 'Persil'
10 *R.* 'Spek's Orange' **11** *R.* 'Strawberry Ice'

LARGE RHODODENDRONS

Large, woodland-type rhododendrons, which can reach tree-like proportions, are grown primarily for their bright, sometimes fragrant, mostly spring flowers which are available in a wide spectrum of shapes and colours. They are ideal for adding colour to shaded areas or woodland gardens. Most leaves are oval and dark green, although the attractive young foliage of *R. bureaui* is light brown. Some cultivars, like 'Cynthia' or 'Purple Splendour', are tolerant of direct sun (in reliably moist soil), making them more versatile than others; they make glorious, spring-flowering screens or hedges for a large garden.

CULTIVATION *Grow in moist but well-drained, humus-rich, acid soil. Most prefer dappled shade in sheltered woodland. Shallow planting is essential. Little formative pruning is necessary, although most can be renovated after flowering to leave a balanced framework of old wood.*

☼ ◑ ◊ ♦ ♈H4

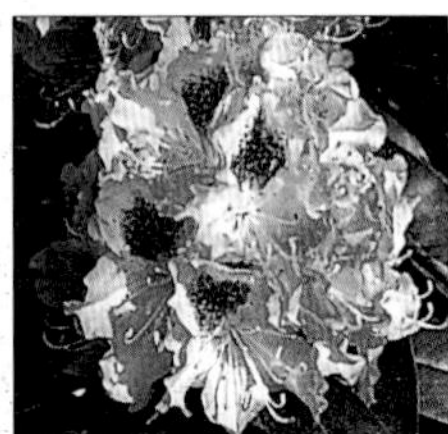

1 ↕↔ 3m (10ft)

2 ↕↔ 3m (10ft)

3 ↕↔ 3.5m (11ft)

1 *R.* 'Blue Peter' **2** *R. bureaui* **3** *R.* 'Crest'

4 ↕↔ 6m (20ft)
5 ↕ to 12m (40ft) ↔ 5m (15ft)
6 ↕↔ 4m (12ft)
7 ↕↔ 3m (10ft)
8 ↕↔ 4m (12ft)
9 ↕↔ 3m (10ft)
10 ↕↔ 3m (10ft)
11 ↕↔ 3m (10ft)

4 *R.* 'Cynthia' **5** *R. falconeri* **6** *R.* 'Fastuosum Flore Pleno' **7** *R.* 'Furnivall's Daughter' **8** *R.* 'Loderi King George' **9** *R.* 'Purple Splendour' **10** *R.* 'Sappho' **11** *R.* 'Susan'

Medium-sized Rhododendrons

These evergreen rhododendrons, between 1.5 and 3m (5–10ft) tall, are much-valued for their attractive, often scented blooms; the flowers are carried amid dark green foliage throughout spring. 'Yellow Hammer' will often produce an early show of flowers in autumn, and the foliage of 'Winsome' is unusual for its bronze tints when young. A vast number of different medium-sized rhododendrons are available, all suitable for shrub borders or grouped together in mass plantings. Some sun-tolerant varieties, especially low-growing forms like 'May Day', are suitable for informal hedging.

CULTIVATION *Grow in moist but well-drained, humus-rich, acid soil. Most prefer light dappled shade. Shallow planting is essential. 'Fragrantissimum' requires extra care in cold areas; provide a thick winter mulch and avoid siting in a frost-pocket. Trim after flowering, if necessary.*

☼ ◑ ◊ ♦ ♀H2–3, H4

1 ↕↔ 2m (6ft)

2 ↕↔ 1.5m (5ft)

1 *R.* 'Fabia' (H4) **2** *R.* 'Golden Torch' (H4)

3 *R.* 'Fragrantissimum' (H2–3) **4** *R.* 'Hydon Dawn' (H4) **5** *R.* 'May Day' (H4)
6 *R.* 'Titian Beauty' (H4) **7** *R.* 'Yellow Hammer' (H4) **8** *R.* 'Winsome' (H4)

Dwarf Rhododendrons

Dwarf rhododendrons are low-lying, evergreen shrubs with mid- to dark green, lance-shaped leaves. They flower throughout spring, in a wide variety of showy colours and flower forms. If soil conditions are too alkaline for growing rhododendrons in the open garden, these compact shrubs are ideal in containers or tubs on shaded patios; 'Ptarmigan' is is particularly suited to this kind of planting since it is able to tolerate periods without water. Dwarf rhododendrons are also effective in rock gardens. In areas with cold winters, the earliest spring flowers may be vulnerable to frost.

CULTIVATION *Grow in moist but well-drained, leafy, acid soil that is enriched with well-rotted organic matter. Site in sun or partial shade, but avoid the deep shade directly beneath a tree canopy. Best planted in spring or autumn; shallow planting is essential. No pruning is necessary.*

☼☀ ◊♦ ♈H3–4, H4

1 ↕↔ 1.1m (3½ft)

2 ↕↔ 1.2m (4ft)

3 ↕↔ 60cm (24in)

4 ↕↔ 45–90cm (18–36in)

1 *R.* 'Cilpinense' (H4) **2** *R.* 'Doc' (H4) **3** *R.* 'Dora Amateis' (H4) **4** *R.* 'Ptarmigan' (H3–4)

RHUS TYPHINA 'DISSECTA'

Rhus typhina, stag's horn sumach (also seen as *R. hirta*), is an upright, deciduous shrub with velvety-red shoots that resemble antlers. This form has long leaves, divided into many finely cut leaflets, which turn a brilliant orange-red in autumn. Upright clusters of less significant, yellow-green flowers are produced in summer, followed by velvety clusters of deep crimson-red fruits.

CULTIVATION *Grow in moist but well-drained, fairly fertile soil, in full sun to obtain best autumn colour. Remove any suckering shoots arising from the ground around the base of the plant.*

☼ ◊◐ ♀H4 ↕2m (6ft) ↔3m (10ft)

RIBES SANGUINEUM 'BROCKLEBANKII'

This slow-growing flowering currant is an upright, deciduous shrub with rounded, aromatic, yellow leaves, bright when young and fading in summer. The tubular, pale pink flowers, borne in hanging clusters in spring, are followed by small, blue-black fruits. 'Tydeman's White' is a very similar shrub, sometimes a little taller, with pure white flowers.

CULTIVATION *Grow in well-drained, fairly fertile soil. Site in sun, with shade during the hottest part of the day. Prune out some older stems after flowering. Cut back overgrown specimens in winter.*

☼☀ ◊ ♀H4 ↕↔ 1.2m (4ft)

RIBES SANGUINEUM 'PULBOROUGH SCARLET'

This vigorous flowering currant, larger than 'Brocklebankii', (see previous page, bottom) is an upright, deciduous shrub, bearing hanging clusters of tubular, dark red flowers with white centres in spring. The aromatic, dark green leaves are rounded with toothed lobes. Small, berry-like, blue-black fruits develop during the summer.

CULTIVATION *Grow in well-drained, moderately fertile soil, in full sun. Cut out some older stems after flowering; overgrown specimens can be pruned hard in winter or early spring.*

☼ ◊♦ ♀H4 ↕2m (6ft) ↔ 2.5m (8ft)

ROBINIA HISPIDA

The bristly locust is an upright and arching, deciduous shrub with spiny shoots, useful for shrub borders on poor, dry soils. Deep rose-pink, pea-like flowers appear in hanging spikes during late spring and early summer; these are followed by brown seed pods. The large, dark green leaves are divided into many oval leaflets.

CULTIVATION *Grow in any but water-logged soil, in sun. Provide shelter from wind to avoid damage to the brittle branches. No pruning is necessary.*

☼ ◊ ♀H4 ↕2.5m (8ft) ↔3m (10ft)

ROBINIA PSEUDOACACIA 'FRISIA'

The black locust, *R. pseudoacacia*, is a fast-growing, broadly columnar, deciduous tree, in this cultivar bearing gentle, yellow-green foliage which is golden-yellow when young, turning orange-yellow in autumn. Usually sparse, hanging clusters of fragrant, pea-like, small white flowers appear in mid-summer. The stems are normally spiny.

CULTIVATION *Grow in moist but well-drained, fertile soil, in full sun. When young, maintain a single trunk by removing competing stems as soon as possible. Do not prune once established.*

☼ 💧💧 🏆H4 ↕15m (50ft) ↔8m (25ft)

RODGERSIA PINNATA 'SUPERBA'

A clump-forming perennial that bears upright clusters of star-shaped, bright pink flowers. These are borne in mid- to late summer above bold, heavily veined, dark green foliage. The divided leaves, up to 90cm (36in) long, are purplish-bronze when young. Good near water, in a bog garden, or for naturalizing at a woodland margin. For creamy-white flowers, look for *R. podophylla*.

CULTIVATION *Best in moist, humus-rich soil, in full sun or semi-shade. Provide shelter from cold, drying winds. Will not tolerate drought.*

☼ ◑ 💧 🏆H4 ↕to 1.2m (4ft) ↔75cm (30in)

LARGE-FLOWERED BUSH ROSES (*ROSA*)

Also known as hybrid tea roses, these deciduous shrubs are commonly grown in formal bedding displays, laid out with neat paths and edging. They are distinguished from other roses in that they carry their large flowers either singly or in clusters of two or three. The first blooms appear in early summer and repeat flushes continue into early autumn. In a formal bed, group five or six of the same cultivar together, and interplant with some standard roses to add some variation in height. These bush roses also combine well with herbaceous perennials and other shrubs in mixed borders.

CULTIVATION *Grow in moist but well-drained, fertile soil in full sun. Cut spent flower stems back to the first leaf for repeat blooms. Prune main stems to about 25cm (10in) above ground level in early spring, and remove any dead or diseased wood as necessary at the base.*

☼ ◊◊ ♀H4

1 ↕↔ 60cm (24in)

2 ↕ to 2m (6ft) ↔ 80cm (32in)

3 ↕ 1.1m (3½ft) ↔ 75cm (30in)

4 ↕ 1.1m (3½ft) ↔ 75cm (30in)

5 ↕ 75cm (30in) ↔ 60cm (24in)

1 *Rosa* ABBEYFIELD ROSE 'Cocbrose' **2** *R.* ALEXANDER 'Harlex' **3** *R.* BLESSINGS **4** *R.* ELINA 'Dicjana' **5** *R.* FREEDOM 'Dicjem'

6 ↕55cm (22in) ↔ 60cm (24in)
7 ↕80cm (32in) ↔ 65cm (26in)
8 ↕75cm (30in) ↔ 70cm (28in)
9 ↕75cm (30in) ↔ 60cm (24in)
10 ↕1m (3ft) ↔ 75cm (30in)
11 ↕1.2m (4ft) ↔ 1m (3ft)
12 ↕1m (3ft) ↔ 60cm (24in)
13 ↕1m (3ft) ↔ 75cm (30in)
14 ↕80cm (32in) ↔ 60cm (24in)
15 ↕1.1m (3½ft) ↔ 60cm (24in)
16 ↕1m (3ft) ↔ 75cm (30in)

6 *R.* 'Indian Summer' **7** *R.* 'Ingrid Bergman' **8** *R.* JUST JOEY **9** *R.* LOVELY LADY 'Dicjubell' **10** *R.* PAUL SHIRVILLE 'Harqueterwife' **11** *R.* PEACE 'Madame A. Meilland' **12** *R.* REMEMBER ME 'Cocdestin' **13** *R.* ROYAL WILLIAM 'Korzaun' **14** *R.* SAVOY HOTEL 'Harvintage' **15** *R.* SILVER JUBILEE **16** *R.* 'Troika'

CLUSTER-FLOWERED BUSH ROSES (*ROSA*)

These very free-flowering bush roses are also known as floribunda roses. Like other roses, they come in a huge range of flower colours but are set apart from the large-flowered bush roses by their large, many-flowered trusses of relatively small blooms. Nearly all are fragrant, some more so than others. They lend themselves well to informal or cottage garden displays, mixing well with herbaceous perennials and other shrubs; remember to consider the colour of the flowers when choosing all neighbouring plants, since blooms will continue to appear from early summer to early autumn.

CULTIVATION *Best in moist but well-drained, fairly fertile soil in full sun. Deadhead for repeat blooms, unless the hips are wanted. Prune main stems to about 30cm (12in) above ground level in early spring, and remove any dead or diseased wood as necessary.*

☼ ◊◊ 🏆H4

1 ↕50cm (20in) ↔ 60cm (24in)

2 ↕1m (3ft) ↔ 75cm (30in)

3 ↕75cm (30in) ↔ 60cm (24in)

4 ↕1m (3ft) ↔ 60cm (24in)

5 ↕1.2m (4ft) ↔ 1m (3ft)

6 ↕80cm (32in) ↔ 75cm (30in)

1 *Rosa* AMBER QUEEN 'Harroony' **2** *R.* ANISLEY DICKSON 'Dickimono' **3** *R.* ANNA LIVIA 'Kormetter' **4** *R.* 'Arthur Bell' **5** *R.* CHINATOWN **6** *R.* CITY OF LONDON 'Harukfore'

7 ↕75cm (30in) ↔ 60cm (24in)
8 ↕1m (3ft) ↔ 75cm (30in)
9 ↕80cm (32in) ↔ 65cm (26in)
10 ↕↔ 75cm (30in)
11 ↕80cm (32in) ↔ 60cm (24in)
12 ↕1.2m (4ft) ↔ 1m (3ft)
13 ↕70cm (28in) ↔ 60cm (24in)
14 ↕75cm (30in) ↔ 60cm (24in)
15 ↕ to 2.2m (8ft) ↔ 1m (3ft)

7 *R.* ESCAPADE 'Harpade' **8** *R.* 'Fragrant Delight' **9** *R.* ICEBERG 'Korbin' **10** *R.* MANY HAPPY RETURNS 'Harwanted' **11** *R.* MARGARET MERRIL 'Harkuly' **12** *R.* MOUNTBATTEN 'Harmantelle' **13** *R.* SEXY REXY 'Macrexy' **14** *R.* TANGO 'Macfirwal' **15** *R.* 'The Queen Elizabeth'

CLIMBING ROSES (*ROSA*)

Climbing roses are often vigorous plants that will reach varying heights depending on the cultivar. All types have stiff, arching stems, usually with dense, glossy leaves divided into small leaflets. The frequently scented flowers are borne in summer, some in one exuberant flush, others having a lesser repeat flowering. They can be trained against walls or fences as decorative features in their own right, planted as a complement to other climbers, such as clematis, or allowed to scramble up into other wall-trained shrubs or even old trees. They are invaluable for disguising unsightly garden buildings, or as a backdrop to a summer border.

CULTIVATION *Best in moist but well-drained, fairly fertile soil, in sun. Deadhead unless hips are wanted. As plants mature, prune back to within the allowed area, after flowering. Occasionally cut an old main stem back to the base to renew growth. Never prune in the first two years.*

☼ ◊◆ ♀H4

1 ↕↔ to 6m (20ft)

2 ↕3m (10ft) ↔ 2.5m (8ft)

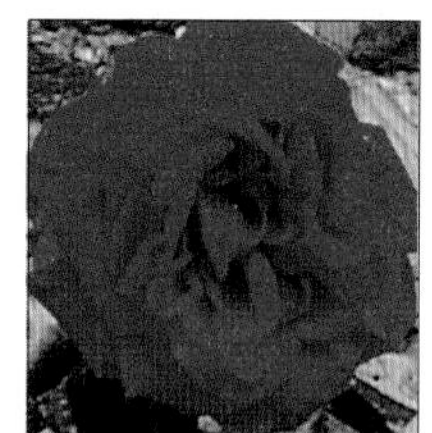

3 ↕↔ 2.2m (7ft)

4 ↕to 10m (30ft) ↔ 6m (20ft)

5 ↕to 5m (15ft) ↔ 4m (12ft)

1 *Rosa banksiae* 'Lutea' **2** *R.* COMPASSION **3** *R.* DUBLIN BAY 'Macdub' **4** *R. filipes* 'Kiftsgate' **5** *R.* 'Gloire de Dijon'

6 ↕ to 3m (10ft) ↔ 2m (6ft)

7 ↕ 3m (10ft) ↔ 2.2m (7ft)

8 ↕↔ 2.5m (8ft)

9 ↕ 5m (15ft) ↔ 3m (10ft)

10 ↕ 3m (10ft) ↔ 2.5m (8ft)

11 ↕ to 6m (20ft) ↔ 4m (12ft)

12 ↕ to 3m (10ft) ↔ 2m (6ft)

6 *R.* GOLDEN SHOWERS **7** *R.* HANDEL 'Macha' **8** *R.* 'Maigold' **9** *R.* 'Madame Alfred Carrière' **10** *R.* 'New Dawn' **11** *R.* 'Madame Grégoire Staechelin' **12** *R.* 'Zéphirine Drouhin'

RAMBLING ROSES (*ROSA*)

Rambling roses are very similar to climbers (see page 380), but with more lax, flexible stems. These are easier to train on to complex structures such as arches, tunnels and pergolas, or ropes and chains suspended between rigid uprights, provided they are solidly built; most ramblers are vigorous. Unlike climbers, they can succumb to mildew if trained flat against walls. All available cultivars have divided, glossy green leaves, borne on thorny or prickly stems. Flowers are often scented, arranged singly or in clusters, and are borne during summer. Some bloom only once, others having a lesser repeat flowering later on.

CULTIVATION *Best in moist but well-drained, fertile soil, in full sun. Train stems of young plants on to a support, to establish a permanent framework; prune back to this each year after flowering has finished, and remove any damaged wood as necessary.*

☼ ◊♦ ♀H4

1 ↕to 5m (15ft) ↔ 3m (10ft)

2 ↕to 5m (15ft) ↔ 4m (12ft)

1 *Rosa* 'Albéric Barbier' **2** *R.* 'Albertine'

3 *R.* 'Bobbie James' **4** *R.* 'Félicité Perpétue' **5** *R.* 'Rambling Rector' **6** *R.* 'Seagull'
7 *R.* 'Sanders' White Rambler' **8** *R.* 'Veilchenblau'

PATIO AND MINIATURE ROSES (*ROSA*)

These small or miniature shrub roses, bred especially for their compact habit, greatly extend the range of garden situations in which roses can be grown. All have very attractive, scented flowers in a wide range of colours, blooming over long periods from summer to autumn amid deciduous, glossy green leaves. With the exception of 'Ballerina', which can grow to a height of about 1.5m (5ft), most are under 1m (3ft) tall, making them invaluable for confined, sunny spaces; planted in large tubs, containers or raised beds, they are also excellent for decorating patios and other paved areas.

CULTIVATION *Grow in well-drained but moist, moderately fertile soil that is rich in well-rotted organic matter. Choose an open, sunny site. Remove all but the strongest shoots in late winter, then reduce these by about one-third of their height. Cut out any dead or damaged wood as necessary.*

☼ ◊◆ ♀H4

1 ↕45cm (18in) ↔ 40cm (16in)

2 ↕to 1.5m (5ft) ↔ 1.2m (4ft)

3 ↕75cm (30in) ↔ 60cm (24in)

4 ↕45cm (18in) ↔ 30cm (12in)

5 ↕50cm (20in) ↔ 40cm (16in)

6 ↕75cm (30in) ↔ 60cm (30in)

1 *Rosa* ANNA FORD 'Harpiccolo' **2** *R.* 'Ballerina' **3** *R.* 'Cecile Brunner' **4** *R.* CIDER CUP 'Dicladida' **5** *R.* GENTLE TOUCH 'Diclulu' **6** *R.* 'Mevrouw Nathalie Nypels'

7 ↕1.2m (4ft) ↔ 1m (3ft)

8 ↕40cm (16in) ↔ 60cm (24in)

9 ↕25cm (10in) ↔ 30cm (12in)

10 ↕40cm (16in) ↔ 35cm (14in)

11 ↕↔ 35cm (14in)

12 ↕↔ 60–90cm (24–30in)

13 ↕↔ 1–1.5m (3–5ft)

7 *R.* 'Perle d'Or' **8** *R.* QUEEN MOTHER 'Korquemu' **9** *R.* 'Stacey Sue'
10 *R.* SWEET DREAM 'Fryminicot' **11** *R.* SWEET MAGIC 'Dicmagic' **12** *R.* 'The Fairy'
13 *R.* 'Yesterday'

ROSES FOR GROUND COVER (*ROSA*)

Ground cover roses are low-growing, spreading, deciduous shrubs, ideal for the front of a border, in both formal and informal situations. They produce beautiful, fragrant flowers over long periods from summer into autumn, amid divided, glossy green leaves, on thorny or prickly, sometimes trailing stems. Only those of really dense habit, like SWANY 'Meiburenac', will provide weed-smothering cover, and even these are only effective if the ground is weed-free to begin with. Most give their best cascading over a low wall, or when used to clothe a steep bank which is otherwise difficult to plant.

CULTIVATION *Best in moist but well-drained, reasonably fertile soil that is rich in humus, in full sun. Prune shoots back after flowering each year to well within the intended area of spread, removing any dead or damaged wood. Annual pruning will enhance flowering performance.*

☼ ◊♦ ♈H4

1 ↕85cm (34in) ↔ 1.1m (3½ft)

2 ↕45cm (18in) ↔ 1.2m (4ft)

3 ↕75cm (30in) ↔ 1.2m (4ft)

4 ↕1m (3ft) ↔ 1.2m (4ft)

1 *Rosa* BONICA 'Meidomonac' **2** *R.* 'Nozomi' **3** *R.* RED BLANKET 'Intercell' **4** *R.* ROSY CUSHION 'Interall'

5 ↕15cm (6in) ↔ 45cm (18in)

6 ↕50cm (20in) ↔ 1.5m (5ft)

7 ↕80cm (32in) ↔ 1.2m (4ft)

8 ↕to 75cm (30in) ↔ 1.7m (5½ft)

9 ↕75cm (30in) ↔ 1.2m (4ft)

5 *R.* SNOW CARPET 'Maccarpe' **6** *R.* SUMA 'Harsuma' **7** *R.* SURREY 'Korlanum'
8 *R.* SWANY 'Meiburenac' **9** *R.* TALL STORY 'Dickooky'

OLD GARDEN ROSES (*ROSA*)

The history of old garden roses extends back to Roman times, demonstrating their lasting appeal in garden design. They are deciduous shrubs, comprised of a very large number of cultivars categorized into many groups, such as the Gallica, Damask and Moss roses. As almost all old garden roses flower in a single flush in early summer, they should be mixed with other flowering plants to maintain a lasting display. Try underplanting non-climbing types with spring bulbs, and climbing cultivars can be interwoven with a late-flowering clematis, for example.

CULTIVATION *Best in moist but well-drained, fertile soil in full sun. Little pruning is necessary; occasionally remove an old stem at the base to alleviate congested growth and to promote new shoots. Trim to shape in spring as necessary. Unless hips are wanted, remove spent flowers as they fade.*

☼ ◊◆ 🏆H4

1 ↕2.2m (7ft) ↔ 1.5m (5ft) **2** ↕1.2m (4ft) ↔ 1m (3ft) **3** ↕1m (3ft) ↔ 1.2m (4ft)

4 ↕1.5m (5ft) ↔ 1.2m (4ft) **5** ↕1.5m (5ft) ↔ 1.2m (4ft) **6** ↕1.5m (5ft) ↔ 1.2m (4ft)

1 *Rosa* 'Alba Maxima' **2** *R.* 'Belle de Crècy' **3** *R.* 'Cardinal de Richelieu' **4** *R.* 'Céleste' **5** *R.* x *centifolia* 'Cristata' (*syn.* 'Chapeau de Napoléon') **6** *R.* x *centifolia* 'Muscosa'

7 ↕↔ 1.2m (4ft) or more

8 ↕↔ to 2m (6ft)

9 ↕↔ 1.2m (4ft)

10 ↕1.5m (5ft) ↔ 1.2m (4ft)

11 ↕1.3m (4½ft) ↔ 1.2m (4ft)

12 ↕1.5m (5ft) ↔ 1.2m (4ft)

13 ↕to 2m (6ft) ↔ 1.2m (4ft)

14 ↕1.5m (5ft) ↔ 1.2m (4ft)

15 ↕1.5m (5ft) ↔ 1.2m (4ft)

16 ↕1.5m (5ft) ↔ 1.2m (4ft)

17 ↕↔ 1.2m (4ft)

18 ↕↔ 1m (3ft)

7 *R.* 'Charles de Mills' **8** *R.* 'De Rescht' **9** *R.* 'Duc de Guiche' **10** *R.* 'Fantin-Latour' **11** *R.* 'Félicité Parmentier' **12** *R.* 'Ferdinand Pichard' **13** *R.* 'Henri Martin' **14** *R.* 'Ispahan' **15** *R.* 'Königin von Dänemark' **16** *R.* 'Madame Hardy' **17** *R.* 'Président de Sèze' **18** *R.* 'Tuscany Superb'

MODERN SHRUB ROSES (*ROSA*)

These roses are slightly larger and more spreading than most others, combining the stature of old garden roses with some of the benefits of modern types. Their general good health and vigour makes them easy to grow and they repeat-flower over a long period, making ideal summer-flowering, deciduous shrubs for the back of a low-maintenance shrub border in a large garden. Like other roses, there is a wide choice in flower shape and colour, and several have a superb fragrance. Flowering begins in early summer, with repeat flushes up until early autumn.

CULTIVATION *Grow in well-drained but moist, moderately fertile soil that is rich in organic matter. Choose an open, sunny site. To keep them at a manageable size, prune every year in early spring. It is often better to let them grow naturally, however, as their form is easily spoiled by severe or careless pruning.*

☼ 00 ♀H4

1 ↕1.5m (5ft) ↔ 1.1m (3½ft) **2** ↕↔ 1.2m (4ft) **3** ↕↔ to 3.5m (11ft)

4 ↕2m (6ft) ↔ 1.5m (5ft) **5** ↕↔ 1.5m (5ft) **6** ↕1.5m (5ft) ↔ 1.2m (4ft)

1 *Rosa* 'Blanche Double de Coubert' **2** *R.* 'Buff Beauty' **3** *R.* 'Cerise Bouquet' **4** *R.* CONSTANCE SPRY **5** *R.* 'Cornelia' **6** *R.* 'Felicia'

7 ↕2m (6ft) ↔ 1m (3ft) **8** ↕1.5m (5ft) ↔ 1m (3ft) **9** ↕1.2m (4ft) ↔ 1.5m (5ft)
10 ↕1.5m (5ft) ↔ 1.2m (4ft) **11** ↕↔ 2.2m (7ft) **12** ↕↔ 2.2m (7ft)
13 ↕↔ 1.1m (3½ft) **14** ↕ 2.2m (7ft) ↔ 2m (6ft)
15 ↕2m (6ft) ↔ 1m (3ft) **16** ↕↔ 1.1m (3½ft) **17** ↕2m (6ft) ↔ 1.2m (4ft)

7 *R.* 'Fred Loads' **8** *R.* GERTRUDE JEKYLL 'Ausbord' **9** *R.* GRAHAM THOMAS 'Ausmas'
10 *R.* 'Jacqueline du Pré' **11** *R.* 'Marguerite Hilling' **12** *R.* 'Nevada' **13** *R.* 'Penelope'
14 *R.* 'Roseraie de l'Häy' **15** *R.* 'Sally Holmes' **16** *R.* 'The Lady' **17** *R.* WESTERLAND 'Korwest'

ROSES FOR HEDGEROWS AND WILD AREAS (*ROSA*)

The best types of rose for hedgerows or wild gardens are the species or wild roses, as many naturalize easily. Either shrubs or climbers, most have a natural-looking, scrambling or arching growth habit with single, five-petalled, often fragrant flowers, which appear in early summer on the previous year's growth. Although the main flowering season is fleeting, in many the flowers develop into beautiful rosehips, just as attractive as the flowers. Hips vary in colour from orange to red or black, and often persist into winter, providing valuable food for hungry wildlife.

CULTIVATION *Best in moist but well-drained, reasonably fertile soil, rich in organic matter. In a wild garden, little pruning is needed; to control hedges, trim after flowering each year, removing any dead or damaged wood. The flowers are borne on the previous summer's stems, so do not remove too many older branches.*

☼ 💧 🏆H4

1 ↕2.2m (7ft) ↔ 2.5m (8ft)

2 ↕80cm (32in) ↔ 1m (3ft)

4 ↕80cm (32in) ↔ 1m (3ft)

5 ↕2m (6ft) ↔ 1.5m (5ft)

3 ↕1m (3ft) ↔ 1.2m (4ft)

1 *R.* 'Complicata' **2** *R.* 'Fru Dagmar Hastrup' **3** *R. gallica* var. *officinalis* **4** *R. gallica* var. *officinalis* 'Versicolor' **5** *R. glauca*

6 ↕1.1m (3½ft) ↔ 1.3m (4½ft)

7 ↕1.5–2.5m (5–8ft) ↔ 1.2–2m (4–6ft)

8 ↕to 3m (10ft) ↔ 2m (6ft)

9 ↕1.2m (4ft) ↔ 1m (3ft)

10 ↕2m (6ft) ↔ 1m (3ft)

11 ↕↔ 1–2.5m (3–8ft)

12 ↕2m (6ft) ↔ 1m (3ft)

13 ↕2.5m (8ft) ↔ 1m (3ft)

6 *R.* 'Golden Wings' **7** *R. mulliganii* **8** *R. nutkana* 'Plena' **9** *R.* × *odorata* 'Mutabilis'
10 *R. primula* **11** *R. rugosa* 'Rubra' **12** *R. xanthina* 'Canary Bird'
13 *R. xanthina* var. *hugonis*

ROSMARINUS OFFICINALIS 'MISS JESSOP'S UPRIGHT'

This vigorous, upright rosemary is an evergreen shrub with aromatic foliage which can be used in cooking. From mid-spring to early summer, whorls of small, purple-blue to white flowers are produced amid the narrow, dark green, white-felted leaves, often with a repeat show in autumn. A good hedging plant for a kitchen garden. For white flowers, try 'Sissinghurst White'.

CULTIVATION *Grow in well-drained, poor to moderately fertile soil. Choose a sunny, sheltered site. After flowering, trim any shoots that spoil the symmetry.*

☼ ♢ ♀H4 ↕↔ 2m (6ft)

ROSMARINUS OFFICINALIS PROSTRATUS GROUP

These low-growing types of rosemary are aromatic, evergreen shrubs ideal for a rock garden or the top of a dry wall. Whorls of small, two-lipped, purple-blue to white flowers are produced in late spring, and often again in autumn. The dark green leaves have white-felted undersides, and can be cut in sprigs for culinary use. Plant in a sheltered position and protect in cold winters.

CULTIVATION *Grow in well-drained, poor to moderately fertile soil, in full sun. Trim or lightly cut back shoots that spoil the symmetry, after flowering.*

☼ ♢ ♀H3 ↕15cm (6in) ↔1.5m (5ft)

RUBUS 'BENENDEN'

This flowering raspberry is an ornamental, deciduous shrub, which has spreading, arching, thornless branches and peeling bark. It is valued for its abundance of large, saucer-shaped, rose-like flowers with glistening, pure white petals in late spring and early summer. The lobed leaves are dark green. Suitable for a shrub border.

CULTIVATION *Grow in any rich, fertile soil, in full sun or partial shade. After flowering, occasionally remove some old stems to the base to relieve overcrowding and to promote new growth.*

☼◑ ◊ 🏆H4 ↕↔ 3m (10ft)

RUBUS THIBETANUS

The ghost bramble is an upright, summer-flowering, deciduous shrub so named for its conspicuously white-bloomed, prickly stems in winter. The small, saucer-shaped, red-purple flowers are carried amid fern-like, white-hairy, dark green leaves, followed by spherical black fruits, also with a whitish bloom. *R. cockburnianus* is another bramble valued for its white winter stems.

CULTIVATION *Grow in any fertile soil, in sun or partial shade. Each spring, cut all flowered stems back to the ground, leaving the previous season's new, unflowered shoots unpruned.*

 ↕↔ 2.5m (8ft)

RUDBECKIA FULGIDA VAR. *SULLIVANTII* 'GOLDSTURM'

This black-eyed Susan is a clump-forming perennial valued for its strongly upright form and large, daisy-like, golden-yellow flowerheads with cone-shaped, blackish-brown centres. These appear above the substantial clumps of lance-shaped, mid-green leaves during late summer and autumn. It is a bold addition to a late summer border, and the cut flowers last reasonably well in water.

CULTIVATION *Grow in any moist but well-drained soil that does not dry out, in full sun or light shade.*

☼ ◑ ◊◊ ♀H4 ↕ to 60cm (24in) ↔ 45cm (18in)

RUDBECKIA 'GOLDQUELLE'

A tall but compact perennial that bears large, fully double, bright lemon-yellow flowers from mid-summer to mid-autumn. These are carried above loose clumps of deeply-divided, mid-green leaves. The flowers are good for cutting.

CULTIVATION *Grow in any moist but well-drained soil, in full sun or light dappled shade.*

☼ ◑ ◊◊ ♀H4 ↕ to 90cm (36in) ↔ 45cm (18in)

SALIX BABYLONICA VAR. *PEKINENSIS* 'TORTUOSA'

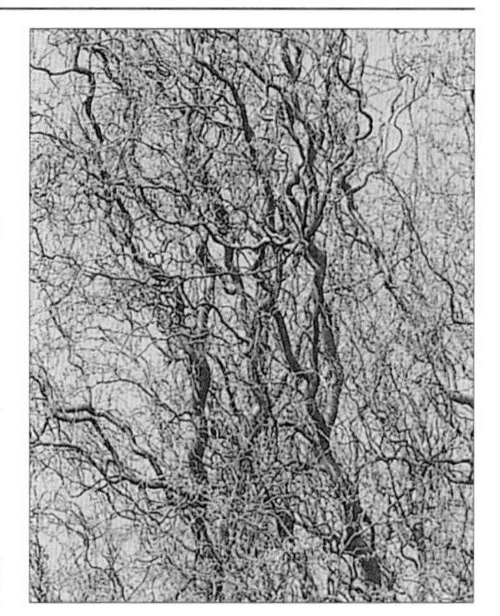

The dragon's claw willow is a fast-growing, upright, deciduous tree with curiously twisted shoots that are striking in winter. In spring, yellow-green catkins appear with the contorted, bright green leaves with grey-green undersides. Plant away from drains, as roots are invasive and water-seeking. Also sold as *S. matsudana* 'Tortuosa'.

CULTIVATION *Grow in any but very dry or shallow, chalky soil. Choose a sunny site. Thin occasionally in late winter to stimulate new growth, which most strongly exhibits the fascinating growth pattern.*

☼ ♦ 🏆H4 ↕15m (50ft) ↔8m (25ft)

SALIX 'BOYDII'

This tiny, very slow-growing, upright, deciduous shrub, with gnarled branches, is suitable for planting in a rock garden or trough. The small, almost rounded leaves are rough-textured, prominently veined and greyish-green. Catkins are only produced occasionally, in early spring.

CULTIVATION *Grow in any deep, moist but well-drained soil, in full sun; willows dislike shallow, chalky soil. When necessary, prune in late winter to maintain a healthy framework.*

☼ ◊♦ 🏆H4 ↕30cm (12in) ↔20cm (8in)

SALIX CAPREA 'KILMARNOCK'

The Kilmarnock willow is a small, weeping, deciduous tree ideal for a small garden. It forms a dense, umbrella-like crown of yellow-brown shoots studded with silvery catkins in mid- and late spring, before the foliage appears. The broad, toothed leaves are dark green on top, and grey-green beneath.

CULTIVATION *Grow in any deep, moist but well-drained soil, in full sun. Prune annually in late winter to prevent the crown becoming congested. Remove shoots that arise on the clear trunk.*

☼ ◊◑ 🏆H4 ↕1.5–2m (5–6ft) ↔2m (6ft)

SALIX HASTATA 'WEHRHAHNII'

This small, slow-growing, upright, deciduous shrub, with dark purple-brown stems and contrasting silvery-grey, early spring catkins, makes a beautiful specimen for winter colour displays. The leaves are oval and bright green.

CULTIVATION *Grow in any moist soil, in sun; does not tolerate shallow, chalky soils. Prune in spring to maintain a balance between young stems, which usually have the best winter colour, and older wood with catkins.*

☼ ◑ 🏆H4 ↕↔ 1m (3ft)

SALIX LANATA

The woolly willow is a rounded, slow-growing, deciduous shrub with stout shoots which have an attractive white-woolly texture when young. Large, upright, golden to grey-yellow catkins emerge on older wood in late spring among the dark green, broadly oval, silvery-grey-woolly leaves.

CULTIVATION *Grow in moist but well-drained soil, in sun. Tolerates semi-shade, but dislikes shallow, chalky soil. Prune occasionally in late winter or early spring to maintain a balance between old and young stems.*

☼ ◑ ◊ ♦ ♀H4 ↕1m (3ft) ↔1.5m (5ft)

SALPIGLOSSIS CASINO SERIES

These upright and compact, weather-resistant annuals, ideal for summer bedding, freely produce a contrasting display of funnel-shaped flowers throughout summer and autumn. Colours range from blue and purple to red, yellow or orange, heavily veined in deeper tints. The lance-shaped, mid-green leaves have wavy margins.

CULTIVATION *Grow in moist but well-drained, fertile soil, in sun. Stems need support. Deadhead regularly. Plant out only after any risk of frost has passed.*

☼ ◊ ♀H3 ↕to 60cm (24in) ↔to 30cm (12in)

SALVIA ARGENTEA

This short-lived perennial forms large clumps of soft, felty-grey leaves around the base of the plant. Spikes of hooded, two-lipped, white or pinkish-white flowers are borne in mid- and late summer, on strong, upright stems. Suits a Mediterranean-style border, but will need shelter in areas with cold winters.

CULTIVATION *Grow in light, very well-drained soil, in a warm, sunny site. Use a cloche or glass panel to protect from excessive winter wet and cold winds.*

☼ ◊ ♈H3 ↕90cm (36in) ↔60cm (24in)

SALVIA CACALIIFOLIA

Usually grown as an annual in cool climates, this upright and hairy, herbaceous perennial bears spikes of deep blue flowers in early summer, held above mid-green foliage. A distinctive sage with a brilliant flower colour, for bedding, summer infilling, or containers.

CULTIVATION *Grow in light, moderately fertile, moist but well-drained soil enriched with organic matter. Choose a site in full sun to light dappled shade.*

☼ ◐ ◊◆ ♈H2 ↕90cm (36in) ↔30cm (12in)

SALVIA COCCINEA 'PSEUDOCOCCINEA'

A bushy, short-lived perennial often grown as an annual in frost-prone climates, with toothed, dark green, hairy leaves and loose, slender spikes of soft cherry-red flowers from summer to autumn. An exotic-looking addition to a summer display, and good in containers.

CULTIVATION *Grow in light, moderately fertile well-drained soil, rich in organic matter. Site in full sun.*

☼ ◊ H3 ↕60cm (24in) ↔30cm (12in)

SALVIA DISCOLOR

This upright, herbaceous perennial, which is normally treated as a summer annual in cool climates, is valued for both its flowers and foliage. Its green leaves have a densely white-woolly surface, forming an unusual display in themselves until the long spikes of deep purplish-black flowers extend above them in late summer and early autumn.

CULTIVATION *Thrives in light, moderately fertile, moist but well-drained, humus-rich soil. A position in full sun or light shade is best.*

☼ ◐ ◊◊ H1 ↕45cm (18in) ↔30cm (12in)

SALVIA FULGENS

An upright, evergreen, summer-flowering subshrub bearing spikes of tubular, two-lipped red flowers. The oval, toothed or notched leaves are rich green above and densely white-woolly beneath. Provides brilliant colour for bedding or containers. May not survive winter in frost-prone climates.

CULTIVATION *Grow in light, moist but well-drained, moderately fertile, humus-rich soil. Site in full sun or semi-shade.*

☼ ◑ ○ ♀H3 ↕15cm (6in) ↔20cm (8in)

SALVIA GUARANITICA 'BLUE ENIGMA'

A subshrubby perennial that grows well as a summer annual for bedding displays in cool climates. It is admired for its deep blue flowers, more fragrant than those of the species, *S. guaranitica*, which tower above the mid-green foliage from the end of summer until late autumn.

CULTIVATION *Grow in light, moderately fertile, moist but well-drained soil enriched with organic matter. Choose a site in full sun to light dappled shade.*

☼ ◑ ○● ♀H3–4 ↕1.5m (5ft) ↔90cm (36in)

SALVIA LEUCANTHA

The Mexican bush is a small evergreen shrub that must be grown under glass in cool climates for its winter flowers to be seen. The leaves are mid-green, white-downy beneath, and long spikes of white flowers with purple to lavender-blue calyces appear in winter and spring. Beautiful year-round in a greenhouse border or large container.

CULTIVATION *Under glass, grow in well-drained potting compost in full light with shade from hot sun, and water moderately while in flower. Otherwise, grow in moist but well-drained, fertile soil in sun or partial shade.*

☼☀ 💧💧 ♈H1 ↕60–100cm (24–39in) ↔40–90cm (16–36in)

SALVIA MICROPHYLLA 'PINK BLUSH'

An unusual flower colour for this shrubby perennial, with mid-green leaves and tall, slender spires of intense fuchsia-pink flowers. Lovely against a sunny, sheltered wall or fence. May not survive harsh winters. 'Kew Red' and 'Newby Hall' are also recommended, both with red flowers.

CULTIVATION *Grow in light, moist but well-drained, moderately fertile soil that is rich in organic matter, in full sun. Plants damaged by frost may be trimmed, but do not cut into old wood.*

☼ 💧 ♈H3–4 ↕90cm (3ft) ↔60cm (2ft)

SALVIA OFFICINALIS 'ICTERINA'

A very attractive, yellow and green variegated form of sage that has a mound-forming, subshrubby habit. The aromatic, evergreen, velvety leaves can be used in cooking. Less significant spikes of small, lilac-blue flowers appear in early summer. Ideal for a herb or kitchen garden. 'Kew Gold' is a very similar plant, although its leaves are often completely yellow.

CULTIVATION *Grow in moist but well-drained, fairly fertile, humus-rich soil. Site in full sun or partial shade.*

☼ ◊ ♀H3 ↕to 80cm (32in) ↔1m (3ft)

SALVIA OFFICINALIS PURPURASCENS GROUP

Purple sage is an upright, evergreen subshrub, suitable for a sunny border. Its red-purple young leaves and spikes of lilac-blue flowers are an attractive combination, the latter appearing during the first half of summer. The foliage is aromatic and can be used for culinary purposes. A useful plant to add colour to a herb garden.

CULTIVATION *Grow in light, moderately fertile, humus-rich, moist but well-drained soil in full sun or light shade. Trim to shape each year after flowering.*

☼ ☀ ◊◊ ♀H4 ↕to 80cm (32in) ↔1m (3ft)

SALVIA OFFICINALIS 'TRICOLOR'

This variegated sage is an upright, evergreen perennial with grey-green woolly, aromatic leaves with cream and pink to beetroot purple marking. In early to mid-summer, it carries spikes of lilac-blue flowers, attractive to butterflies. Less hardy than the species, so choose a warm, sheltered site, or grow for summer display.

CULTIVATION *Grow in moist but well-drained, reasonably fertile soil enriched with organic matter, in full sun or light dappled shade. Shelter from winter wet and cold, drying winds, and trim back untidy growth each year after flowering.*

☼ ◑ ◊♦ 🏆H3 ↕80cm (32in) ↔1m (3ft)

SALVIA PATENS 'CAMBRIDGE BLUE'

This lovely cultivar of *S. patens* is an upright perennial, with tall, loose spikes of pale blue flowers. It is a striking addition to a herbaceous or mixed border, or bedding and patio tubs. The flowers are borne during mid-summer to mid-autumn, above the oval, hairy, mid-green leaves. In frost-prone climates, shelter at the base of a warm wall.

CULTIVATION *Grow in well-drained soil, in full sun. Overwinter young plants in frost-free conditions.*

☼ ◊ 🏆H3 ↕45–60cm (18–24in) ↔45cm (18in)

SALVIA PRATENSIS HAEMATODES GROUP

A short-lived perennial, sometimes sold as *S. haematodes*, forming basal clumps of large, dark green leaves. In early and mid-summer, spreading spikes of massed blue-violet flowers, with paler throats, emerge from the centre of the clumps. Provides colour for bedding, infilling in beds and borders, or containers. 'Indigo' is another attractive recommended cultivar.

CULTIVATION *Grow in moist but well-drained, moderately fertile, humus-rich soil. Site in full sun or light shade.*

☼ ◊ ♀H4 ↕to 90cm (36in) ↔30cm (12in)

SALVIA SPLENDENS 'SCARLET KING'

This compact, bushy perennial, with dense spikes of long-tubed, scarlet flowers, is usually grown as an annual, especially in cold climates. The flowers are borne from summer to autumn, above dark green toothed leaves. Its long-lasting, brilliant flowers are an invaluable addition to any bedding or container display. 'Red Riches' and 'Vanguard' also make fine bedding salvias.

CULTIVATION *Grow in moist but well-drained, moderately fertile soil that is rich in humus. Choose a site in full sun.*

☼ ◊ ♀H3 ↕to 25cm (10in) ↔23–35cm (9–14in)

SALVIA X *SYLVESTRIS* 'MAINACHT'

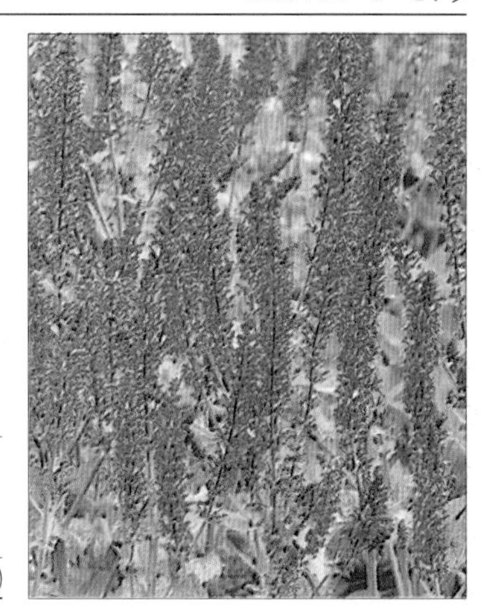

A neat, clump-forming, pleasantly aromatic perennial bearing tall, dense, upright spikes of indigo-blue flowers during early and mid-summer. The narrow, softly hairy, mid-green leaves are scalloped at the edges. Provides strong contrast for silver-leaved plants in a herbaceous border. 'Blauhugel' and 'Tänzerin' are other recommended cultivars.

CULTIVATION *Grow in well-drained, fertile soil, in sun. Tolerates drought. Cut back after the first flush of flowers to encourage a further set of blooms.*

☼ ◊ 🏆H4 ↕70cm (28in) ↔45cm (18in)

SALVIA ULIGINOSA

The bog sage is a graceful, upright perennial bearing spikes of clear blue flowers from late summer to mid-autumn. These are carried above lance-shaped, toothed, mid-green leaves, on branched stems. Good for moist borders; in frost-prone areas, it can be container-grown in a cool conservatory.

CULTIVATION *Needs moist but well-drained, fertile soil or compost. Choose a sunny, sheltered position. Taller plants will need support.*

☼ ♦ 🏆H3–4 ↕to 2m (6ft) ↔90cm (36in)

SAMBUCUS NIGRA 'GUINCHO PURPLE'

This popular cultivar of the common elder is an upright shrub with dark green, divided leaves; these turn black-purple then red in autumn. In early summer, musk-scented, pink-tinged white flowers are borne in large, flattened clusters, followed by small black fruits. Elders are ideal in new gardens, because they establish themselves in a short space of time.

CULTIVATION *Grow in any fertile soil, in sun or partial shade. For the best foliage effect, either cut all stems to the ground in winter, or prune out old stems and reduce length of young shoots by half.*

☼ ◑ ♦ ♈H4 ↕↔ 6m (20ft)

SANTOLINA CHAMAECYPARISSUS

Cotton lavender is a rounded, evergreen shrub grown for its foliage. The slender, white-woolly stems are densely covered with narrow, grey-white, finely cut leaves. The small, yellow flowerheads in summer can be removed to enhance the foliage effect. Suitable for a mixed border, or as low, informal hedging. For a dwarf version, look for var. *nana.*

CULTIVATION *Grow in well drained, poor to moderately fertile soil, in full sun. Remove old flowerheads and trim long shoots in autumn. Cut old, straggly plants back hard each spring.*

☼ ◊ ♈H4 ↕50cm (20in) ↔1m (3ft)

SANTOLINA ROSMARINIFOLIA 'PRIMROSE GEM'

A dense, rounded, evergreen shrub, similar to *S. chamaecyparissus* (see facing page, below), but with bright green leaves and paler flowers. These are borne at the tips of slender stems in mid-summer, above the finely cut aromatic leaves. Useful for filling gaps in a sunny border.

CULTIVATION *Grow in well drained, poor to moderately fertile soil, in full sun. In autumn, remove old flower-heads and prune long shoots. Cut old, straggly plants back hard each spring.*

☼ ◊ ♕H4 ↕60cm (24in) ↔1m (3ft)

SAPONARIA OCYMOIDES

Tumbling Ted is a sprawling, mat-forming perennial that carries a profusion of tiny, flat pink flowers in summer. The hairy, bright green leaves are small and oval. Excellent as part of a dry bank, scree or rock garden, although it may swamp smaller plants. 'Rubra Compacta' is a neater version of this plant, with dark red flowers.

CULTIVATION *Grow in gritty, sharply drained soil, in full sun. Cut back hard after flowering to keep compact.*

☼ ◊ ♕H4 ↕8cm (3in) ↔45cm (18in) or more

SAPONARIA x *OLIVANA*

A cushion-forming, summer-flowering perennial that produces abundant clusters of small, pale pink flowers around the edge of a mound of small and narrow, mid-green leaves. Good for rock gardens, scree slopes and dry banks.

CULTIVATION *Grow in sharply drained soil, in a sunny site. Top dress the soil around the plant with grit or gravel.*

☼ ◊ ♈H4 ↕5cm (2in) ↔15cm (6in)

SARCOCOCCA CONFUSA

Christmas box is a dense, evergreen shrub, giving an unparalleled winter fragrance. Clusters of small white flowers appear in mid-winter, followed by small, glossy black fruits. The tiny, oval leaves are glossy and dark green. Excellent as a low hedge near a door or entrance. Tolerates atmospheric pollution, dry shade and neglect.

CULTIVATION *Grow in moist but well-drained, fertile, humus-rich soil. Site in deep or semi-shade with protection from cold, drying winds. Remove dead and damaged growth each year in spring.*

◐ ● ♦ ♈H4 ↕2m (6ft) ↔1m (3ft)

SARCOCOCCA HOOKERIANA VAR. *DIGYNA*

This perfumed, evergreen shrub is very similar to *S. confusa* (see facing page, below), but with a more compact and spreading habit. The tiny, fragrant white flowers, which are followed by small, black or blue-black fruits, have pink anthers; they are borne amid glossy leaves, more slender and pointed than those of *S. hookeriana*, in winter. The flowers are good for cutting.

CULTIVATION *Grow in moist but well-drained, fertile, humus-rich soil, in shade. Dig up spreading roots in spring to confine to its allotted space.*

◐● 💧 🏆H4 ↕1.5m (5ft) ↔2m (6ft)

SARRACENIA FLAVA

Yellow trumpet is a carnivorous perennial bearing nodding yellow flowers in spring. Some leaves are modified into large, upright, nectar-secreting, insect-catching pitchers. These are yellow-green and red-marked, with round mouths and hooded tops. In frost-prone areas, grow outdoors as an annual, or in a cool greenhouse or conservatory.

CULTIVATION *Grow in wet but sharply drained, rich, acid soil or compost, in sun. Irrigate plentifully with lime-free water, but keep slightly drier in winter. Miniumum temperature 2°C (35°F).*

↕50–100cm (20–39in) ↔to 1m (3ft)

SAXIFRAGA 'JENKINSIAE'

This neat and slow-growing, evergreen perennial forms very dense cushions of grey-green foliage. It produces an abundance of solitary, cup-shaped, pale pink flowers with dark centres in early spring, on short, slender red stems. Good for rock gardens or troughs.

CULTIVATION *Best in moist but sharply drained, moderately fertile, neutral to alkaline soil, in full sun. Provide shade from the hottest summer sun.*

☼ ◊ ♈H4 ↕5cm (2in) ↔20cm (8in)

SAXIFRAGA 'SOUTHSIDE SEEDLING'

A mat-forming, evergreen perennial that is suitable for a rock garden. Open sprays of small, cup-shaped white flowers, spotted heavily with red, are borne in late spring and early summer. The oblong to spoon-shaped, pale green leaves form large rosettes close to soil level.

CULTIVATION *Grow in very sharply drained, moderately fertile, alkaline soil. Choose a position in full sun.*

☼ ◊ ♈H4 ↕30cm (12in) ↔20cm (8in)

SCABIOSA CAUCASICA 'CLIVE GREAVES'

This delicate, perennial scabious, with solitary, lavender-blue flowerheads, is ideal for a cottage garden. The blooms have pincushion-like centres, and are borne above the clumps of grey-green leaves during mid- to late summer. The flowerheads cut well for indoor arrangements.

CULTIVATION *Grow in well-drained, moderately fertile, neutral to slightly alkaline soil, in full sun. Deadhead to prolong flowering.*

☼ ◊ H4 ↕↔ 60cm (24in)

SCABIOSA CAUCASICA 'MISS WILLMOTT'

A clump-forming perennial that is very similar to 'Clive Greaves' (above), but with white flowerheads. These are borne in mid- to late summer and are good for cutting. The lance-shaped leaves are grey-green and arranged around the base of the plant. Suits a cottage garden.

CULTIVATION *Grow in well-drained, moderately fertile, neutral to slightly alkaline soil, in full sun. Deadhead to prolong flowering.*

 ↕ 90cm (36in) ↔ 60cm (24in)

SCHIZOSTYLIS COCCINEA 'MAJOR'

A vigorous, robust, clump-forming perennial, bearing gladiolus-like spikes of large red flowers. These appear in late summer on stiff and upright stems above the narrow, almost floppy, mid-green leaves. Good in sheltered spots, for a border front, or above water level in a waterside planting. 'Jennifer' is very similar, with pink flowers.

CULTIVATION *Grow in moist but well-drained, fertile soil, with a site in full sun. Plants rapidly become congested, but are easily lifted and divided every few years in spring.*

☼ ♦ ♀H4 ↕to 60cm (24in) ↔30cm (12in)

SCHIZOSTYLIS COCCINEA 'SUNRISE'

This clump-forming, vigorous perennial has the same general appearance, demands and usage as 'Major' (above), but with upright spikes of salmon-pink flowers which open in autumn. The long leaves are sword-shaped and ribbed. When cut, the flowers last well in water.

CULTIVATION *Best in moist but well-drained, fertile soil, in sun. Naturally forms congested clumps, but these are easily lifted and divided in spring.*

☼ ♦ ♀H4 ↕to 60cm (24in) ↔30cm (12in)

SCILLA BIFOLIA

A small, bulbous perennial bearing early-spring flowers which naturalizes well under trees and shrubs, or in grass. The slightly one-sided spikes of many star-shaped, blue to purple-blue flowers are carried above the clumps of narrow, basal leaves.

CULTIVATION *Grow in well-drained, moderately fertile, humus-rich soil, in full sun or partial shade.*

☼ ◊ 🏆H4 ↕8–15cm (3–6in) ↔5cm (2in)

SCILLA MISCHTSCHENKOANA 'TUBERGENIANA'

This dwarf, bulbous perennial has slightly earlier flowers than *S. bifolia* (above), which are silvery-blue with darker stripes. They are grouped together in elongating spikes, appearing at the same time as the semi-upright, narrow, mid-green leaves. Naturalizes in open grass. Also known as *S. tubergeniana*.

CULTIVATION *Grow in well-drained, moderately fertile soil that is rich in well-rotted organic matter, in full sun.*

☼ ◊ 🏆H4 ↕10–15cm (4–6in) ↔5cm (2in)

SEDUM KAMTSCHATICUM 'VARIEGATUM'

This clump-forming, semi-evergreen perennial has eye-catching, fleshy leaves, mid-green with pink tints and cream margins. During late summer, these contrast nicely with flat-topped clusters of small, star-shaped, yellow flowers which age to crimson later in the season. Suitable for rock gardens and borders.

CULTIVATION *Grow in well-drained, gritty, fertile soil. Choose a site in full sun, but will tolerate light shade.*

☼ ◊ ♀H4 ↕10cm (4in) ↔25cm (10in)

SEDUM 'RUBY GLOW'

This low-growing perennial is an ideal choice for softening the front of a mixed border. It bears masses of small, star-shaped, ruby-red flowers from mid-summer to early autumn, above clumps of fleshy, green-purple leaves. The nectar-rich flowers will attract bees, butterflies and other beneficial insects.

CULTIVATION *Grow in well-drained, fertile soil that has adequate moisture in summer. Position in full sun.*

☼ ◊ ♀H4 ↕25cm (10in) ↔45cm (18in)

SEDUM SPATHULIFOLIUM 'CAPE BLANCO'

A vigorous, evergreen perennial that forms a mat of silvery-green foliage, often tinted bronze-purple, with a heavy bloom of white powder over the innermost leaves. Small clusters of star-shaped, bright yellow flowers are borne just above the leaves in summer. A very attractive addition to a trough or raised bed.

CULTIVATION *Grow in well-drained, moderately fertile, gritty soil. Position in full sun, but tolerates light shade.*

☼ ◊ ♆H4 ↕10cm (4in) ↔60cm (24in)

SEDUM SPATHULIFOLIUM 'PURPUREUM'

This fast-growing, summer-flowering perennial forms tight, evergreen mats of purple-leaved rosettes; the central leaves are covered with a thick, silvery bloom. Flat clusters of small, star-shaped, bright yellow flowers appear throughout summer. Suitable for a rock garden, or the front of a sunny, well-drained border.

CULTIVATION *Grow in gritty, moderately fertile soil with good drainage, in sun or partial shade. Trim occasionally to prevent encroachment on other plants.*

☼ ◑ ◊ ♆H4 ↕10cm (4in) ↔60cm (24in)

SEDUM SPECTABILE 'BRILLIANT'

This cultivar of *S. spectabile*, the ice plant, is a clump-forming, deciduous perennial with brilliant pink flowerheads, excellent for the front of a border. The small, star-shaped flowers, packed into dense, flat clusters on fleshy stems, appear in late summer above the succulent, grey-green leaves. The flowerheads are attractive to bees and butterflies, and dry well on or off the plant.

CULTIVATION *Grow in well-drained, fertile soil with adequate moisture during summer, in full sun.*

☼ ◊ ♡H4 ↕↔ 45cm (18in)

SEDUM SPURIUM 'SCHORBUSER BLUT'

This vigorous, evergreen perennial forms mats of succulent, mid-green leaves which become purple-tinted when mature. Rounded clusters of star-shaped, deep pink flowers are borne during late summer. Suitable for a rock garden.

CULTIVATION *Grow in well-drained, moderately fertile, neutral to slightly alkaline soil, in full sun. Tolerates light shade. To improve flowering, divide the clumps or mats every 3 or 4 years.*

☼ ☀ ◊ ♡H4 ↕10cm (4in) ↔60cm (24in)

SEDUM TELEPHIUM SUBSP. *MAXIMUM* 'ATROPURPUREUM'

This clump-forming, deciduous perennial is valued for its very dark purple foliage which contrasts well with other plants. During summer and early autumn, attractive pink flowers with orange-red centres are clustered above the oval, slightly scalloped leaves.

CULTIVATION *Grow in well-drained, moderately fertile, neutral to slightly alkaline soil, in full sun. Divide clumps every 3 or 4 years to improve flowering.*

☼ ◊ 🏆H4 ↕45–60cm (18–24in) ↔30cm (12in)

SELAGINELLA KRAUSSIANA

Trailing spikemoss is a mat-forming, evergreen perennial with a moss-like appearance. The trailing stems are clothed in tiny, scale-like, bright green leaves. An excellent foliage plant that can easily be grown in a conservatory in frost-prone climates. Also recommended are 'Variegata', with cream-splashed leaves, and 'Brownii', which forms neat mounds, 15cm (6in) across.

CULTIVATION *Grow in moist, peaty soil, in semi-shade. Keep just moist in winter. Needs a humid atmosphere under glass. Minimum temperature 2°C (35°F).*

 🏆H1 ↕2.5cm (1in) ↔indefinite

SEMPERVIVUM ARACHNOIDEUM

The cobweb houseleek is a mat-forming, evergreen succulent, so-named because the foliage is webbed with white hairs. The small, fleshy, mid-green to red leaves are arranged in tight rosettes. In summer, flat clusters of star-shaped, reddish-pink flowers appear on leafy stems. Suitable for growing in a scree bed, wall crevice or trough.

CULTIVATION *Grow in gritty, sharply drained, poor to moderately fertile soil. Choose a position in full sun.*

☼ ◊ ♆H4 ↕8cm (3in) ↔30cm (12in)

SEMPERVIVUM CILIOSUM

A mat-forming, evergreen succulent carrying very hairy, dense rosettes of incurved, lance-shaped, grey-green leaves. It bears flat, compact heads of star-shaped, greenish-yellow flowers throughout summer. The rosettes of leaves die after flowering, but are rapidly replaced. Best in an alpine house in areas prone to very damp winters.

CULTIVATION *Grow in gritty, sharply drained, poor to moderately fertile soil, in sun. Tolerates drought conditions, but dislikes winter wet or climates that are warm and humid.*

☼ ◊ ♆H4 ↕8cm (3in) ↔30cm (12in)

SEMPERVIVUM TECTORUM

The common houseleek is a vigorous, mat-forming, evergreen succulent with large, open rosettes of thick, oval, bristle-tipped, blue-green leaves, often suffused red-purple. In summer, dense clusters of star-shaped, red-purple flowers appear on upright, hairy stems. Very attractive growing on old roof tiles or among terracotta fragments.

CULTIVATION *Grow in gritty, sharply drained, poor to moderately fertile soil. Choose a site in full sun.*

☼ ◊ 🏆H4 ↕15cm (6in) ↔50cm (20in)

SENECIO CINERARIA 'SILVER DUST'

This mound-forming, moderately fast-growing, evergreen shrub is usually grown as an annual for its attractive, lacy foliage. The almost white leaves are deeply cut and densely hairy. Plants kept into the second season bear loose heads of coarse, daisy-like, mustard-yellow flowerheads in mid-summer; many gardeners prefer to remove them. Ideal for creating massed foliage effects in summer bedding schemes.

CULTIVATION *Grow in well-drained, fertile soil, in sun. Nip out flower buds if desired, or deadhead regularly.*

 ↕↔ 30cm (12in)

SILENE SCHAFTA

A clump-forming, spreading, semi-evergreen perennial with floppy stems bearing small, bright green leaves. Profuse sprays of long-tubed, deep magenta flowers with notched petals are borne from late summer to autumn. Suitable for a raised bed or rock garden.

CULTIVATION *Grow in well-drained, neutral to slightly alkaline soil, in full sun or light dappled shade.*

☼ ◐ ◊ 🏆H4 ↕25cm (10in) ↔30cm (12in)

SKIMMIA × *CONFUSA* 'KEW GREEN' (MALE)

A compact, dome-shaped, evergreen shrub carrying aromatic, pointed, mid-green leaves. Conical spikes of fragrant, creamy-white flowers open in spring. There are no berries, but it will pollinate female skimmias if they are planted nearby. Good in a shrub border or woodland garden.

CULTIVATION *Grow in moist but well-drained, moderately fertile, humus-rich soil. Tolerates full sun to deep shade, atmospheric pollution and neglect. Requires little or no pruning.*

☼ ● ♦ 🏆H4 ↕0.5–3m (1½–10ft) ↔1.5m (5ft)

SKIMMIA JAPONICA 'RUBELLA' (MALE)

This tough, dome-shaped, evergreen shrub bears dark red flower buds in autumn and winter, opening in spring as fragrant heads of white flowers. The oval leaves have red rims. No berries are produced, but it will pollinate female skimmias nearby. (Where space permits only one plant, 'Robert Fortune' produces both flowers and berries.) Tolerates pollution and coastal conditions.

CULTIVATION *Grow in moist, fertile, neutral to slightly acid soil, in partial or full shade. Requires little pruning, but cut back any shoots that spoil the shape.*

☼☀ 💧 🏆H4 ↕↔ to 6m (20ft)

SMILACINA RACEMOSA

False spikenard is a clump-forming perennial bearing dense, feathery spikes of creamy-white, often green-tinged flowers in mid- to late spring. These are occasionally followed by red berries. The lance-shaped, pale green, luxuriant leaves turn yellow in autumn. A beautiful specimen for a woodland garden or shady border.

CULTIVATION *Grow in moist, neutral to acid, fertile soil that is rich in humus, in light or deep shade.*

☼☀ 💧 🏆H4 ↕to 90cm (36in) ↔60cm (24in)

SOLANUM CRISPUM 'GLASNEVIN'

This long-flowering Chilean potato tree is a fast-growing, scrambling, woody-stemmed, evergreen climber. Fragrant, deep purple-blue flowers, borne in clusters at the tips of the stems during summer and autumn, are followed by small, yellow-white fruits. The leaves are oval and dark green. Slightly tender in cold areas, so grow on a warm, sunny wall.

CULTIVATION *Grow in any moist but well-drained, moderately fertile soil, in full sun or semi-shade. Cut back weak and badly placed growth in spring. Tie to a support as growth proceeds.*

H3 ↕6m (20ft)

SOLANUM JASMINOIDES 'ALBUM'

This white-flowered potato vine is a scrambling, woody-stemmed, semi-evergreen climber. It produces broad clusters of fragrant, star-shaped, milk-white flowers, with prominent, lemon-yellow anthers, from summer into autumn. They are followed by black fruits. The leaves are dark green and oval. May not survive in cold areas.

CULTIVATION *Grow in any moist but well-drained, fertile soil, in full sun or semi-shade. Thin out shoots in spring. The climbing stems need support.*

H3 ↕6m (20ft)

SOLIDAGO 'GOLDENMOSA'

This compact, vigorous goldenrod is a bushy perennial topped with bright golden-yellow flowerheads in late summer and early autumn. The leaves are wrinkled and mid-green. Valuable in a wild garden or for late summer colour; the flowers are good for cutting. Can be invasive.

CULTIVATION *Grow in well-drained, poor to moderately fertile, preferably sandy soil, in full sun. Remove flowered stems to prevent self-seeding.*

☼ ◊ 🏆H4 ↕to 75cm (30in) ↔45cm (18in)

SOLLYA HETEROPHYLLA

The bluebell creeper is a twining, evergreen climber which must be grown in a cool conservatory in frost-prone climates. Clusters of nodding, bell-shaped blue flowers, followed by blue berries, are borne over a long period from early summer to autumn. The leaves are lance-shaped and deep green. Train over an arch or pergola, or into a host shrub.

CULTIVATION *Grow in moist but well-drained, moderately fertile, humus-rich soil or compost, in a sunny, sheltered site. Water sparingly during winter. Minimum temperature 2°C (35°F).*

☼ ◊♦ 🏆H1 ↕1.5–2m (5–6ft)

SORBUS ARIA 'LUTESCENS'

This compact whitebeam is a broadly columnar, deciduous tree, bearing oval and toothed, silvery-grey foliage which turns russet and gold in autumn. Clusters of white flowers appear in late spring, followed by brown-speckled, dark red berries. It makes a beautiful specimen tree, tolerating a wide range of conditions. 'Majestica' is similar but taller, with larger leaves.

CULTIVATION *Grow in moist but well-drained, fertile soil, in sun. Tolerates heavy clay soils, semi-shade, urban pollution and exposed conditions. Remove any dead wood in summer.*

☼ ◑ ◊◆ ♈H4 ↕10m (30ft) ↔8m (25ft)

SORBUS HUPEHENSIS VAR. *OBTUSA*

This variety of the Hubei rowan, *S. hupehensis*, is an open and spreading tree which gives a fine display of autumn colour. Broad clusters of white flowers in late spring are followed by round white berries; these ripen to dark pink later in the season. The blue-green leaves, divided into many leaflets, turn scarlet before they fall.

CULTIVATION *Grow in any moist but well-drained soil, preferably in full sun, but tolerates light shade. Remove any dead or diseased wood in summer.*

☼ ◑ ◊◆ ♈H4 ↕↔ to 8m (25ft)

SORBUS 'JOSEPH ROCK'

This broadly columnar, upright, deciduous tree has bright green leaves which are divided into many sharply-toothed leaflets. These colour attractively to orange, red and purple in autumn. In late spring, white flowers appear in broad clusters, followed by round, pale yellow berries which ripen to orange-yellow.

CULTIVATION *Grow in moist but well-drained, fertile soil, in sun. Very prone to fireblight, the main sign of which is blackened leaves; affected growth must be pruned back in summer to at least 60cm (24in) below the diseased area.*

☼ ◊◊ ♀H4 ↕10m (30ft) ↔7m (22ft)

SORBUS REDUCTA

A deciduous shrub that forms a low thicket of upright branches. Much-valued for its ornamental, dark green foliage which turns a rich red in autumn. Small, open clusters of white flowers appear in late spring, followed by white, crimson-flushed berries. Tolerates pollution.

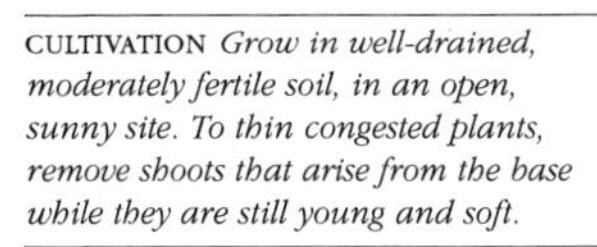

CULTIVATION *Grow in well-drained, moderately fertile soil, in an open, sunny site. To thin congested plants, remove shoots that arise from the base while they are still young and soft.*

☼ ◊ ♀H4 ↕1–1.5m (3–5ft) ↔2m (6ft)

SORBUS VILMORINII

A spreading shrub or small tree with elegant, arching branches bearing dark green leaves divided into many leaflets. The deciduous foliage gives a lovely display in autumn, turning orange- or bronze-red. Clusters of white flowers appear in late spring and early summer, followed later in the season by dark red berries which age to pink then white.

CULTIVATION *Grow in well-drained, moderately fertile, humus-rich soil, in full sun or dappled shade. Remove any dead or diseased wood in summer.*

☼ ◐ ◊ ♀H4 ↕↔ 5m (15ft)

SPARTIUM JUNCEUM

Spanish broom is an upright shrub with slender, dark green shoots which are almost leafless. A profusion of fragrant, pea-like, rich golden-yellow flowers appear at the end of the stems from early summer to early autumn. These are followed by flattened, dark brown seed pods. Particularly useful on poor soils.

CULTIVATION *Grow in any but water-logged soil, in a warm, sunny site. When young, cut back main stems by half each spring to promote a bushy habit. Once established, trim every few years, but do not cut into old wood.*

☼ ◊◊ ♀H4 ↕15cm (6in) ↔20cm (8in)

SPIRAEA JAPONICA 'ANTHONY WATERER'

A compact, deciduous shrub that makes a good informal flowering hedge. The lance-shaped, dark green leaves, usually margined with creamy-white, are red when young. Dense heads of tiny pink flowers are borne amid the foliage in mid- to late summer.

CULTIVATION *Grow in any well-drained, fairly fertile soil that does not dry out, in full sun. On planting, cut back stems to leave a framework 15cm (6in) high; prune back close to this every year in spring. Deadhead after flowering.*

☼ ◊ 🏆H4 ↕↔ to 1.5m (5ft)

SPIRAEA JAPONICA 'GOLDFLAME'

This compact, deciduous, flowering shrub bears pretty, bright yellow leaves which are bronze-red when young. Dense, flattened heads of tiny, dark pink flowers appear at the tips of slightly arching stems during mid- and late summer. Ideal for a rock garden. 'Nana' is even smaller, to 45cm (18in) tall.

CULTIVATION *Grow in well-drained soil that does not dry out completely, in full sun. On planting, cut back stems to a framework 15cm (6in) high; prune back close to this each year in spring. Deadhead after flowering.*

 ↕↔ 75cm (30in)

SPIRAEA NIPPONICA 'SNOWMOUND'

This fast-growing and spreading, deciduous shrub has arching, reddish-green stems. The dense clusters of small white flowers in mid-summer make an invaluable contribution to any shrub border. The rounded leaves are bright green when young, darkening as they age.

CULTIVATION *Grow in any moderately fertile soil that does not dry out too much during the growing season, in full sun. Cut back flowered stems in autumn, and remove any weak growth.*

☼ ◊ 🏆H4 ↕↔ 1.2–2.5m (4–8ft)

SPIRAEA X *VANHOUTTEI*

Bridal wreath is a fast-growing, deciduous shrub, more compact in habit than *S. nipponica* 'Snowmound' (above), but with similar mounds of white flowers during early summer. The diamond-shaped leaves are dark green above with blue-green undersides. Grow as an informal hedge, or in a mixed border.

CULTIVATION *Grow in any well-drained, fertile soil that does not dry out, in sun. In autumn, cut back flowered stems, removing any weak or damaged growth.*

☼ ◊ 🏆H4 ↕2m (6ft) ↔1.5m (5ft)

STACHYURUS PRAECOX

This spreading, deciduous shrub bears oval, mid-green leaves on arching, red-purple shoots. Hanging spikes of tiny, bell-shaped, pale yellow-green flowers appear on the bare stems in late winter and early spring. Suitable for a shrub border, and lovely in a woodland garden.

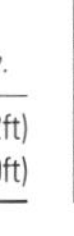

CULTIVATION *Grow in moist but well-drained, humus-rich, fertile, neutral to acid soil. Prefers partial shade, but will tolerate full sun if soil is kept reliably moist. Regular pruning is unnecessary.*

☼◐ ◊♦ ♀H4 ↕1–4m (3–12ft) ↔3m (10ft)

STIPA GIGANTEA

Golden oats is a fluffy, evergreen, perennial grass forming dense tufts of narrow, mid-green leaves. In summer, these are topped by silvery- to purplish-green flowerheads which turn gold when mature, and persist well into winter. Makes an imposing feature at the back of a border.

CULTIVATION *Grow in well-drained, fertile soil, in full sun. Remove dead leaves and flowerheads in early spring.*

☼ ◊ ♀H4 ↕to 2.5m (8ft) ↔1.2m (4ft)

STYRAX JAPONICUS

Japanese snowbell is a gracefully spreading, deciduous tree bearing hanging clusters of fragrant, bell-shaped, dainty white flowers which are often tinged with pink. These appear during early to mid-summer, amid oval, rich green leaves which turn yellow or red in autumn. Ideal for a woodland garden.

CULTIVATION *Grow in moist but well-drained, neutral to acid soil, in full sun with shelter from cold, drying winds. Tolerates dappled shade. Leave to develop naturally without pruning.*

☼☀ ◊♦ ♀H4 ↕10m (30ft) ↔8m (25ft)

STYRAX OBASSIA

The fragrant snowbell is a broadly columnar, deciduous tree bearing beautifully rounded, dark green leaves which turn yellow in autumn. Fragrant, bell-shaped white flowers are produced in long, spreading clusters in early and mid-summer.

CULTIVATION *Grow in moist but well-drained, fertile, humus-rich, neutral to acid soil, in full sun or partial shade. Shelter from cold, drying winds. Dislikes pruning; leave to develop naturally.*

☼☀ ◊♦ ♀H4 ↕12m (40ft) ↔7m (22ft)

SYMPHYTUM X *UPLANDICUM* 'VARIEGATUM'

This upright, clump-forming, bristly perennial has large, lance-shaped, mid-green leaves with broad cream margins. Drooping clusters of pink-blue buds open to blue-purple flowers from late spring to late summer. Best suited to a wild garden or shady border. Less invasive than green-leaved types.

CULTIVATION *Grow in any moist soil, in sun or partial shade. For the best foliage effect, remove flowering stems before they bloom. Liable to form plain green leaves if grown in poor or infertile soil.*

H4 ↕90cm (36in) ↔60cm (24in)

SYRINGA MEYERI 'PALIBIN'

This compact, slow-growing, deciduous shrub with a rounded shape is much valued for its abundant clusters of fragrant, lavender-pink flowers in late spring and early summer. The leaves are dark green and oval. Makes a bold contribution to any shrub border. Sometimes seen as *S. palibiniana.*

CULTIVATION *Grow in deep, moist but well-drained, fertile, preferably alkaline soil, in full sun. Deadhead for the first few years until established. Prune out weak and damaged growth in winter.*

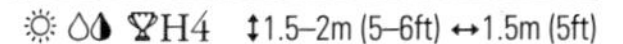

H4 ↕1.5–2m (5–6ft) ↔1.5m (5ft)

SYRINGA PUBESCENS SUBSP. *MICROPHYLLA* 'SUPERBA'

This upright to spreading, conical, deciduous shrub bears spikes of fragrant, rose-pink flowers at the tips of slender branches. First appearing in spring, they continue to open at irregular intervals until autumn. The oval, mid-green leaves are red-green when young. Makes a good screen or informal hedge. 'Miss Kim' is a smaller, more compact cultivar.

CULTIVATION *Grow in moist but well-drained, fertile, humus-rich, neutral to alkaline soil, in full sun. Prune out any weak or damaged growth in winter.*

☼ ◊♦ ♀H4 ↕to 6m (20ft) ↔6m (20ft)

SYRINGA VULGARIS 'CHARLES JOLY'

This dark-purple-flowered form of common lilac is a spreading shrub or small tree. Very fragrant, double flowers appear during late spring and early summer in dense, conical clusters. The deciduous leaves are heart-shaped to oval and dark green. Use as a backdrop in a shrub or mixed border.

CULTIVATION *Grow in moist but well-drained, fertile, humus-rich, neutral to alkaline soil, in full sun. Young shrubs require minimal pruning; old, lanky stems can be cut back hard in winter.*

☼ ◊♦ ♀H4 ↕↔ 7m (22ft)

SYRINGA VULGARIS 'KATHERINE HAVEMEYER'

A spreading lilac, forming a large shrub or small tree, producing dense clusters of very fragrant, double, lavender-blue flowers. These open from purple buds in late spring and early summer. The deciduous, mid-green leaves are heart-shaped. 'Madame Antoine Buchner' is a very similar recommended lilac, with slightly pinker flowers.

CULTIVATION *Grow in moist but well-drained, fertile, neutral to alkaline soil that is rich in humus, in sun. Mulch regularly. Little pruning is necessary, but tolerates hard pruning to renovate.*

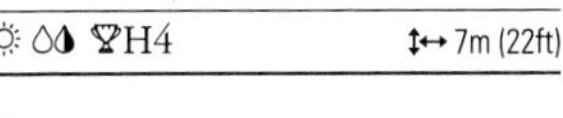

☼ ◊◊ 🏆H4 ↕↔ 7m (22ft)

SYRINGA VULGARIS 'MADAME LEMOINE'

This lilac is very similar in form to 'Katherine Havemeyer' (above), but has compact spikes of large, very fragrant, double white flowers. These are borne in late spring and early summer, amid the deciduous, heart-shaped to oval, mid-green leaves. 'Vestale' is another recommended white lilac, with single flowers.

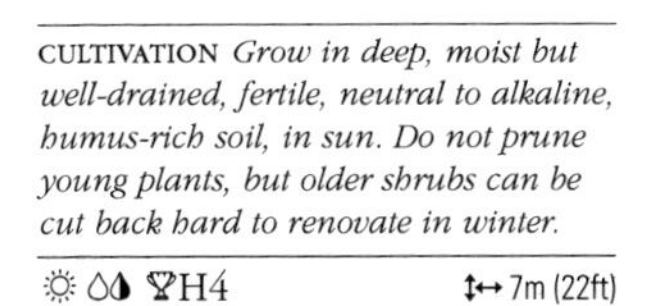

CULTIVATION *Grow in deep, moist but well-drained, fertile, neutral to alkaline, humus-rich soil, in sun. Do not prune young plants, but older shrubs can be cut back hard to renovate in winter.*

☼ ◊◊ 🏆H4 ↕↔ 7m (22ft)

TAMARIX TETRANDRA

This large, arching shrub, with feathery foliage on purple-brown shoots, bears plumes of light pink flowers in mid- to late spring. The leaves are reduced to tiny, needle-like scales. Particularly useful on light, sandy soils; in mild coastal gardens, it makes a good windbreak or hedge. For a similar shrub that flowers in late summer, look for *T. ramossissima* 'Rubra'.

CULTIVATION *Grow in well-drained soil, in full sun. Cut back young plants by almost half after planting. Prune each year after flowering, or the shrub may become top-heavy and unstable.*

☼ ♢ ♀H4 ↕↔ 3m (10ft)

TANACETUM COCCINEUM 'BRENDA'

A bushy, herbaceous perennial, grown for its daisy-like, magenta-pink, yellow-centred flowerheads in early summer, are borne on upright stems above aromatic, finely divided, grey-green foliage. The cut flowers last well in water. 'James Kelway' is very similar, while 'Eileen May Robinson' has pale pink flowers.

CULTIVATION *Grow in well-drained, fertile, neutral to slightly acid soil, in an open, sunny site. Cut back after the first flush of flowers to encourage a second flowering later in the season.*

☼ ♢ ♀H4 ↕70–80cm (28–32in) ↔45cm (18in)

TAXUS BACCATA

Yew is a slow-growing, broadly conical, evergreen conifer. The needle-like, dark green leaves are arranged in two ranks along the shoots. Male plants bear yellow cones in spring, and female plants produce cup-shaped, fleshy, bright red fruits in autumn. Excellent as a dense hedge, which can be clipped to shape, and as a backdrop to colourful plants. All parts are toxic.

CULTIVATION *Grow in any well-drained, fertile soil, in sun to deep shade. Tolerates chalky or acidic soils. Plant both sexes together, for berries. Trim or cut back to renovate, in summer or early autumn.*

☼☀ ◊ ♈H4 ↕to 20m (70ft) ↔10m (30ft)

TAXUS BACCATA 'DOVASTONII AUREA' (FEMALE)

This slow-growing, evergreen conifer has wide-spreading, horizontally tiered branches which weep at the tips. It is smaller than the common yew (above), and has yellow-margined to golden-yellow foliage. Fleshy, bright red fruits appear in autumn. All parts of this plant are poisonous if eaten.

CULTIVATION *Grow in any well-drained, fertile soil, in sun or deep shade. Tolerates chalky or acidic conditions. Plant close to male yews for a reliable display of autumn berries. Trim or cut back to renovate, in summer or early autumn.*

☼☀ ◊ ♈H4 ↕3–5m (10–15ft) ↔2m (6ft)

TAXUS BACCATA 'FASTIGIATA' (FEMALE)

Irish yew is a dense, strongly upright, evergreen conifer which becomes columnar with age. The dark green leaves are not two-ranked like other yews, but stand out all around the shoots. All are female, bearing fleshy, berry-like, bright red fruits in late summer. All parts are toxic if eaten.

CULTIVATION *Grow in any reliably moist soil, but tolerates most conditions including very dry, chalky soils, in full sun or deep shade. Plant with male yews for a reliable crop of berries. Trim or cut back to renovate, in summer or early autumn, if necessary.*

☼☀ ◊♦ ♀H4 ↕to 10m (30ft) ↔4m (12ft)

THALICTRUM DELAVAYI 'HEWITT'S DOUBLE'

An upright, clump-forming perennial, more floriferous than *T. delavayi*, that produces upright sprays of long-lasting, pompon-like, rich mauve flowers from mid-summer to early autumn. Fnely divided, mid-green leaves are carried on slender stems shaded dark purple. An excellent foil in a herbaceous border to plants with bolder leaves and flowers.

CULTIVATION *Grow in moist but well-drained, humus-rich soil, in sun or light shade. Divide clumps and replant every few years to maintain vigour.*

☼☀ ◊♦ ♀H4 ↕1.2m (4ft) or more ↔60cm (24in)

THALICTRUM FLAVUM SUBSP. *GLAUCUM*

This subspecies of yellow meadow rue is a summer-flowering, clump-forming perennial that bears large, upright heads of fragrant, sulphur-yellow flowers. These are carried above mid-green, divided leaves. Good at the margins of woodland.

CULTIVATION *Best in moist, humus-rich soil, in partial shade. Tolerates sun and dry soil. Flower stems may need staking.*

H4 ↕to 1m (3ft) ↔60cm (24in)

THUJA OCCIDENTALIS 'HOLMSTRUP'

This shrub-like form of white cedar is a slow-growing conifer with a conical shape. The dense, mid-green leaves are apple-scented and arranged in vertical sprays. Small oval cones appear amid the foliage. Plant alone as a specimen tree, or use as a hedge.

CULTIVATION *Grow in deep, moist but well-drained soil, in full sun. Shelter from cold, drying winds. Trim as necessary in spring and late summer.*

H4 ↕to 4m (12ft) ↔3–5m (10–15ft)

THUJA OCCIDENTALIS 'RHEINGOLD'

This bushy, spreading, slow-growing conifer is valued for its golden-yellow foliage, which is pink-tinted when young, and turns bronze in winter. Small, oval cones are carried amid the billowing sprays of apple-scented, scale-like leaves. Good as a specimen tree.

CULTIVATION *Grow in deep, moist but well-drained soil, in a sheltered, sunny site. Trim in spring and late summer, but be careful not to spoil the form.*

☼ ◊♦ ♈H4 ↕1–2m (3–6ft) ↔3–5m (10–15ft)

THUJA ORIENTALIS 'AUREA NANA'

This dwarf Chinese thuja is an oval-shaped conifer with fibrous, red-brown bark. The yellow-green foliage, which fades to bronze over winter, is arranged in flat, vertical sprays. Flask-shaped cones are borne amid the foliage. Good in a rock garden.

CULTIVATION *Grow in deep, moist but well-drained soil, in sun with shelter from cold, drying winds. Trim in spring and again in late summer as necessary.*

☼ ◊♦ ♈H4 ↕↔ to 60cm (24in)

THUJA PLICATA 'STONEHAM GOLD'

This slow-growing, dwarf form of western red cedar is a conical conifer with fissured, red-brown bark and flattened, irregularly arranged sprays of bright gold, aromatic foliage; the tiny, scale-like leaves are very dark green within the bush. The cones are small and elliptic. Ideal for a rock garden.

CULTIVATION *Grow in deep, moist but well-drained soil, in full sun with shelter from cold, drying winds. Trim in spring and again in late summer.*

☼ ◊◊ H4 ↕↔ to 2m (6ft)

THUNBERGIA GRANDIFLORA

The blue trumpet vine is a vigorous, woody-stemmed, evergreen climber which can be grown as an annual in cold climates. Lavender- to violet-blue, sometimes white, trumpet-shaped flowers with yellow throats appear in hanging clusters during summer. The oval to heart-shaped, dark green leaves are softly hairy.

CULTIVATION *Grow in moist but well-drained, fertile soil or compost, in sun. Provide shade during the hottest part of the day. Give the climbing stems support. Minimum temperature 10°C (50°F).*

 ↕5–10m (15–30ft)

THUNBERGIA MYSORENSIS

A spring-flowering, fast-growing, woody-stemmed climber bearing hanging spikes of large yellow flowers with brownish-red to purple tubes. The narrow, evergreen leaves are dark green with prominent veins. Must be grown in a warm conservatory or heated greenhouse in climates with cool winters.

CULTIVATION *Grow in moist but well-drained, fertile, humus-rich soil or compost with shade from midday sun. Give the climbing stems support. Minimum temperature 15°C (59°F).*

☼ ◊◊ ♈H1 ↕to 6m (20ft)

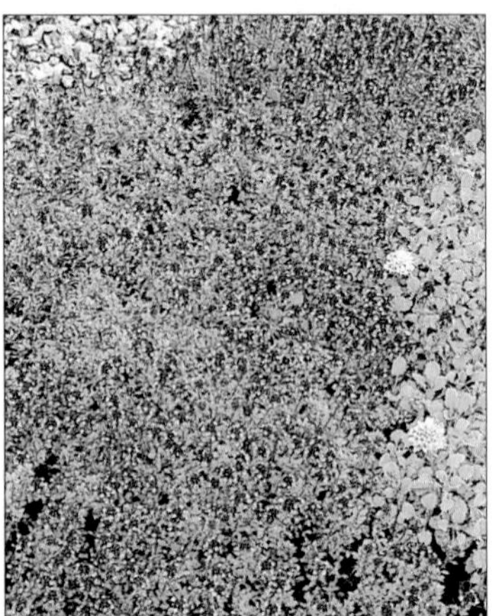

THYMUS x *CITRIODORUS* 'BERTRAM ANDERSON'

A low-growing, rounded, evergreen shrub carrying small, narrow, grey-green leaves, strongly suffused with yellow. They are aromatic and can be used in cooking. Heads of pale lavender-pink flowers are borne above the foliage in summer. Lovely in a herb garden. Sometimes sold as 'Anderson's Gold'.

CULTIVATION *Best in well-drained, neutral to alkaline soil, in full sun. Trim after flowering, and remove sprigs for cooking as they are needed.*

☼ ◊ ♈H4 ↕to 30cm (12in) ↔to 25cm (10in)

THYMUS x *CITRIODORUS* 'SILVER QUEEN'

This rounded, evergreen shrub is similar to 'Bertram Anderson' (see facing page, below), but with silver-white foliage. Masses of oblong, lavender-pink flowerheads are borne throughout summer. Plant in a herb garden; the aromatic leaves can be used in cooking.

CULTIVATION *Grow in well-drained, neutral to alkaline soil, in full sun. Trim after flowering, and remove sprigs for cooking as needed.*

☼ ◊ 🏆H4 ↕to 30cm (12in) ↔to 25cm (10in)

THYMUS SERPYLLUM VAR. *COCCINEUS*

A mat-forming, evergreen subshrub with finely hairy, trailing stems bearing tiny, aromatic, mid-green leaves. Crimson-pink flowers are borne in congested whorls during summer. Suitable for planting in paving crevices, where the foliage will release its fragrance when stepped on. May also be seen as *T. praecox* 'Coccineus'.

CULTIVATION *Grow in well-drained, neutral to alkaline, gritty soil. Choose a position in full sun. Trim lightly after flowering to keep the plant neat.*

☼ ◊ 🏆H4 ↕25cm (10in) ↔45cm (18in)

TIARELLA CORDIFOLIA

This vigorous, summer-flowering, evergreen perennial is commonly known as foam flower, and gets its name from the tiny, star-shaped, creamy-white flowers. These are borne in a profusion of upright sprays, above lobed, pale green leaves which turn bronze-red in autumn. Ideal as ground cover in a woodland garden, as is the similar *T. wherryi.*

CULTIVATION *Best in cool, moist soil that is rich in humus, in deep or light shade. Tolerates a wide range of soil types.*

◑ ● ◆ ♀H4 ↕10–30cm (4–12in) ↔to 30cm (12in)

TOLMIEA MENZIESII 'TAFF'S GOLD'

A spreading, clump-forming, semi-evergreen perennial carrying ivy-like, long-stalked, pale lime-green leaves, mottled with cream and pale yellow. An abundance of tiny, nodding, slightly scented, green and chocolate-brown flowers appear in slender, upright spikes during late spring and early summer. Plant in groups to cover the ground in a woodland garden.

CULTIVATION *Grow in moist but well-drained, humus-rich soil, in partial or deep shade. Sun will scorch the leaves.*

◑ ● ◇◆ ♀H4 ↕30–60cm (12–24in) ↔1m (3ft)

TRACHELOSPERMUM JASMINOIDES

Star jasmine is an evergreen, woody-stemmed climber with attractive, oval, glossy dark green leaves. The very fragrant flowers, creamy-white aging to yellow, have five twisted petal lobes. They are borne during mid- to late summer, followed by long seed pods. In cold areas, grow in the shelter of a warm wall with a deep mulch around the base of the plant. The variegated-leaved form, 'Variegatum', is also recommended.

CULTIVATION *Grow in any well-drained, moderately fertile soil, in full sun or partial shade. Tie in young growth.*

☼☀ ◊ ♈H2–3 ↕9m (28ft)

TRADESCANTIA x *ANDERSONIANA* 'J.C. WEGUELIN'

This tufted, clump-forming perennial bears large, pale blue flowers with three wide-open, triangular petals. These appear from early summer to early autumn in paired clusters at the tips of branching stems. The slightly fleshy, mid-green leaves are long, pointed and arching. Effective in a mixed or herbaceous border.

CULTIVATION *Grow in moist, fertile soil, in sun or partial shade. Deadhead to encourage repeat flowering.*

☼☀ ◑ ♈H4 ↕60cm (24in) ↔45cm (18in)

TRADESCANTIA × *ANDERSONIANA* 'OSPREY'

This clump-forming perennial bears clusters of large white flowers on the tips of the upright stems from early summer to early autumn. Each flower has three triangular petals, surrounded by two leaf-like bracts. The mid-green leaves are narrow and often purple-tinted. A long-flowering plant for a mixed or herbaceous border, lovely mixed with dark blue-flowered 'Isis'.

CULTIVATION *Grow in moist but well-drained, fertile soil, in sun or partial shade. Deadhead to prevent self-seeding.*

☼☀ 💧💧 ♈H4 ↕60cm (24in) ↔45cm (18in)

TRICYRTIS FORMOSANA

An upright, herbaceous perennial grown for its white, purple-spotted, star-shaped flowers on zig-zagging, softly hairy stems. These appear in early autumn above lance-shaped, dark green leaves which clasp the stems. An unusual plant for a shady border or open woodland garden.

CULTIVATION *Grow in moist soil that is rich in humus. Choose a sheltered site in deep or partial shade. In cold areas, provide a deep winter mulch where there is unlikely to be deep snow cover.*

☀● 💧 ♈H4 ↕to 80cm (32in) ↔45cm (18in)

TRILLIUM GRANDIFLORUM

Wake robin is a vigorous, clump-forming perennial grown for its large, three-petalled, pure white flowers which often fade to pink. These are carried during spring and summer on slender stems, above a whorl of three large, dark green, almost circular leaves. Effective in the company of hostas. The cultivar 'Flore Pleno' has double flowers.

CULTIVATION *Grow in moist but well-drained, leafy, neutral to acid soil, in deep or light shade. Provide an annual mulch of leaf mould in autumn.*

◐● ◊♦ ♀H4 ↕to 40cm (16in) ↔30cm (12in)

TRILLIUM LUTEUM

An upright, clump-forming perennial valued for its sweet-scented, golden- or bronze-green flowers in spring. These are produced above a whorl of oval, pointed, mid-green leaves which are heavily marked with paler green. Suits a moist, shady border.

CULTIVATION *Grow in moist but well-drained, humus-rich, preferably acid to neutral soil, in deep or partial shade. Mulch with leaf mould each autumn.*

◐● ◊♦ ♀H4 ↕to 40cm (16in) ↔to 30cm (12in)

TROLLIUS x *CULTORUM* 'ORANGE PRINCESS'

This globeflower is a robust, clump-forming perennial with orange-gold flowers. These are held above the mid-green foliage in late spring and early summer. The leaves are deeply cut with five rounded lobes. Good for bright colour beside a pond or stream, or in a damp border. For bright yellow flowers, choose the otherwise similar 'Goldquelle'.

CULTIVATION *Best in heavy, moist, fertile soil, in full sun or partial shade. Cut stems back hard after the first flush of flowers to encourage further blooms.*

☼ ◐ ♦ ♀H4 ↕to 90cm (36in) ↔45cm (18in)

TROPAEOLUM MAJUS 'HERMINE GRASHOFF'

This double-flowered nasturtium is a strong-growing, often scrambling, annual climber. Long-spurred, bright red flowers appear during summer and autumn, above the light green, wavy-margined leaves. Excellent for hanging baskets and other containers.

CULTIVATION *Grow in moist but well-drained, fairly poor soil, in full sun. The climbing stems need support. Minimum temperature 2°C (35°F).*

☼ ◊ ♀H1+3 ↕1–3m (3–10ft) ↔1.5–5m (5–15ft)

TROPAEOLUM SPECIOSUM

The flame nasturtium is a slender, herbaceous climber, producing long-spurred, bright vermilion flowers throughout summer and autumn. These are followed by small, bright blue fruits. The mid-green leaves are divided into several leaflets. Effective growing through dark-leaved hedging plants, which contrast well with its flowers.

CULTIVATION *Grow in moist, humus-rich, neutral to acid soil, in full sun or partial shade. Provide shade at the roots, and support the climbing stems.*

☼◑ ◊♦ 🏆H4 ↕to 3m (10ft)

TSUGA CANADENSIS 'JEDDELOH'

This dwarf form of the eastern hemlock is a small, vase-shaped conifer with deeply furrowed, purplish-grey bark. The bright green foliage is made up of needle-like leaves which are arranged in two ranks along the stems. An excellent small specimen tree for shady places; also popular for bonsai training.

CULTIVATION *Grow in moist but well-drained, humus-rich soil, in full sun or partial shade. Provide shelter from cold, drying winds. Trim during summer.*

☼◑ ◊♦ 🏆H4 ↕1.5m (5ft) ↔2m (6ft)

TULIPA CLUSIANA VAR. *CHRYSANTHA*

This yellow-flowered lady tulip is a bulbous perennial which flowers in early- to mid-spring. The bowl- to star-shaped flowers, tinged red or brownish-purple on the outsides, are produced in clusters of up to three per stem, above the linear, grey-green leaves. Suitable for a raised bed or rock garden.

CULTIVATION *Grow in well-drained, fertile soil, in full sun with shelter from strong winds. Deadhead and remove any fallen petals after flowering.*

☼ ◊ ♀H4 ↕30cm (12in)

TULIPA LINIFOLIA

This slender, variable, bulbous perennial bears bowl-shaped red flowers in early and mid-spring. These are carried above the linear, grey-green leaves with wavy red margins. The petals have yellow margins and black-purple marks at the base. Good for a rock garden.

CULTIVATION *Grow in sharply drained, fertile soil, in full sun with shelter from strong winds. Deadhead and remove any fallen petals after flowering.*

☼ ◊ ♀H4 ↕20cm (8in)

TULIPA LINIFOLIA BATALINII GROUP

Slender, bulbous perennials, often sold as *T. batalinii*, bearing solitary, bowl-shaped, pale yellow flowers with dark yellow or bronze marks on the insides. These appear from early to mid-spring above linear, grey-green leaves with wavy red margins. Use in spring bedding; the flowers are good for cutting.

CULTIVATION *Grow in sharply drained, fertile soil, in full sun with shelter from strong winds. Deadhead and remove any fallen petals after flowering.*

☼ ◊ ♈H4 ↕35cm (14in)

TULIPA TURKESTANICA

This bulbous perennial produces up to 12 star-shaped white flowers per stem in early and mid-spring. They are flushed with greenish-grey on the outsides and have yellow or orange centres. The linear, grey-green leaves are arranged beneath the flowers. Grow in a rock garden or sunny border, away from paths or seating areas as the flowers have an unpleasant scent.

CULTIVATION *Grow in well-drained, fertile soil, in full sun with shelter from strong winds. Deadhead and remove any fallen petals after flowering.*

☼ ◊ ♈H4 ↕30cm (12in)

TULIPA CULTIVARS

These cultivated varieties of tulip are spring-flowering, bulbous perennials, with a much wider range of flower colour than any other spring bulbs, from the buttercup-yellow 'Hamilton' through to the violet-purple 'Blue Heron' and the multi-coloured, red, white and blue 'Union Jack'. This diversity makes them invaluable for bringing variety into the garden, either massed together in large containers or beds, or planted in a mixed border. Flower shape is also varied; as well as the familiar cup-shaped blooms, as in 'Dreamland', there are also conical, goblet- and star-shaped forms. The flowers are good for cutting.

CULTIVATION *Grow in well-drained, fertile soil, in sun with shelter from strong winds and excessive wet. Remove spent flowers. Lift bulbs once the leaves have died down, and store over summer in a cold greenhouse to ripen. Replant the largest bulbs in autumn.*

☼ ◊ ♈H4

1 ↕15cm (6in)　2 ↕60cm (24in)　3 ↕50cm (20in)

4 ↕40cm (16in)　5 ↕50cm (20in)　6 ↕30cm (12in)

1 *T.* 'Ancilla' **2** *T.* 'Blue Heron' **3** *T.* 'China Pink' **4** *T.* 'Don Quichotte' **5** *T.* 'Hamilton' **6** *T.* 'Oriental Splendour'

7 *T.* 'Dreamland' **8** *T.* 'Keizerskroon' **9** *T.* 'Prinses Irene' **10** *T.* 'Queen of Sheba' **11** *T.* 'Red Riding Hood' **12** *T.* 'Spring Green' **13** *T.* 'Union Jack' **14** *T.* 'West Point'

UVULARIA GRANDIFLORA

Large merrybells is a slow-spreading, clump-forming perennial bearing solitary or paired, narrowly bell-shaped, sometimes green-tinted yellow flowers. They hang gracefully from slender, upright stems during mid- to late spring, above the downward-pointing, lance-shaped, mid-green leaves. Excellent for a shady border or woodland garden.

CULTIVATION *Grow in moist but well-drained, fertile soil that is rich in humus, in deep or partial shade.*

◐● ◊◊ ♈H4 ↕to 75cm (30in) ↔30cm (12in)

VACCINIUM CORYMBOSUM

The highbush blueberry is a dense, deciduous, acid-soil-loving shrub with slightly arching shoots. The oval leaves are mid-green, turning yellow or red in autumn. Hanging clusters of small, often pink-tinged white flowers appear in late spring and early summer, followed by sweet, edible, blue-black berries. Best suited to a woodland garden on lime-free soil.

CULTIVATION *Grow in moist but well-drained, peaty or sandy, acid soil, in sun or light shade. Trim in winter.*

☼◐ ◊◊ ♈H4 ↕↔ 1.5m (5ft)

VACCINIUM GLAUCOALBUM

A mound-forming, dense, evergreen shrub bearing elliptic, leathery, dark green leaves with bright bluish-white undersides. Very small, pink-tinged white flowers appear in hanging clusters during late spring and early summer, followed by edible, white-bloomed, blue-black berries. Suits a lime-free woodland garden.

CULTIVATION *Grow in open, moist but well-drained, peaty or sandy, acid soil, in sun or partial shade. Trim in spring.*

☼ ◐ ◊♦ ♕H3–4 ↕50–120cm (20–48in) ↔1m (3ft)

VACCINIUM VITIS-IDAEA KORALLE GROUP

These heavy-fruiting cowberries are creeping, evergreen shrubs with oval, glossy dark green leaves, shallowly notched at the tips. In late spring and early summer, small, bell-shaped, white to deep pink flowers appear in dense, nodding clusters. These are followed by a profusion of round, bright red berries which are edible, but taste acidic. Good ground cover for a lime-free, woodland garden.

CULTIVATION *Best in peaty or sandy, moist but well-drained, acid soil, in full sun or partial shade. Trim in spring.*

☼ ◐ ◊♦ ♕H4 ↕25cm (10in) ↔indefinite

VELTHEIMIA BRACTEATA

A bulbous perennial with basal rosettes of thick and waxy, strap-like, glossy, dark green leaves, from which upright flowering stems grow in spring, to be topped by a dense cluster of tubular, pink-purple flowers with yellow spots. An unusual house or conservatory plant where winter protection is needed.

CULTIVATION *Plant bulbs in autumn with the neck just above soil level, in loam-based potting compost with added sharp sand. Site in full sun, reducing watering as the leaves fade. Keep the soil just moist during dormancy. Minimum temperature 2–5°C (35–41°F).*

☼ ◊ ♈H1 ↕45cm (18in) ↔30cm (12in)

VERATRUM NIGRUM

An imposing, rhizomatous perennial that produces a tall and upright, branching spike of many reddish-brown to black flowers from the centre of a basal rosette of pleated, mid-green leaves. The small, star-shaped flowers that open in late summer have an unpleasant scent, so choose a moist, shady site not too close to paths or patios.

CULTIVATION *Grow in deep, fertile, moist but well-drained soil with added organic matter. If in full sun, make sure the soil remains moist. Shelter from cold, drying winds. Divide congested clumps in autumn or early spring.*

☼ ◐ ◊ ♦ ♈H4 ↕60–120cm (24–48in) ↔60cm (24in)

VERBASCUM BOMBYCIFERUM

This mullein is a very tall perennial which forms basal rosettes of densely packed leaves covered in silky silver hairs. Short-lived, it dies back after the magnificent display of its tall, upright flower spike in summer, which is covered in silver hairs and sulphur-yellow flowers. For a large border, or may naturalize by self-seeding in a wild garden.

CULTIVATION *Grow in alkaline, poor, well-drained soil in full sun. Support may be needed in fertile soil because of the resultant more vigorous growth. Divide plants in spring, if necessary.*

☼ ◊ 🏆H4 ↕1.8m (8ft) ↔to 60cm (24in)

VERBASCUM 'COTSWOLD BEAUTY'

A tall, evergreen, short-lived perennial that makes a bold addition to any large border. It bears upright spikes of saucer-shaped, peachy-pink flowers, with darker centres, over a long period from early to late summer. These spikes tower over the wrinkled, grey-green foliage, most of which is clumped near the base of the plant. It may naturalize in a wild or lightly wooded garden.

CULTIVATION *Best in well-drained, poor, alkaline soil in full sun. In fertile soil, it grows taller and will need support. Divide in spring, if necessary.*

☼ ◊ 🏆H4 ↕1.2m (4ft) ↔45cm (18in)

VERBASCUM 'GAINSBOROUGH'

This short-lived, semi-evergreen perennial is valued for its spires of saucer-shaped, soft yellow flowers, borne throughout summer. Most of the oval and grey-green leaves are arranged in rosettes around the base of the stems. A very beautiful, long-flowering plant for a herbaceous or mixed border.

CULTIVATION *Grow in well-drained, fertile soil, in an open, sunny site. Often short-lived, but easily propagated by root cuttings taken in winter.*

☼ ♢ ♀H4 ↕to 1.2m (4ft) ↔30cm (12in)

VERBASCUM DUMULOSUM

This evergreen subshrub forms small, spreading domes of densely felted, grey or grey-green leaves on white-downy stems. In late spring and early summer, clusters of small, saucer-shaped yellow flowers with red-purple eyes appear amid the foliage. In cold areas, grow in small crevices of a warm, sunny wall.

CULTIVATION *Best in gritty, sharply drained, moderately fertile, preferably alkaline soil, in full sun. Shelter from excessive winter wet.*

☼ ♢ ♀H2–3 ↕to 25cm (10in) ↔to 40cm (16in)

VERBASCUM 'HELEN JOHNSON'

This evergreen perennial produces spikes of saucer-shaped flowers in an unusual light pink-brown during early to late summer. The oval, wrinkled, finely downy, grey-green leaves are arranged in rosettes around the base of the stems. Naturalizes well in a wild garden.

CULTIVATION *Grow in well-drained, poor, alkaline soil, in full sun. Flower stems will need staking in fertile soil, where it grows larger.*

☼ ◊ H4 ↕90cm (36in) ↔30cm (12in) or more

VERBASCUM 'LETITIA'

This dense, rounded, evergreen subshrub produces a continuous abundance of small, clear yellow flowers with reddish-purple centres throughout summer. The lance-shaped, irregularly toothed, grey-green leaves are carried beneath the clustered flowers. Suitable for a raised bed or rock garden. In cold climates, grow in crevices of a warm and protected drystone wall.

CULTIVATION *Best in sharply drained, fairly fertile, alkaline soil, in sun. Choose a site not prone to excessive winter wet.*

☼ ◊ H3 ↕to 25cm (10in) ↔to 30cm (12in)

VERBENA 'SISSINGHURST'

A mat-forming perennial bearing rounded heads of small, brilliant magenta-pink flowers, which appear from late spring to autumn – most prolifically in summer. The dark green leaves are cut and toothed. Excellent for edging a path or growing in a tub. In cold areas, overwinter under glass.

CULTIVATION *Grow in moist but well-drained, moderately fertile soil or compost. Choose a site in full sun.*

☼ ◊ ♈H2–3 ↕to 20cm (8in) ↔to 1m (3ft)

VERONICA GENTIANOIDES

This early-summer-flowering, mat-forming perennial bears shallowly cup-shaped, pale blue or white flowers. These are carried in upright spikes which arise from rosettes of glossy, broadly lance-shaped, dark green leaves at the base of the plant. Excellent for the edge of a border.

CULTIVATION *Grow in moist but well-drained, moderately fertile soil, in full sun or light shade.*

☼ ◐ ◊ ♈H4 ↕↔ 45cm (18in)

VERONICA PROSTRATA

The prostrate speedwell is a dense, mat-forming perennial. In early summer, it produces upright spikes of saucer-shaped, pale to deep blue flowers at the tips of sprawling stems. The small, bright to mid-green leaves are narrow and toothed. Grow in a rock garden. The cultivar 'Spode Blue' is also recommended.

CULTIVATION *Best in moist but well-drained, poor to moderately fertile soil. Choose a position in full sun.*

☼ ◊ ♈H4 ↕to 15cm (6in) ↔to 40cm (16in)

VERONICA SPICATA SUBSP. *INCANA*

The silver speedwell is an entirely silver-hairy, mat-forming perennial with upright flowering stems. These are tall spikes of star-shaped, purple-blue flowers, borne from early to late summer. The narrow leaves are silver-hairy and toothed. Ideal for a rock garden. The cultivar 'Wendy', with bright blue flowers, is also recommended.

CULTIVATION *Grow in well-drained, poor to moderately fertile soil, in full sun. Protect from winter wet.*

☼ ◊ ♈H4 ↕↔ 30cm (12in)

VIBURNUM X *BODNANTENSE* 'DAWN'

A strongly upright, deciduous shrub carrying toothed, dark green leaves, bronze when young. From late autumn to spring, when branches are bare, small, tubular, heavily scented, dark pink flowers, which age to white, are borne in clustered heads, followed by small, blue-black fruits. 'Charles Lamont' is another lovely cultivar of this shrub, as is 'Deben', with almost white flowers.

CULTIVATION *Grow in any deep, moist but well-drained, fertile soil, in full sun. On old crowded plants, cut the oldest stems back to the base after flowering.*

☼ ◊♦ ♆H4 ↕3m (10ft) ↔2m (6ft)

VIBURNUM X *CARLCEPHALUM*

This vigorous, rounded, deciduous shrub has broadly heart-shaped, irregularly toothed, dark green leaves which turn red in autumn. Rounded heads of small, fragrant white flowers appear amid the foliage during late spring. Suits a shrub border or a woodland garden.

CULTIVATION *Grow in deep, moisture-retentive, fertile soil. Tolerates sun or semi-shade. Little pruning is necesary.*

☼ ◑ ♦ ♆H4 ↕↔ 3m (10ft)

VIBURNUM DAVIDII

A compact, evergreen shrub that forms a dome of dark green foliage consisting of oval leaves with three distinct veins. In late spring, tiny, tubular white flowers appear in flattened heads; female plants bear tiny but decorative, oval, metallic-blue fruits, later in the season. Looks good planted in groups.

CULTIVATION *Best in deep, moist but well-drained, fertile soil, in sun or semi-shade. Grow male and female shrubs together for reliable fruiting. Keep neat, if desired, by cutting wayward stems back to strong shoots, or to the base of the plant, in spring.*

☼☀ 00 ♈H4 ↕↔ 1–1.5m (3–5ft)

VIBURNUM FARRERI

This strongly upright, deciduous shrub has oval, toothed, dark green leaves which are bronze when young and turn red-purple in autumn. During mild periods in winter and early spring, small, fragrant, white or pink-tinged flowers are borne in dense clusters on the bare stems. These are occasionally followed by tiny, bright red fruits.

CULTIVATION *Grow in any reliably moist but well-drained, fertile soil, in full sun or partial shade. Thin out old shoots after flowering.*

☼☀ 00 ♈H4 ↕3m (10ft) ↔2.5m (8ft)

VIBURNUM OPULUS 'XANTHOCARPUM'

A vigorous, deciduous shrub bearing maple-like, lobed, mid-green leaves which turn yellow in autumn. Flat heads of showy white flowers are produced in late spring and early summer, followed by large bunches of spherical, bright yellow berries. 'Notcutt's Variety' and 'Compactum' are recommended red-berried cultivars, the latter much smaller, to only 1.5m (5ft) tall.

CULTIVATION *Grow in any moist but well-drained soil, in sun or semi-shade. Cut out older stems after flowering to relieve overcrowding.*

☼☀ ◊♦ ♉H4 ↕5m (15ft) ↔4m (12ft)

VIBURNUM PLICATUM 'MARIESII'

A spreading, deciduous shrub bearing distinctly tiered branches. These are clothed in heart-shaped, toothed, dark green leaves which turn red-purple in autumn. In late spring, saucer-shaped white flowers appear in spherical, lacecap-like heads. Few berries are produced; for a heavily fruiting cultivar, look for 'Rowallane'.

CULTIVATION *Grow in any well-drained, fairly fertile soil, in sun or semi-shade. Requires little pruning, other than to remove damaged wood after flowering; be careful not to spoil the form.*

☼☀ ◊ ♉H4 ↕3m (10ft) ↔4m (12ft)

VIBURNUM TINUS 'EVE PRICE'

This very compact, evergreen shrub has dense, dark green foliage. Over a long period from late winter to spring, pink flower buds open to tiny, star-shaped white flowers, carried in flattened heads; they are followed by small, dark blue-black fruits. Can be grown as an informal hedge.

CULTIVATION *Grow in any moist but well-drained, moderately fertile soil, in sun or partial shade. Train or clip after flowering to maintain desired shape.*

☼☀ ◊♦ ♀H4 ↕↔ 3m (10ft)

VINCA MAJOR 'VARIEGATA'

This variegated form of the greater periwinkle, also known as 'Elegantissima', is an evergreen subshrub with long, slender shoots bearing oval, dark green leaves which have creamy-white margins. Dark violet flowers are produced over a long period from mid-spring to autumn. Useful as ground cover for a shady bank, but may be invasive.

CULTIVATION *Grow in any moist but well-drained soil. Tolerates deep shade, but flowers best with part-day sun.*

☀● ◊♦ ♀H4 ↕45cm (18in) ↔indefinite

VINCA MINOR 'ATROPURPUREA'

This dark-flowered lesser periwinkle is a mat-forming, ground cover shrub with long trailing shoots. Dark plum-purple flowers are produced over a long period from mid-spring to autumn, amid the oval, dark green leaves. For light blue flowers, choose 'Gertrude Jekyll'.

CULTIVATION *Grow in any but very dry soil, in full sun for best flowering, but tolerates partial shade. Restrict growth by cutting back hard in early spring.*

☼☀ ◊◊ ♈H4 ↕10–20cm (4–8in) ↔indefinite

VIOLA CORNUTA

The horned violet is a spreading, evergreen perennial which produces an abundance of slightly scented, spurred, violet to lilac-blue flowers; the petals are widely separated, with white markings on the lower ones. The flowers are borne amid the mid-green, oval leaves from spring to summer. Suitable for a rock garden. The Alba Group have white flowers.

CULTIVATION *Grow in moist but well-drained, poor to moderately fertile soil, in sun or partial shade. Cut back after flowering to keep compact.*

☼☀ ◊◊ ♈H4 ↕to 15cm (6in) ↔to 40cm (16in)

VIOLA 'JACKANAPES'

A robust, clump-forming, evergreen perennial bearing spreading stems with oval and toothed, bright green leaves. Spurred, golden-yellow flowers with purple streaks appear in late spring and summer; the upper petals are deep brownish-purple. Good for containers or summer bedding.

CULTIVATION *Grow in moist but well-drained, fairly fertile soil, in full sun or semi-shade. Often short-lived, but easily grown from seed sown in spring.*

☼ ◐ ◊◆ 🏆H4 ↕to 12cm (5in) ↔to 30cm (12in)

VIOLA 'NELLIE BRITTON'

This clump-forming, evergreen perennial with spreading stems produces an abundance of spurred, pinkish-mauve flowers over long periods in summer. The oval, mid-green leaves are toothed and glossy. Suitable for the front of a border.

CULTIVATION *Best in well-drained but moist, moderately fertile soil, in full sun or partial shade. Deadhead frequently to prolong flowering.*

☼ ◐ ◊◆ 🏆H4 ↕to 15cm (6in) ↔to 30cm (12in)

VITIS COIGNETIAE

The crimson glory vine is a fast-growing, deciduous climber with large, heart-shaped, shallowly lobed, dark green leaves which turn bright red in autumn. Small, blue-black grape-like fruits appear in autumn. Train against a wall or over a trellis.

CULTIVATION *Grow in well-drained, neutral to alkaline soil, in sun or semi-shade. Autumn colour is best on poor soils. Pinch out the growing tips after planting and allow the strongest shoots to form a permanent framework. Prune back to this each year in mid-winter.*

☼ ◑ 💧 ♀H4 ↕15m (50ft)

VITIS VINIFERA 'PURPUREA'

This purple-leaved grape vine is a woody, deciduous climber bearing rounded, lobed, toothed leaves; these are white-hairy when young, turning plum-purple, then dark purple, before they fall. Tiny, pale green summer flowers are followed by small purple grapes in autumn. Grow over a robust fence or pergola, or through a large shrub or tree.

CULTIVATION *Grow in well-drained, slightly alkaline soil, in sun or semi-shade. Autumn colour is best on poor soils. Prune back to an established framework each year in mid-winter.*

☼ ◑ 💧 ♀H4 ↕7m (22ft)

WEIGELA FLORIDA 'FOLIIS PURPUREIS'

A compact, deciduous shrub with arching shoots that produces clusters of funnel-shaped, dark pink flowers with pale insides in late spring and early summer. The bronze-green foliage is made up of oval, tapered leaves. Pollution-tolerant, so is ideal for urban gardens.

CULTIVATION *Best in well-drained, fertile, humus-rich soil, in full sun. Prune out some older branches at ground level each year after flowering.*

☼ ◊ 🏆H4 ↕1m (3ft) ↔1.5m (5ft)

WEIGELA 'FLORIDA VARIEGATA'

A dense, deciduous shrub that produces abundant clusters of funnel-shaped, dark pink flowers with pale insides. These are borne in late spring and early summer amid attractive, grey-green leaves with white margins. Suitable for a mixed border or open wood-land. Tolerates urban pollution. 'Praecox Variegata' is another pretty variegated weigela.

CULTIVATION *Grow in any well-drained, fertile, humus-rich soil, in full sun. Prune out some of the oldest branches each year after flowering.*

 ↕2–2.5m (6–8ft)

WISTERIA FLORIBUNDA 'ALBA'

This white-flowered Japanese wisteria is a fast-growing, woody climber with bright green, divided leaves. The fragrant, pea-like flowers appear during early summer in very long, drooping spikes; bean-like, velvety-green seed pods usually follow. Train against a wall, up into a tree or over a well-built arch.

CULTIVATION *Grow in moist but well-drained, fertile soil, in sun or partial shade. Prune back new growth in summer and in late winter to control spread and promote flowering.*

☼☀ ♢♦ ♈H4 ↕9m (28ft) or more

WISTERIA FLORIBUNDA 'MULTIJUGA'

This lilac-blue-flowered Japanese wisteria is a vigorous, woody-stemmed climber. Long, hanging flower clusters open during early summer, usually followed by bean-like, velvety-green seed pods. Each mid-green leaf is composed of many oval leaflets. Train against a wall or up into a tree.

CULTIVATION *Grow in moist but well-drained, fertile soil, in full sun or partial shade. Trim in mid-winter, and again 2 months after flowering.*

☼☀ ♢♦ ♈H4 ↕9m (28ft) or more

WISTERIA SINENSIS

Chinese wisteria is a vigorous, deciduous climber that produces long, hanging spikes of fragrant, pea-like, lilac-blue to white flowers. These appear amid the bright green, divided leaves in late spring and early summer, usually followed by bean-like, velvety-green seed pods. 'Alba' is a white-flowered form.

CULTIVATION *Grow in moist but well-drained, fertile soil. Provide a sheltered site in sun or partial shade. Cut back long shoots to 2 or 3 buds, in late winter.*

☼ ◐ 💧💧 🏆H4 ↕9m (28ft) or more

YUCCA FILAMENTOSA 'BRIGHT EDGE'

An almost stemless, clump-forming shrub with basal rosettes of rigid, lance-shaped, dark green leaves, to 75cm (30in) long, with broad yellow margins. Tall spikes, to 2m (6ft) or more, of nodding, bell-shaped white flowers, tinged with green or cream, are borne in mid- to late summer. An architectural specimen for a border or courtyard. The leaves of 'Variegata' are edged white.

CULTIVATION *Grow in any well-drained soil, in full sun. Remove old flowers at the end of the season. In cold or frost-prone areas, mulch over winter.*

☼ 💧 🏆H3 ↕75cm (30in) ↔1.5m (5ft)

YUCCA FLACCIDA 'IVORY'

An almost stemless, evergreen shrub that forms a dense, basal clump of sword-like leaves. Tall spikes, to 1.5m (5ft) or more, of nodding, bell-shaped white flowers are borne in mid- and late summer. The lance-shaped, dark blue-green leaves, fringed with curly or straight threads, are arranged in basal rosettes. Thrives in coastal gardens and on sandy soils.

CULTIVATION *Grow in any well-drained soil, but needs a hot, dry position in full sun to flower well. In cold or frost-prone areas, mulch over winter.*

☼ ◊ ♀H3–4 ↕55cm (22in) ↔1.5m (5ft)

ZANTEDESCHIA AETHIOPICA

This relatively small-flowered arum lily is a clump-forming perennial. Almost upright, cream-yellow flowers are borne in succession from late spring to mid-summer, followed by long, arrow-shaped leaves, evergreen in mild areas. May be used as a waterside plant in shallow water. May not survive in cold areas.

CULTIVATION *Best in moist but well-drained, fertile soil that is rich in humus. Choose a site in full sun or partial shade. Mulch deeply over winter in cold, frost-prone areas.*

☼ ◐ ◊ ♦ ♀H3 ↕90cm (36in) ↔60cm (24in)

ZANTEDESCHIA AETHIOPICA 'GREEN GODDESS'

This green-flowered arum lily is a clump-forming, robust perennial which is evergreen in mild climates. Upright and white-centred flowers appear from late spring to mid-summer, above the dull green, arrow-shaped leaves. Good in shallow water, in a site sheltered from heavy frosts.

CULTIVATION *Grow in moist, humus-rich, fertile soil. Choose a site in full sun or partial shade. Mulch deeply over winter in cold, frost-prone areas.*

☼ ◑ ◊ ♦ ♀H3 ↕90cm (36in) ↔60cm (24in)

ZAUSCHNERIA CALIFORNICA 'DUBLIN'

This deciduous Californian fuchsia, sometimes called 'Glasnevin', is a clump-forming perennial bearing a profusion of tubular, bright red flowers during late summer and early autumn. The grey-green leaves are lance-shaped and hairy. Provides spectacular, late-season colour for a dry stone wall or border. In climates with cold winters, grow at the base of a warm wall.

CULTIVATION *Best in well-drained, moderately fertile soil, in full sun. Provide shelter from cold, drying winds.*

☼ ◊ ♀H3 ↕to 25cm (10in) ↔to 30cm (12in)

THE PLANTING GUIDE

CONTAINER PLANTS FOR SPRING

A tub or pot of evergreens and bright spring flowers and bulbs brings cheer to a front door or patio on a chilly early spring day. Tailor the soil to suit the plants: camellias and rhododendrons can be grown in ericaceous compost, for example. Keep containers in a sheltered place for early flowers, water through the winter if necessary, and give liquid feed as growth begins in spring.

AGAPANTHUS CAMPANULATUS 'ALBOVITTATUS'
Perennial, not hardy Soft blue flowers in large heads above white-striped leaves.
↕90cm (3ft) ↔ 45cm (18in)

AJUGA REPTANS 'CATLIN'S GIANT'
♀ **Perennial** Clumps of green leaves and spikes of blue flowers.
↕20cm (8in) ↔ 60–90cm (24–36in)

BERGENIA PURPURASCENS 'BALLAWLEY'
♀ **Perennial** page **88**

CAMELLIA JAPONICA CULTIVARS
♀ **Evergreen shrubs** pages **100–101**

CAMELLIA X *WILLIAMSII* 'DONATION'
♀ **Evergreen shrub** page **104**

CHAENOMELES 'GEISHA GIRL'
Shrub Soft apricot, semi-double flowers.
↕1m (3ft) ↔ 1.2m (4ft)

CHAMAECYPARIS LAWSONIANA 'MINIMA GLAUCA'
♀ **Evergreen shrub** A small conifer of neat habit with blue-green leaves.
↕60cm (24in)

CHIONODOXA FORBESII 'PINK GIANT'
Bulb Star-shaped, white-centred pink flowers.
↕10–20cm (4–8in) ↔ 3cm (1¼in)

CHIONODOXA SARDENSIS
♀ **Bulb** Bright blue, starry flowers in early spring.
↕10–20cm (4–8in) ↔ 3cm (1¼in)

CROCUS CHRYSANTHUS 'LADYKILLER'
♀ **Bulb** White, scented flowers with purple stripes on buds.
↕7cm (3in) ↔ 5cm (2in)

ERICA CARNEA 'MYRETOUN RUBY'
♀ **Evergreen shrub** Spreading habit, with pink flowers that deepen as they age.
↕15cm (6in) ↔ 45cm (18in)

HEDERA HELIX 'KOLIBRI'
♀ **Evergreen shrub** Trailing stems with leaves variegated with white and green.
↕45cm (18in)

HYACINTHUS ORIENTALIS 'CITY OF HAARLEM'
♀ **Bulb** page **263**

HYACINTHUS ORIENTALIS 'GIPSY QUEEN'
♀ **Bulb** Orange/pink scented flowers.
↕25cm (10in)

HYACINTHUS ORIENTALIS 'HOLLYHOCK'
Bulb Dense spikes of double red flowers.
↕20cm (8in)

HYACINTHUS ORIENTALIS 'VIOLET PEARL'
Bulb Spikes of scented, amethyst-violet flowers with paler petal edges.
↕25cm (10in)

ILEX X *ALTACLERENSIS* 'GOLDEN KING'
♀ **Evergreen shrub** page **271**

MYOSOTIS 'BLUE BALL'
♀ **Biennial** Compact plant with azure-blue flowers.
↕ 15cm (6in) ↔ 20cm (8in)

NARCISSUS 'CEYLON'
Bulb page **348**

NARCISSUS 'CHEERFULNESS'
Bulb page **348**

NARCISSUS SMALL DAFFODILS
Bulbs pages **346–347**

PIERIS 'FOREST FLAME'
♀ **Evergreen shrub** page **402**

PIERIS FORMOSA VAR. *FORRESTII* 'WAKEHURST'
♀ **Evergreen shrub** page **402**

PIERIS JAPONICA 'PURITY'
♀ **Evergreen shrub** A compact variety with white flowers and pale young growth.
↕↔ 1m (3ft)

PRIMULA GOLD-LACED GROUP
♀ **Perennial** page **415**

PRIMULA 'GUINEVERE'
♀ **Perennial** page **415**

RHODODENDRON CILPINENSE GROUP
♀ **Evergreen shrub** page **444**

RHODODENDRON 'CURLEW'
♀ **Evergreen shrub** Small trusses of bright yellow flowers.
↕↔ 60cm (24in)

RHODODENDRON 'DOC'
♀ **Evergreen shrub** page **444**

RHODODENDRON 'HOMEBUSH'
♀ **Shrub** page **438**

RHODODENDRON 'MOTHER'S DAY'
♀ **Evergreen shrub** Bright red-flowered evergreen azalea.
↕↔ 1.5m (5ft)

RHODODENDRON 'PTARMIGAN'
♀ **Evergreen shrub** page **444**

RHODODENDRON 'SUSAN'
♀ **Evergreen shrub** page **441**

TULIPA 'APRICOT BEAUTY'
Bulb Pastel, salmon-pink flowers of great beauty.
↕ 35cm (14in)

TULIPA 'CAPE COD'
Bulb Yellow flowers striped with red above prettily marked leaves.
↕ 20cm (8in)

TULIPA 'CARNAVAL DE NICE'
Bulb Red and white-striped blooms like double peonies.
↕ 40cm (16in)

TULIPA CLUSIANA VAR. *CHRYSANTHA*
♀ **Bulb** page **522**

TULIPA CULTIVARS
♀ **Bulbs** pages **524–525**

TULIPA PRAESTANS 'UNICUM'
Bulb Cream-edged leaves and one to four scarlet flowers per stem.
↕ 30cm (12in)

TULIPA 'TORONTO'
♀ **Bulb** Multi-headed tulips with three to five flowers of deep coral-pink.
↕ 25cm (10in)

VIOLA UNIVERSAL SERIES
♀ **Biennial** Wide range of colours with flowers throughout spring.
↕ 15cm (6in) ↔ to 30cm (12in)

VIOLA 'VELOUR BLUE'
Biennial Compact plants with masses of small light blue flowers.
↕ 15cm (6in) ↔ 20cm (8in)

CONTAINER PLANTS FOR SUMMER

Containers can bring colour to areas that have no soil; they can be filled with permanent plants, or used to add extra colour when filled with annuals and tender perennials. In a large tub, combine the two, with a shrub augmented by lower, temporary plants. There is a huge choice, but every container will need regular watering and feeding if it is to look at its best.

ABUTILON 'CANARY BIRD'
♈ **Shrub, not hardy** Bushy plant with pendulous yellow flowers.
↕↔ to 3m (10ft)

ABUTILON MEGAPOTAMICUM
♈ **Shrub, not hardy** page **30**

AGAPANTHUS CAMPANULATUS SUBSP. *PATENS*
♈ **Perennial, not hardy** page **47**

ARGYRANTHEMUM 'VANCOUVER'
♈ **Perennial, not hardy** page **67**

CLEMATIS 'DOCTOR RUPPEL'
♈ **Climber** page **129**

CROCOSMIA 'LUCIFER'
♈ **Perennial** page **152**

CUPHEA IGNEA
♈ **Perennial, not hardy** page **155**

DIASCIA BARBERAE 'BLACKTHORN APRICOT'
♈ **Perennial, borderline hardy** page **174**

DIASCIA VIGILIS
♈ **Perennial, not hardy** The creeping stems carry loose spikes of pink flowers with yellow centres.
↕30cm (12in) ↔ 60cm (24in)

FUCHSIA 'ANNABEL'
♈ **Shrub, not hardy** page **213**

FUCHSIA 'CELIA SMEDLEY'
♈ **Shrub, not hardy** page **214**

FUCHSIA 'DISPLAY'
♈ **Shrub, not hardy** Upright plant with carmine and pink flowers.
↕60–75cm (24–30in) ↔ 45–60cm (18–24in)

FUCHSIA 'NELLIE NUTTALL'
♈ **Shrub, not hardy** page **215**

FUCHSIA 'SNOWCAP'
♈ **Shrub, not hardy** page **215**

FUCHSIA 'SWINGTIME'
♈ **Shrub, not hardy** page **215**

FUCHSIA 'THALIA'
♈ **Shrub, not hardy** page **215**

FUCHSIA 'WINSTON CHURCHILL'
♈ **Shrub, not hardy** Pink and lavender, double flowers.
↕↔ 45–75cm (18–30in)

GAZANIAS
♈ **Perennials, not hardy** page **222**

HOSTA 'FRANCEE'
♈ **Perennial** page **259**

HYDRANGEA MACROPHYLLA 'AYESHA'
♈ **Shrub** Unusual because the blooms are cupped, resembling *Syringa* (lilac) flowers.
↕1.5m (5ft) ↔ 2m (6ft)

LILIUM FORMOSANUM VAR. *PRICEI*
♈ **Perennial** page **315**

LILIUM LONGIFLORUM
♈ **Perennial, not hardy** page **316**

LILIUM MONADELPHUM
♈ **Perennial** page **317**

LOTUS MACULATUS
♀ **Perennial, not hardy** Narrow, silvery leaves contrast with orange, claw-like flowers on trailing shoots.
↕ 20cm (8in) ↔ indefinite

NICOTIANA LANGSDORFII
♀ **Annual** Airy spikes of small, green, tubular flowers with flaring trumpets.
↕ to 1.5m (5ft) ↔ to 35cm (14in)

OSTEOSPERMUM CULTIVARS
♀ **Perennials, not hardy** page **361**

PELARGONIUM 'ATTAR OF ROSES'
♀ **Perennial, not hardy** page **372**

PELARGONIUMS, FLOWERING
♀ **Perennials, not hardy** pages **374–376**

PENSTEMON 'ANDENKEN AN FRIEDRICH HAHN'
♀ **Perennial** page **378**

PENSTEMON 'HEWELL PINK BEDDER'
♀ **Perennial, not hardy** A free-flowering, bushy plant with tubular, pink flowers.
↕ 45cm (18in) ↔ 30cm (12in)

PETUNIA 'LAVENDER STORM'
Annual Compact, ground-covering plants with large blooms.
↕ 30cm (12in) ↔ 40cm (16in)

PETUNIA 'MILLION BELLS BLUE'
Annual Masses of small flowers on bushy, compact plants.
↕ 25cm (10in)

PETUNIA 'PRISM SUNSHINE'
Annual The most reliable, large-flowered yellow petunia.
↕ 38cm (15in)

PLUMBAGO AURICULATA
♀ **Scrambling shrub, not hardy** page **407**

ROSA 'ANNA FORD'
♀ **Patio rose** page **456**

ROSA 'GENTLE TOUCH'
♀ **Patio rose** page **456**

ROSA 'QUEEN MOTHER'
♀ **Patio rose** page **457**

ROSA 'SWEET MAGIC'
♀ **Patio rose** page **457**

ROSA 'THE FAIRY'
♀ **Patio rose** page **457**

SALVIA PATENS 'CAMBRIDGE BLUE'
♀ **Perennial, not hardy** page **477**

SCAEVOLA AEMULA 'BLUE WONDER'
Perennial, not hardy Blue, fan-shaped flowers on a sprawling, vigorous bush.
↕ 15cm (6in) ↔ 1.5m (5ft)

TROPAEOLUM MAJUS 'HERMINE GRASHOFF'
♀ **Climber, not hardy** page **520**

VERBENA 'LAWRENCE JOHNSTON'
♀ **Perennial, not hardy** Bright green foliage with intense, fiery red flowers.
↕ 45cm (18in) ↔ 60cm (24in)

VERBENA 'SISSINGHURST'
Annual Spreading plant with fine leaves and magenta-pink flowers.
↕ to 20cm (8in) ↔ to 1m (3ft)

CONTAINER PLANTS FOR AUTUMN

Autumn is a season of great change, as the leaves of deciduous plants become brilliant before they fall. Most annuals have finished their display but some plants, especially tender perennials like fuchsias, dwarf chrysanthemums and dahlias, continue to provide splashes of colourful flowers. Evergreens form a strong and solid background to all garden displays.

ARGYRANTHEMUM 'JAMAICA PRIMROSE'
♀ **Perennial, not hardy** page 67

CABBAGE 'NORTHERN LIGHTS'
Biennial Frilled ornamental cabbage in shades of white, pink and red.
↕↔ 30cm (12in)

CANNA 'LUCIFER'
Tender perennial Dwarf cultivar with green leaves and red, yellow-edged flowers.
↕ 60cm (24in) ↔ 50cm (20in)

CHAMAECYPARIS LAWSONIANA 'ELLWOOD'S GOLD'
♀ **Conifer** page 117

CLEMATIS 'MADAME JULIA CORREVON'
♀ **Climber** page 133

ESCALLONIA LAEVIS 'GOLD BRIAN'
Evergreen shrub Pink flowers held against bright yellow leaves.
↕↔ 1m (3ft)

GAULTHERIA MUCRONATA 'CRIMSONIA'
♀ **Evergreen shrub** Acid-loving fine-leaved shrub with showy, deep pink berries.
↕↔ 1.2m (4ft)

HIBISCUS SYRIACUS 'WOODBRIDGE'
♀ **Shrub** page 257

HYPERICUM X *MOSERIANUM* 'TRICOLOR'
Shrub Variegated narrow leaves, flushed pink, and yellow flowers.
↕ 30cm (12in) ↔ 60cm (24in)

JUNIPERUS COMMUNIS 'COMPRESSA'
♀ **Conifer** page 293

MYRTUS COMMUNIS
♀ **Evergreen shrub** page 344

NANDINA DOMESTICA 'FIREPOWER'
Shrub Compact plant with red autumn colour and berries.
↕ 45cm (18in) ↔ 60cm (24in)

OSMANTHUS HETEROPHYLLUS 'VARIEGATUS'
♀ **Evergreen shrub** Holly-like cream-edged leaves and fragrant, white flowers in autumn.
↕↔ 5m (15ft)

RUDBECKIA HIRTA 'TOTO'
Annual Orange flowers with black centres on very compact plants.
↕↔ 20cm (8in)

THUJA ORIENTALIS 'AUREA NANA'
♀ **Conifer** page 512

VIBURNUM TINUS 'VARIEGATUM'
Evergreen shrub The leaves are broadly margined with creamy yellow.
↕↔ 3m (10ft)

CONTAINER PLANTS FOR WINTER

Few plants flower during cold winter months; use those that do in containers by doors and windows, where they can be seen from the comfort of the home. Many evergreens, especially variegated ones, look cheery when other plants are bare. Plant shrubs out in the garden after a few years in pots. To protect roots from freezing, wrap pots with bubble polythene in very cold spells.

AUCUBA JAPONICA 'CROTONIFOLIA'
♈ **Evergreen shrub** page **78**

CYCLAMEN COUM
♈ **Bulb** Deep green heart-shaped leaves and pink flowers.
↕5–8cm (2–3in) ↔ 10cm (4in)

ERICA CARNEA 'SPRINGWOOD WHITE'
♈ **Evergreen shrub** page **187**

ERICA X *DARLEYENSIS* 'ARTHUR JOHNSON'
♈ **Evergreen shrub** The upright stems have deep green leaves and bear pink flowers in winter.
↕75cm (30in) ↔ 60cm (42in)

EUONYMUS FORTUNEI 'HARLEQUIN'
Evergreen shrub Compact shrub, young leaves heavily splashed with white.
↕↔ 40cm (16in)

FATSIA JAPONICA 'VARIEGATA'
♈ **Evergreen shrub** Large, lobed leaves, edged with white; creamy-white flowers on mature plants.
↕↔ 2m (6ft)

FESTUCA GLAUCA 'BLAUFUCHS'
♈ **Ornamental grass** page **207**

GAULTHERIA MUCRONATA 'MULBERRY WINE'
♈ **Evergreen shrub** page **220**

HEDERA HELIX 'EVA'
♈ **Evergreen climber** Small grey-green leaves edged white.
↕1.2m (4ft)

ILEX AQUIFOLIUM 'FEROX ARGENTEA'
♈ **Evergreen shrub** page **273**

JUNIPERUS COMMUNIS 'HIBERNICA'
♈ **Conifer** Compact, upright plant with grey-green needles.
↕3–5m (10–15ft) ↔ 30cm (12in)

LAMIUM MACULATUM 'BEACON SILVER'
Perennial Bright silver-grey foliage and magenta-pink flowers.
↕20cm (8in) ↔ 1m (3ft)

LAURUS NOBILIS
♈ **Evergreen shrub or small tree** page **304**

LIRIOPE MUSCARI 'JOHN BURCH'
Perennial Grassy plant with gold-striped leaves and violet flowers.
↕30cm (12in) ↔ 45cm (18in)

SKIMMIA JAPONICA 'RUBELLA'
♈ **Evergreen shrub** page **495**

VIBURNUM DAVIDII
♈ **Evergreen shrub** page **535**

VIOLA 'FLORAL DANCE'
Biennial Winter pansies in mixed colours.
↕15cm (6in) ↔ to 30cm (12in)

VIOLA 'MELLO 21'
Biennial Pansy mixture of 21 colours.
↕15cm (6in) ↔ to 30cm (12in)

YUCCA GLORIOSA 'VARIEGATA'
♈ **Evergreen shrub** Erect shrub with sharp, pointed leaves edged with yellow.
↕↔ 2m (6ft)

CONTAINERS IN SUNNY SITES

A sunny position where pots and tubs can be placed allows the cultivation of a huge range of plants. Apart from most bedding, many tender plants that must spend cold winters under glass should thrive in a sheltered sunny position and bring a touch of the exotic. Specimen plants are best grown in their own containers; simply group them together to create masterful associations.

ALOE VARIEGATA
♀ **Succulent, not hardy** page **53**

BRACHYCOME 'STRAWBERRY MIST'
Perennial, not hardy Low, spreading plant with feathery foliage and pink daisy-like flowers. Sometimes sold as *Brachyscome.*
↕25cm (10in) ↔ 45cm (18in)

CORDYLINE AUSTRALIS 'ALBERTII'
♀ **Evergreen shrub, not hardy** page **137**

CORREA BACKHOUSEANA
♀ **Evergreen shrub, not hardy** page **140**

ERICA ERIGENA 'IRISH DUSK'
♀ **Evergreen shrub** Deep pink flowers are produced from autumn to spring on plants with greyish-green leaves.
↕60cm (24in) ↔ 45cm (18in)

FELICIA AMELLOIDES 'SANTA ANITA'
♀ **Subshrub, not hardy** page **206**

FUSCHIA 'MADAME CORNÉLLISSEN'
♀ **Shrub** Upright, with masses of red and white single flowers.
↕↔ 60cm (24in)

GAZANIA 'DAYBREAK RED STRIPE'
Perennial, not hardy Raise from seed for showy yellow blooms striped with red.
↕ to 20cm (8in) ↔ to 25cm (10in)

GERANIUM 'ANN FOLKARD'
♀ **Perennial** page **226**

GERANIUM CINEREUM VAR. *SUBCAULESCENS*
♀ **Perennial** page **227**

GERANIUM PALMATUM
♀ **Perennial, not hardy** Bold clumps of foliage with large mauve flowers in summer.
↕↔ to 1.2m (4ft)

HEBE ALBICANS
♀ **Evergreen shrub** page **240**

HEBE 'MRS WINDER'
♀ **Evergreen shrub** Dark foliage, flushed purple, and violet flowers in late summer.
↕1m (3ft) ↔ 1.2m (4ft)

HOSTA 'KROSSA REGAL'
♀ **Perennial** A magnificent cultivar with upright grey leaves and tall spikes of lilac flowers.
↕70cm (28in) ↔ 75cm (30in)

IMPATIENS SUPER ELFIN SERIES
♀ **Annuals** page **275**

IPOMOEA 'HEAVENLY BLUE'
♀ **Annual climber** page **277**

JASMINUM MESNYI
♀ **Climber, not fully hardy** page **291**

LATHYRUS ODORATUS 'PATIO MIXED'
♀ **Annual climber** page **303**

LEPTOSPERMUM SCOPARIUM 'RED DAMASK'
♀ **Evergreen shrub** Rather tender plant with tiny leaves and double, deep pink flowers.
↕↔ 3m (10ft)

LOBELIA ERINUS 'KATHLEEN MALLARD'
Perennial, not hardy Neat, double-flowered blue lobelia.
↕10cm (4in) ↔ 30cm (12in)

MIMULUS AURANTIACUS
🏆 **Evergreen shrub, not hardy** page 341

MYRTUS COMMUNIS SUBSP. *TARENTINA*
🏆 **Evergreen shrub, not hardy** page 345

OSTEOSPERMUM 'SILVER SPARKLER'
🏆 **Perennial, not hardy** White daisy-like flowers and bright, variegated leaves.
↕↔ 45cm (18in)

PASSIFLORA 'AMETHYST'
🏆 **Climber, not hardy** Fast-growing climber, produces beautiful lavender flowers all summer.
↕ 4m (12ft)

PELARGONIUM 'CLORINDA'
Perennial, not hardy Rough, scented foliage and quite large, bright pink flowers.
↕ 45–50cm (18–20in) ↔ 20–25cm (8–10in)

PELARGONIUM 'DEACON MOONLIGHT'
Perennial, not hardy Pale lilac double flowers above neat foliage on compact plants.
↕ 20cm (8in) ↔ 25cm (10in)

PELARGONIUM 'DOLLY VARDEN'
Perennial, not hardy page 375

PELARGONIUM 'VISTA DEEP ROSE'
Perennial, not hardy Single colour in this seed-raised F2 series.
↕↔ 30cm (12in)

PENSTEMON 'APPLE BLOSSOM'
🏆 **Perennial, not hardy** page 378

PENSTEMON 'OSPREY'
🏆 **Perennial** Spikes of white flowers edged with pink on an upright plant.
↕↔ 45cm (18in)

PETUNIA 'DUO PEPPERMINT'
🏆 **Annual** Double, pink flowers, quite weather-resistant in poor summers.
↕ 20cm (8in) ↔ 30cm (12in)

PHORMIUM 'SUNDOWNER'
🏆 **Perennial** page 393

REHMANNIA GLUTINOSA
🏆 **Perennial** Slightly sticky leaves on tall stems that carry dusky pink, foxglove-like flowers.
↕ 15–30cm (6–12in) ↔ to 30cm (12in)

RHODOCHITON ATROSANGUINEUS
🏆 **Evergreen climber, not hardy** page 435

SEMPERVIVUM ARACHNOIDEUM
🏆 **Succulent** page 492

SOLENOPSIS AXILLARIS
Perennial, not hardy Dome-shaped, feathery, with delicate but showy star-shaped, blue flowers.
↕↔ 30cm (12in)

THYMUS × *CITRIODORUS* 'AUREUS'
🏆 **Evergreen shrub** Low-growing bush, scented of lemon, with gold leaves and pink flowers.
↕ 30cm (12in) ↔ to 25cm (10in)

TROPAEOLUM MAJUS 'MARGARET LONG'
Perennial, not hardy Compact, trailing plant with double soft orange flowers.
↕↔ 30cm (12in)

TWEEDIA CAERULEA
🏆 **Perennial, not hardy** A straggly twining plant with grey leaves and starry sky-blue flowers.
↕ 60cm–1m (24–36in)

CONTAINERS IN SHADE

A shady site can be a difficult place to grow plants – the soil is often dry – but a wide range of interesting plants can be grown in containers. Remember that rain may not reach plants under trees; regular watering is needed all the year, plus feeding in summer. Add bulbs for spring colour and flowering plants in summer. Combine plants for form and texture as well as colour.

ACER PALMATUM 'CRIMSON QUEEN'
♀ **Shrub** Arching shoots with finely toothed purple leaves.
↕ 3m (10ft) ↔ 4m (12ft)

AGERATUM 'SOUTHERN CROSS'
Annual Neat, bushy plants with white and blue fluffy flowers.
↕↔ 25cm (10in)

AJUGA REPTANS 'RAINBOW'
Perennial Mats of bronze leaves spotted with pink and yellow, and blue flowers.
↕ 15cm (6in) ↔ 45cm (18in)

BEGONIA NONSTOP SERIES
♀ **Perennials, not hardy** Compact plants with double flowers in many colours.
↕↔ 30cm (12in)

CAMELLIA JAPONICA 'LAVINIA MAGGI'
♀ **Evergreen shrub** page **101**

DRYOPTERIS ERYTHROSORA
♀ **Hardy fern** Colourful fern with copper-red young fronds.
↕ 60cm (24in) ↔ 38cm (15in)

DRYOPTERIS WALLICHIANA
♀ **Hardy fern** page **180**

x *FATSHEDERA LIZEI* 'ANNEMIEKE'
♀ **Evergreen shrub** Procumbent stems and leaves with bright gold centres. Sometimes sold as 'Lemon and Lime'.
↕ 1.2–2m (4–6ft) or more ↔ 3m (10ft)

FATSIA JAPONICA
♀ **Evergreen shrub** page **206**

FUCHSIA 'DOLLAR PRINCESS'
♀ **Shrub, not hardy** A robust plant with small, red and purple, double flowers.
↕ 30–45cm (12–18in) ↔ 45–60cm (18–24in)

HEDERA HELIX 'IVALACE'
♀ **Evergreen climber** page **246**

HEDERA HELIX 'MANDA'S CRESTED'
♀ **Evergreen climber** Uneven, curled and twisted, fingered leaves, coppery in winter.
↕ 1.2m (4ft)

HOSTA 'ROYAL STANDARD'
♀ **Perennial** page **260**

ILEX CRENATA 'CONVEXA'
♀ **Evergreen shrub** page **274**

IMPATIENS 'BLACKBERRY ICE'
Annual Double purple flowers and white-variegated foliage.
↕ to 70cm (28in)

IMPATIENS SUPER ELFIN SERIES
♀ **Annuals** page **275**

IPHEION UNIFLORUM 'WISLEY BLUE'
♀ **Bulbous perennial** page **276**

PIERIS JAPONICA 'LITTLE HEATH'
♀ **Evergreen shrub** Compact, acid soil-loving, with white-edged leaves flushed pink in spring.
↕↔ 60cm (24in)

RHODODENDRON 'VUYK'S SCARLET'
♀ **Evergreen shrub** page **437**

VIOLA 'IMPERIAL ANTIQUE SHADES'
Biennial Large flowers in pink, cream and parchment colours.
↕ 15cm (6in) ↔ 25cm (8in)

FOLIAGE FOR HANGING BASKETS

Although flowering plants are the most popular choice for hanging baskets, they can often be enhanced with attractive foliage plants. Silver-leaved plants are the most widely grown foil for flowers, but there are trailing plants that have gold, green and red foliage, and it is not difficult to plant a beautiful basket using solely plants chosen for their colourful and contrasting leaves.

AJUGA REPTANS 'BURGUNDY GLOW'
♀ **Perennial** page **49**

ASPARAGUS DENSIFLORUS 'MYERSII'
♀ **Perennial** page **70**

BEGONIAS, FOLIAGE
♀ **Perennials, not hardy** pages **80–81**

CHLOROPHYTUM COMOSUM 'VITTATUM'
♀ **Perennial, not fully hardy** Variegated narrow leaves; arching stems bearing plantlets.
↕ 15–20cm (6–8in) ↔ 15–30cm (6–12in)

GLECHOMA HEDERACEA 'VARIEGATA'
Perennial Long, pendent stems with grey-green leaves edged with white.
↔ trailing to 2m (6ft)

HEDERA HELIX 'GOLDCHILD'
♀ **Evergreen climber** page **246**

HELICHRYSUM PETIOLARE
♀ **Evergreen shrub, not hardy** page **250**

HELICHRYSUM PETIOLARE 'LIMELIGHT'
♀ **Shrub, not hardy** Arching stems with felted pale green leaves, yellow in sun.
↕ 1m (3ft) ↔ indefinite

HELICHRYSUM PETIOLARE 'VARIEGATUM'
♀ **Evergreen shrub, not hardy** page **251**

LAMIUM MACULATUM 'AUREUM'
Perennial Creeping, with bright yellow leaves marked white, and pink flowers.
↕ 20cm (8in) ↔ 1m (3ft)

LOTUS BERTHELOTII
♀ **Subshrub, not hardy** page **326**

LYSIMACHIA CONGESTIFLORA 'OUTBACK SUNSET'
Perennial, not hardy Semi-trailing, with green, cream and bronze leaves and clusters of yellow flowers.
↕ 15cm (6in) ↔ 30cm (12in)

LYSIMACHIA NUMMULARIA 'AUREA'
♀ **Perennial** page **329**

PELARGONIUM 'SWANLAND LACE'
Perennial, not hardy Trailing plant with pink flowers and leaves veined with yellow.
↕↔ 30cm (12in)

PLECTRANTHUS FORSTERI 'MARGINATUS'
Perennial, not hardy Scented foliage, edged with white. New growth arches strongly.
↕ 25cm (10in) ↔ 1m (3ft)

SAXIFRAGA STOLONIFERA 'TRICOLOR'
♀ **Perennial** Round leaves edged white and flushed pink; long strings of plantlets.
↕↔ to 30cm (12in)

SELAGINELLA KRAUSSIANA
♀ **Perennial, not hardy** page **491**

TROPAEOLUM ALASKA SERIES
♀ **Annuals** Easily raised, bushy nasturtiums with white-splashed leaves.
↕ to 30cm (12in) ↔ to 45cm (18in)

FLOWERS FOR HANGING BASKETS

Most gardens, and houses or flats without gardens, can be brightened by a few hanging baskets. They are easy to plant, and there are lots of cheerful plants to choose from, all with a long flowering season and a low or trailing habit. Baskets require regular watering and feeding to grow and flower well all summer. Automatic irrigation systems take the worry out of holiday watering.

ABUTILON MEGAPOTAMICUM
♀ **Shrub, not hardy** page **30**

ACALYPHA REPTANS
Perennial, not hardy Spreading plant with short red-hot cats'-tail flowers.
↕15cm (6in) ↔ 60cm (24in)

ANAGALLIS MONELLII
♀ **Perennial** Spreading plant with small leaves and bright, intense, deep blue flowers.
↕10–20cm (4–8in) ↔ 40cm (16in)

ANTIRRHINUM 'CANDELABRA LEMON BLUSH'
Perennial, not hardy Bushy habit with trailing flower stems carrying pale yellow flowers, flushed with pink.
↕↔ 38cm (15in)

BEGONIA FUCHSIOIDES
♀ **Perennial, not hardy** Arching stems set with tiny, glossy leaves bear pendent pink or red flowers.
↕75cm (30in) ↔ 45cm (18in)

BEGONIA 'ILLUMINATION ORANGE'
♀ **Perennial, not hardy** page **82**

BEGONIA 'IRENE NUSS'
♀ **Perennial, not hardy** page **83**

BEGONIA SUTHERLANDII
♀ **Perennial, not hardy** page **83**

BIDENS FERULIFOLIA
♀ **Perennial, not hardy** page **90**

CONVOLVULUS SABATIUS
♀ **Perennial, not hardy** page **136**

DIASCIA 'LILAC BELLE'
♀ **Perennial** Loose spikes of mauve-pink flowers on low-growing plants.
↕20cm (8in) ↔ 30cm (12in)

FUCHSIA 'GOLDEN MARINKA'
♀ **Shrub, not hardy** page **216**

FUCHSIA 'JACK SHAHAN'
♀ **Shrub, not hardy** page **216**

FUCHSIA 'LA CAMPANELLA'
♀ **Shrub, not hardy** page **216**

FUCHSIA 'LENA'
♀ **Shrub, not hardy** page **216**

IMPATIENS TEMPO SERIES
♀ **Annuals** page **275**

LOBELIA 'COLOUR CASCADE'
♀ **Annual** Trailing stems with flowers in shades of blue, pink and white.
↕15cm (6in) ↔ 45cm (18in)

LOBELIA RICHARDSONII
♀ **Perennial, not hardy** Bushy but rather sparse, trailing stems with small leaves and blue flowers.
↕15cm (6in) ↔ 30cm (12in)

MIMULUS AURANTIACUS
♀ **Evergreen shrub, not hardy** page **341**

PAROCHETUS COMMUNIS
♀ **Perennial, not hardy** A fast-growing trailer with clover-like leaves and pale blue, pea-like flowers.
↕10cm (4in) ↔ 30cm (12in)

PELARGONIUM 'AMETHYST'
♀ **Perennial, not hardy** page **374**

PELARGONIUM 'LILA MINI CASCADE'
Perennial, not hardy Single, pink flowers on a compact, but vigorous plant.
↕ 45–50cm (18–20in) ↔ 15–20cm (6–8in)

PELARGONIUM 'MADAME CROUSSE'
♀ **Perennial, not hardy** Trailing plant with semi-double, pink flowers.
↕ 50–60cm (20–24in) ↔ 15–20cm (6–8in)

PELARGONIUM 'OLDBURY CASCADE'
Perennial, not hardy Rich, red flowers on a compact plant with cream-variegated leaves.
↕ 45cm (18in) ↔ 30cm (12in)

PELARGONIUM 'ROULETTA'
Perennial, not hardy Double white flowers, heavily margined with red.
↕ 50–60cm (20–24in) ↔ 15–20cm (6–8in)

PELARGONIUM 'THE BOAR'
♀ **Perennial, not hardy** page **376**

PELARGONIUM 'VILLE DE PARIS'
Perennial, not hardy Single, pink flowers produced in profusion on a pendulous plant.
↕ 60cm (24in) ↔ 45cm (18in)

PELARGONIUM 'YALE'
♀ **Perennial, not hardy** The trailing stems carry clusters of semi-double bright red flowers.
↕ 20–25cm (8–10in) ↔ 15–20cm (6–8in)

PETUNIA DADDY SERIES
Annuals Large flowers, heavily veined with darker shades.
↕ to 35cm (14in) ↔ 30–90cm (12–36in)

PETUNIA 'MARCO POLO ADVENTURER'
Annual Large, double flowers of bright rose pink on trailing plants.
↕↔ 38cm (15in)

PETUNIA 'SURFINIA PASTEL PINK'
Annual Large flowers of mid-pink on a strong plant that trails well.
↕ 23–40cm (9–16in) ↔ 30–90cm (12–36in)

SUTERA CORDATA 'KNYSNA HILLS'
Perennial, not hardy Bushy plant with massed heads of tiny, pale pink flowers.
↕ 20cm (8in) ↔ 60cm (24in)

SUTERA CORDATA 'SNOWFLAKE'
Perennial, not hardy Spreading, small-leafed plant with tiny, five-petalled white flowers.
↕ 10cm (4in) ↔ 60cm (24in)

VERBENA 'IMAGINATION'
♀ **Perennial, not hardy** Seed-raised, loosely-branched plants with small, deep violet flowers.
↕↔ 38cm (15in)

VERBENA 'TAPIEN PINK'
Perennial, not hardy Trailing stems with clusters of small pink flowers.
↕↔ 38cm (15in)

VIOLA 'SUNBEAM'
Biennial Semi-trailing plant with small yellow flowers.
↕↔ 30cm (12in)

SPRING-FLOWERING BULBS

The first signs that winter is giving way to spring are usually the flowers of spring bulbs. From tiny winter aconites that open as the leaves push through the soil to majestic crown imperials, there are bulbs for every situation. Most are tolerant of a wide range of soils and many can be grown in the shade of trees. Grown in pots, they can be brought into the home to be enjoyed.

ALLIUM CRISTOPHII
♀ **Borderline hardy** page **51**

ANEMONE BLANDA
'WHITE SPLENDOUR'
♀ **Hardy** page **58**

CAMASSIA CUISICKII
Hardy Sheaves of narrow leaves and tall racemes of blue star-like blooms.
↕ 60–80cm (24–32in) ↔ 10cm (4in)

CHIONODOXA LUCILIAE
♀ **Hardy** page **120**

CHIONODOXA SARDENSIS
♀ **Hardy** Deep blue starry flowers on slender stems.
↕ 10–20cm (4–8in) ↔ 3cm (1¼in)

CORYDALIS FLEXUOSA
Hardy Clumps of delicate foliage and nodding bright blue flowers.
↕ 15–30cm (6–12in) ↔ 20cm (8in)

CORYDALIS SOLIDA 'GEORGE BAKER'
♀ **Hardy** page **142**

CROCUS ANGUSTIFOLIUS
♀ **Hardy** Clusters of orange-yellow flowers in spring.
↕↔ 5cm (2in)

CROCUS CHRYSANTHUS
'BLUE PEARL'
♀ **Hardy** Lilac-blue flowers with white and yellow centres.
↕ 8cm (3in) ↔ 4cm (1½in)

CROCUS CHRYSANTHUS
'E.A. BOWLES'
♀ **Hardy** page **153**

CROCUS SIEBERI 'HUBERT EDELSTEIN'
♀ **Hardy** page **153**

CROCUS SIEBERI 'TRICOLOR'
♀ **Hardy** page **153**

CROCUS TOMMASINIANUS
♀ **Hardy** An early-flowering crocus with slender, silvery-lilac flowers.
↕ 8–10cm (3–4in) ↔ 2.5cm (1in)

ERANTHIS HYEMALIS
♀ **Hardy** page **185**

ERYTHRONIUM CALIFORNICUM
'WHITE BEAUTY'
♀ **Hardy** The recurved, white flowers are held above mottled leaves.
↕ 15–35cm (6–14in) ↔ 10cm (4in)

ERYTHRONIUM 'PAGODA'
♀ **Hardy** page **194**

FRITILLARIA ACMOPETALA
♀ **Hardy** page **210**

FRITILLARIA IMPERIALIS
'AUREOMARGINATA'
Hardy Tall stems carrying heads of orange flowers, with yellow-margined leaves.
↕ 1.5m (5ft) ↔ 25–30cm (10–12in)

FRITILLARIA PALLIDIFLORA
♀ **Hardy** page **211**

GALANTHUS NIVALIS
♀ **Hardy** The common, but easily-grown snowdrop, best planted while in leaf.
↕↔ 10cm (4in)

HERMODACTYLUS TUBEROSUS
Hardy Straggly leaves almost hide the velvety green and black iris-like flowers.
↕20–40cm (8–16in) ↔ 5cm (2in)

HYACINTHUS ORIENTALIS 'BLUE JACKET'
♀ **Hardy** page **262**

HYACINTHUS ORIENTALIS 'ANNA MARIE'
♀ **Hardy** Spikes of scented, pale pink flowers.
↕20cm (8in) ↔ 8cm (3in)

IPHEION UNIFLORUM 'FROYLE MILL'
♀ **Hardy** Clump-forming plant with onion-scented leaves and violet, star-shaped flowers.
↕15–20cm (6–8in)

IRIS BUCHARICA
♀ **Hardy** page **280**

IRIS RETICULATA 'CANTAB'
Hardy Pale blue flowers with deeper falls. Long leaves develop after flowering.
↕10–15cm (4–6in)

MUSCARI LATIFOLIUM
Hardy Broad leaves and bicoloured spikes of flowers in pale and dark blue.
↕20cm (8in) ↔ 5cm (2in)

NARCISSUS 'ACTAEA'
♀ **Hardy** page **348**

NARCISSUS BULBOCODIUM
♀ **Hardy** page **346**

NARCISSUS 'FEBRUARY GOLD'
♀ **Hardy** page **346**

NARCISSUS 'HAWERA'
♀ **Hardy** page **347**

NARCISSUS 'ICE FOLLIES'
♀ **Hardy** page **349**

NARCISSUS 'LITTLE WITCH'
Hardy Golden yellow flowers with reflexed petals.
↕22cm (9in)

NARCISSUS MINOR
♀ **Hardy** page **347**

NARCISSUS 'PASSIONALE'
♀ **Hardy** page **349**

NARCISSUS TRIANDRUS
♀ **Hardy** page **347**

SCILLA BIFOLIA
♀ **Hardy** page **487**

SCILLA MISCHTSCHENKOANA 'TUBERGENIANA'
♀ **Hardy** page **487**

SCILLA SIBERICA
♀ **Hardy** Bell-shaped flowers of cobalt blue above bright green leaves.
↕10–20cm (4–8in) ↔ 5cm (2in)

TULIPA 'APELDOORN'S ELITE'
♀ **Hardy** Rich yellow flowers feathered with red to give a subtle effect.
↕60cm (24in)

TULIPA 'BALLADE'
♀ **Hardy** Elegant flowers of rich pink with white petal edges and tips.
↕50cm (20in)

TULIPA 'CHINA PINK'
♀ **Hardy** page **524**

TULIPA 'MAUREEN'
♀ **Hardy** Oval flowers of ivory white.
↕50cm (20in)

TULIPA 'MRS JOHN T. SCHEEPERS'
♀ **Hardy** Large flowers of crisp, pale yellow.
↕60cm (24in)

TULIPA 'RED SURPRISE'
♀ **Hardy** Bright red flowers that open to produce starry blooms.
↕20cm (8in)

SUMMER-FLOWERING BULBS

Summer-flowering bulbs bring sparkle to gardens with their bright colours and exotic shapes. Unfortunately, many are not hardy in frost-prone climates, but they are not expensive, and most can be lifted in autumn and kept in a frost-free shed. Dahlias and gladioli are familiar to everyone, but modern lilies are easy to grow too, and tigridias and hedychiums are even more exotic.

ALLIUM GIGANTEUM
♀ **Hardy** page **51**

ALLIUM HOLLANDICUM 'PURPLE SENSATION'
♀ **Hardy** page **51**

ALLIUM KARATAVIENSE
♀ **Not hardy** page **52**

BLETILLA STRIATA
Not hardy Slightly tender terrestrial orchid: a leafy plant with spikes of 1–6 small, pink flowers.
↕↔ 30–60cm (12–24in)

CRINUM x *POWELLII*
♀ **Hardy** Huge bulbs produce large leaves and long pink trumpet flowers.
↕ 1.5m (5ft) ↔ 30cm (12in)

CROCOSMIA 'EMILY MCKENZIE'
Hardy Sword-shaped leaves and bright orange flowers marked with bronze.
↕ 60cm (24in) ↔ 8cm (3in)

DAHLIA 'BISHOP OF LLANDAFF'
♀ **Cormous perennial, not hardy** page **162**

DAHLIA 'GLORIE VAN HEEMSTEDE'
♀ **Hardy cormous perennial** Medium-sized bright yellow flowers shaped like waterlilies.
↕ 1.3m (4½ft) ↔ 60cm (2ft)

DAHLIA 'JESCOT JULIE'
Cormous perennial, not hardy Long-petalled flowers of orange and red.
↕ 1m (36in) ↔ 45cm (18in)

DAHLIA 'PEARL OF HEEMSTEDE'
♀ **Hardy** This waterlily dahlia has flowers of silvery pink.
↕ 1m (3ft) ↔ 45cm (18in)

DRACUNCULUS VULGARIS
Hardy Spotted stems support interesting leaves and huge, purple, unpleasantly scented spathes.
↕ to 1.5m (5ft) ↔ 60cm (24in)

EREMURUS 'CLEOPATRA'
Hardy Rosettes of straplike leaves and tall spikes of small soft orange flowers.
↕ 1.5m (5ft) ↔ 60cm (24in)

EUCOMIS BICOLOR
Not hardy Rosettes of broad leaves below spikes of starry, cream, purple-edged blooms.
↕ 30–60cm (12–24in) ↔ 20cm (8in)

GLADIOLUS CALLIANTHUS
♀ **Not hardy** page **231**

GLADIOLUS COMMUNIS SUBSP. *BYZANTINUS*
♀ **Hardy** page **232**

GLADIOLUS 'GREEN WOODPECKER'
♀ **Not hardy** Greenish-yellow flowers with red markings.
↕ 1.5m (5ft) ↔ 12cm (5in)

HEDYCHIUM COCCINEUM
♀ **Hardy** An exotic-looking plant with spikes of tubular, scented, orange, pink or cream flowers.
↕ 3m (10ft) ↔ 1m (3ft)

HEDYCHIUM GARDNERIANUM
♀ **Perennial, not hardy** Large heads of spidery, showy, sweetly scented cream flowers.
↕ 2–2.2m (6–7ft)

HEDYCHIUM 'TARA'
♀ **Hardy** A selection of *H. coccineum* with deeper-coloured flowers.
↕ 3m (10ft) ↔ 1m (3ft)

IXIA VIRIDIFLORA
Not hardy Unusual pale turquoise-green flowers on tall, wiry stems.
↕ 30–60cm (12–24in)

LEUCOJUM AESTIVUM 'GRAVETYE GIANT'
♀ **Hardy** page **311**

LILIUM 'BLACK DRAGON'
♀ **Hardy** Tall stems carry pure white, scented flowers that are deep maroon in bud.
↕ 1.5m (5ft)

LILIUM 'EVEREST'
Hardy Large white flowers, heavily scented, with dark spots, on tall stems.
↕ 1.5m (5ft)

LILIUM HENRYI
♀ **Hardy** page **315**

LILIUM 'KAREN NORTH'
♀ **Hardy** Elegant flowers with reflexed petals of orange-pink with deeper spots.
↕ 1–1.3m (3–4½ft)

LILIUM MARTAGON VAR. *ALBUM*
♀ **Hardy** page **316**

LILIUM PYRENAICUM
♀ **Hardy** page **318**

LILIUM REGALE
♀ **Hardy** page **318**

NECTAROSCORDUM SICULUM
Hardy Plants smell strongly of garlic and produce loose umbels of pendulous cream flowers.
↕ to 1.2m (4ft) ↔ 10cm (4in)

RHODOHYPOXIS BAURII 'TETRA RED'
Not hardy Forms clumps or mats of bright, squat, starry flowers.
↕↔ 10cm (4in)

TIGRIDIA PAVONIA
Not hardy Broad, spotted flowers in shades of red, yellow and white, each lasting a single day.
↕ 1.5m (5ft) ↔ 10cm (4in)

TROPAEOLUM POLYPHYLLUM
Hardy Trailing stems with finely-divided grey leaves, and bright yellow flowers.
↕ 5–8cm (2–3in) ↔ to 1m (3ft)

ZANTEDESCHIA AETHIOPICA 'GREEN GODDESS'
♀ **Hardy** page **545**

AUTUMN-FLOWERING BULBS

Hardy bulbs and corms can provide colour and interest when most plants are dying down. Most require full sun and well-drained soil, if their flowers are to benefit from the last fine days of the year. Many do not fit easily with bulb production cycles, and do not flower well until they are established in the garden, so they may need some effort to find in nurseries.

ALLIUM CALLIMISCHON
Hardy Slender stems with white or pale pink flowers for dry soils.
↕8–35cm (4–14in) ↔ 5cm (2in)

x *AMARYGIA PARKERI*
Hardy This rare hybrid produces large pink trumpet flowers in mild gardens.
↕1m (3ft) ↔ 30cm (12in)

AMARYLLIS BELLADONNA
Hardy In sheltered sites the dark stems support beautiful pink trumpets.
↕60cm (24in) ↔ 10cm (4in)

COLCHICUM SPECIOSUM 'ALBUM'
🏆 **Hardy** page **135**

COLCHICUM 'WATERLILY'
🏆 **Hardy** Beautiful white goblets that appear without leaves.
↕18cm (7in) ↔ 10cm (4in)

CRINUM x *POWELLII* 'ALBUM'
🏆 **Hardy** page **151**

CROCUS BANATICUS
🏆 **Hardy** page **154**

CROCUS BORYI
🏆 **Hardy** Each corm produces up to four pale lilac and cream flowers.
↕8cm (3in) ↔ 5cm (2in)

CROCUS GOULIMYI
🏆 **Hardy** page **154**

CROCUS KOTSCHYANUS
🏆 **Hardy** page **154**

CROCUS MEDIUS
🏆 **Hardy** page **154**

CROCUS OCHROLEUCUS
🏆 **Hardy** page **154**

CROCUS PULCHELLUS
🏆 **Hardy** page **154**

CROCUS SPECIOSUS
🏆 **Hardy** Violet-blue, scented flowers with orange stigmas.
↕15cm (6in)

CYCLAMEN HEDERIFOLIUM
🏆 **Hardy** page **158**

LEUCOJUM AUTUMNALE
🏆 **Hardy** page **311**

MERENDERA MONTANA
Hardy Starry lilac flowers; needs well-drained soil.
↕↔ 5cm (2in)

NERINE BOWDENII
🏆 **Hardy** page **350**

STERNBERGIA LUTEA
Hardy Deep green leaves and bright yellow crocus-like flowers.
↕15cm (6in) ↔ 8cm (3in)

TROPAEOLUM TUBEROSUM 'KEN ASLET'
🏆 **Not hardy** Clambering stems with small leaves and yellow, hooded flowers.
↕2–4m (6–12ft)

ZEPHYRANTHES CANDIDA
Hardy Clumps of narrow leaves and white, crocus-like flowers over many weeks.
↕10–20cm (4–8in) ↔ 8cm (3in)

SPRING BEDDING

When the last bedding plants of the season die down, it is time to replace them with young plants that will survive the winter, then flower in spring. Most of these are biennials, such as Canterbury bells, sweet Williams and wallflowers. Add winter-flowering pansies for early colour, especially in containers. Plant spring bulbs among the bedding plants for extra interest.

BELLIS PERENNIS 'POMPONETTE'
♀ **Biennial** page **84**

CAMPANULA MEDIUM 'CALYCANTHEMA'
Biennial Cup-and-saucer Canterbury bells in pink, blue and white.
↕ to 75cm (30in)

DIANTHUS BARBATUS 'AURICULA-EYED MIXED'
♀ **Biennial** Traditional sweet Williams with flowers zoned in white, pink and maroon.
↕ to 60cm (24in)

ERYSIMUM CHEIRI 'CLOTH OF GOLD'
Biennial Bright yellow, scented flowers.
↕ 45cm (18in)

ERYSIMUM CHEIRI 'FIREKING IMPROVED'
Biennial Established wallflower with orange-red flowers.
↕ 45cm (18in)

ERYSIMUM CHEIRI 'PRINCE PRIMROSE YELLOW'
Biennial Dwarf type with large flowers.
↕ 30cm (12in)

HYACINTHUS ORIENTALIS 'FONDANT'
Bulb Flowers of clean, pure pink with no hint of blue.
↕ 25cm (10in)

HYACINTHUS ORIENTALIS 'OSTARA'
♀ **Bulb** page **263**

HYACINTHUS ORIENTALIS 'PINK PEARL'
♀ **Bulb** page **264**

HYACINTHUS ORIENTALIS 'VIOLET PEARL'
Bulb Amethyst-coloured flowers.
↕ 25cm (10in)

MYOSOTIS 'SPRING SYMPHONY BLUE'
Biennial The blue flowers associate well with most spring bulbs.
↕ 15cm (6in)

POLYANTHUS PRIMROSES
♀ **Perennials** page **416**

TULIPA 'HALCRO'
♀ **Bulb** Large flowers of deep salmon-red.
↕ 70cm (28in)

TULIPA 'ORANJE NASSAU'
♀ **Bulb** Fiery flowers in shades of red and scarlet.
↕ 30cm (12in)

TULIPA 'SPRING GREEN'
♀ **Bulb** page **525**

VIOLA 'FELIX'
Biennial Yellow and purple, "whiskered" flowers on tufted plants.
↕ 15cm (6in)

VIOLA 'RIPPLING WATERS'
Biennial Large flowers of dark purple edged with white are produced in spring.
↕ 15cm (6in)

VIOLA ULTIMA SERIES
♀ **Biennial** Wide range of colours on plants that flower through mild spells.
↕ 15cm (6in)

SUMMER BEDDING

Although bedding out on the grand scale will probably never be as popular as it was in Victorian times, most gardeners find room for plants that grow and flower quickly once planted out. Although most are not tolerant of frost, some are annuals and others are tender perennials. Most also have a long flowering period and, even if only for one season, are excellent value.

AGERATUM HOUSTONIANUM 'PACIFIC'
♀ **Annual** Compact plants with dense heads of purple-blue flowers.
↕ 20cm (8in)

ANTIRRHINUM SONNET SERIES
♀ **Annuals** page **62**

BEGONIA 'PIN UP'
♀ **Perennial, not hardy** page **83**

BEGONIA SEMPERFLORENS OLYMPIA SERIES
♀ **Perennials, not hardy** Compact, large-flowered bedding begonias.
↕↔ 20cm (8in)

DAHLIA 'COLTNESS GEM'
Perennial, not hardy Reliable, single-flowered dahlias in clear, bright colours.
↕↔ 45cm (18in)

DIASCIA RIGESCENS
♀ **Perennial, not hardy** page **174**

HELIOTROPIUM ARBORESCENS 'PRINCESS MARINA'
♀ **Perennial, not hardy** page **252**

IMPATIENS DECO SERIES
Annuals Large, bright flowers on plants with dark green leaves.
↕↔ to 20cm (8in)

LOBELIA 'COMPLIMENT SCARLET'
♀ **Perennial** Bold plant with green foliage and large scarlet flowers on tall spikes.
↕ 75cm (30in) ↔ 30cm (12in)

LOBELIA 'CRYSTAL PALACE'
♀ **Annual** page **322**

LOBELIA ERINUS 'MRS CLIBRAN'
♀ **Annual** A compact plant with white-eyed, bright blue flowers.
↕ 10–15cm (4–6in)

NICOTIANA 'LIME GREEN'
♀ **Annual** page **350**

NICOTIANA x *SANDERAE* 'DOMINO SALMON PINK'
Annual Upward-facing flowers of salmon-pink on compact plants.
↕ 30–45cm (12–18in)

NICOTIANA SYLVESTRIS
♀ **Perennial** page **351**

NIGELLA 'MISS JEKYLL'
♀ **Annual** page **351**

PELARGONIUM MULTIBLOOM SERIES
♀ **Perennials, not hardy** page **376**

PELARGONIUM VIDEO SERIES
♀ **Perennials, not hardy** page **376**

PENSTEMON 'BEECH PARK'
♀ **Perennial** Large pink and white trumpets throughout summer. Sometimes sold as 'Barbara Barker'.
↕ 75cm (30in) ↔ 45cm (18in)

PENSTEMON 'BURGUNDY'
Perennial Deep, wine-red flowers above dark green leaves.
↕ 90cm (36in) ↔ 45cm (18in)

PENSTEMON 'MAURICE GIBBS'
♀ **Perennial, not hardy** Large-leaved plants bear spikes of cerise flowers with white throats.
↕ 75cm (30in) ↔ 45cm (18in)

PENSTEMON 'MYDDELTON GEM'
Perennial Long-flowering plant with pale green leaves and deep pink, tubular flowers.
↕75cm (30in) ↔ 45cm (18in)

PETUNIA CARPET SERIES
♀ **Annuals** page **382**

PETUNIA 'MIRAGE REFLECTIONS'
♀ **Annual** Weather-resistant multiflora petunias in pastel colours with darker veins.
↕30cm (12in) ↔ 60cm (24in)

PHLOX DRUMMONDII 'TAPESTRY'
Annual Wide range of pastel colours and bicolors on bushy plants.
↕50cm (20in) ↔ 38cm (15in)

POLEMONIUM 'LAMBROOK MAUVE'
♀ **Perennial** page **408**

PORTULACA GRANDIFLORA SUNDIAL SERIES
Annuals Sun-loving plants with brilliant flowers, with petals like satin.
↕10cm (4in) ↔ 15cm (6in)

RUDBECKIA HIRTA 'RUSTIC DWARFS'
Annual Mixture of autumn shades on large, dark-eyed flowers.
↕to 60cm (24in)

SALPIGLOSSIS CASINO SERIES
♀ **Annuals** page **471**

SALVIA COCCINEA 'LADY IN RED'
♀ **Annual** Bushy plants bear slender spikes of small but showy red flowers all summer.
↕40cm (16in)

SALVIA FARINACEA 'VICTORIA'
♀ **Annual** The small, deep blue flowers are crowded on spikes held well above the leaves.
↕to 60cm (24in) ↔ to 30cm (12in)

SALVIA FULGENS
♀ **Subshrub** page **474**

SALVIA PRATENSIS HAEMATODES GROUP
♀ **Perennial, not hardy** page **478**

SALVIA SPLENDENS 'SCARLET KING'
♀ **Perennial, not hardy** page **478**

SALVIA SPLENDENS SIZZLER SERIES
Annuals, some ♀
Early-flowering plants with blooms in colours from red to lavender and white.
↕25–30cm (10–12in)

TAGETES 'DISCO ORANGE'
♀ **Annual** Weather-resistant, single flowers.
↕20–25cm (8–10in)

TAGETES 'SAFARI SCARLET'
♀ **Annual** Large, bright double flowers.
↕20–25cm (8–10in)

TAGETES 'ZENITH RED'
Annual Afro-French marigold with double red flowers.
↕30cm (12in)

VERBENA × *HYBRIDA* 'PEACHES AND CREAM'
Annual Seed-raised spreading plants with peach-coloured flowers that fade to cream.
↕30cm (12in) ↔ 45cm (18in)

VERBENA × *HYBRIDA* 'SILVER ANNE'
♀ **Perennial, not hardy** Sweetly scented pink flowers that fade almost to white, above divided foliage.
↕30cm (12in) ↔ 60cm (24in)

VARIEGATED HERBACEOUS PLANTS

Variegation may be restricted to a narrow rim around the edge of a leaf, or it may be more spectacular. Regular patterning with white, cream or gold may help to define the shape of large leaves but random streaks and splashes can make some plants look untidy. Remember that plants may revert back to their plain form: always remove any shoots with plain green leaves.

AQUILEGIA VULGARIS VERVAENEANA GROUP
Perennial Leaves are marbled and splashed with yellow and green, below white, pink or blue flowers.
↕90cm (36in) ↔ 45cm (18in)

ARABIS PROCURRENS 'VARIEGATA'
♀ **Perennial, not fully hardy** page **63**

ARMORACIA RUSTICANA 'VARIEGATA'
Perennial Deep-rooted plants with large leaves heavily splashed with white, especially in spring.
↕1m (3ft) ↔ 45cm (18in)

ASTRANTIA 'SUNNINGDALE VARIEGATED'
♀ **Perennial** page **77**

BRUNNERA MACROPHYLLA 'HADSPEN CREAM'
♀ **Perennial** page **92**

COREOPSIS 'CALYPSO'
Perennial Narrow leaves edged with gold, and yellow flowers with a red zone.
↕ ↔ 38cm (15in)

GAURA LINDHEIMERI 'CORRIE'S GOLD'
Perennial White flowers on delicate thin stems, and small leaves edged with gold.
↕to 1.5m (5ft) ↔ 90cm (36in)

HEMEROCALLIS FULVA 'KWANZO VARIEGATA'
Perennial Arching leaves with bright white stripes, and occasional double orange flowers.
↕75cm (30in)

HOSTA FORTUNEI VAR. *ALBOPICTA*
♀ **Perennial** page **259**

HOSTA 'GOLDEN TIARA'
♀ **Perennial** page **259**

HOSTA 'GREAT EXPECTATIONS'
Perennial Glaucous-green, puckered leaves with broad yellow centres, and greyish white flowers.
↕55cm (22in) ↔ 85cm (34in)

HOSTA 'SHADE FANFARE'
♀ **Perennial** page **260**

HOSTA SIEBOLDII VAR. *ELEGANS*
♀ **Perennial** page **260**

HOSTA TARDIANA GROUP 'HALCYON'
♀ **Perennial** page **261**

HOSTA UNDULATA VAR. *UNIVITTATA*
♀ **Perennial** page **261**

HOSTA VENUSTA
♀ **Perennial** page **261**

HOSTA 'WIDE BRIM'
♀ **Perennial** page **261**

HOUTTUYNIA CORDATA 'CHAMELEON'
Perennial Vivid leaves in shades of cream, green and red, and small, white flowers.
↕to 15–30cm (6–12in) or more ↔ indefinite

IRIS LAEVIGATA 'VARIEGATA'
♀ **Perennial** page **278**

IRIS PALLIDA 'VARIEGATA'
♀ **Perennial** page **288**

IRIS VARIEGATA
♀ **Perennial** page **290**

LYSIMACHIA PUNCTATA 'ALEXANDER'
Perennial Creeping plant with spires of yellow flowers and white-edged foliage, tinged pink in spring.
↕1m (3ft) ↔ 60cm (2ft)

MIMULUS LUTEUS 'VARIEGATUS'
Perennial Creeping plant with yellow flowers and pale green leaves edged with white.
↕30cm (12in) ↔ 45cm (18in)

MOLINIA CAERULEA 'VARIEGATA'
♀ **Ornamental grass** page **342**

PERSICARIA VIRGINIANA 'PAINTER'S PALETTE'
Perennial Bright foliage splashed with white and marked with red and brown, on red stems.
↕40–120cm (16–48in) ↔ 60–140cm (24–56in)

PHALARIS ARUNDINACEA 'PICTA'
♀ **Ornamental grass** page **383**

PHLOX PANICULATA 'PINK POSIE'
Perennial Compact, strong-growing phlox with white-edged leaves and pink flowers.
↕75cm (30in) ↔ 60cm (24in)

PHORMIUM COOKIANUM SUBSP. *HOOKERI* 'TRICOLOR'
♀ **Evergreen perennial** page **392**

PHYSOSTEGIA VIRGINIANA 'VARIEGATA'
Perennial Upright plant with greyish leaves edged with white, and deep pink flowers.
↕75cm (30in) ↔ 60cm (24in)

PLEIOBLASTUS AURICOMUS
♀ **Bamboo** page **406**

PULMONARIA RUBRA 'DAVID WARD'
Perennial Coral-red flowers in spring, and large leaves broadly margined with white.
↕to 40cm (16in) ↔ 90cm (36in)

SAXIFRAGA STOLONIFERA 'TRICOLOR'
♀ **Evergreen perennial** A slightly tender plant with round leaves edged with white and pink.
↕↔ 30cm (12in)

SISYRINCHIUM STRIATUM 'AUNT MAY'
Perennial Upright fans of narrow, grey leaves edged with cream, and spikes of cream flowers.
↕↔ to 50cm (20in)

SYMPHYTUM X *UPLANDICUM* 'VARIEGATUM'
♀ **Perennial** page **505**

VERONICA GENTIANOIDES 'VARIEGATA'
Perennial Mats of deep green leaves margined with white, and spikes of small, pale blue flowers.
↕↔ 45cm (18in)

VINCA MAJOR 'VARIEGATA'
♀ **Perennial** page **537**

VARIEGATED TREES AND SHRUBS

While flowers usually have a short season, foliage provides colour for at least half the year – and all year if evergreen. Variegated shrubs increase that interest with their bright colouring, and some have flowers that complement the foliage. Many variegated evergreens are useful to brighten shady areas and are good in pots and containers; they are also popular with flower arrangers.

ACER CAMPESTRE 'CARNIVAL'
Tree A tree that can be pruned to keep it smaller, with pink-splashed leaves.
↕8m (25ft) ↔ 5m (15ft)

ACER NEGUNDO 'FLAMINGO'
♡ **Tree** page **33**

ACER PALMATUM 'BUTTERFLY'
♡ **Shrub or small tree** page **34**

ACER PLATANOIDES 'DRUMMONDII'
♡ **Tree** page **37**

ACER PSEUDOPLATANUS 'LEOPOLDII'
♡ **Tree** The leaves are pink at first in spring, then speckled with yellow.
↕↔ 10m (30ft)

ARALIA ELATA 'VARIEGATA'
♡ **Tree** page **64**

BERBERIS THUNBERGII 'ROSE GLOW'
♡ **Shrub** page **86**

BUDDLEJA DAVIDII 'HARLEQUIN'
Shrub Striking leaves edged with white, and purple flowers.
↕↔ 2.5m (8ft)

BUDDLEJA DAVIDII 'SANTANA'
Shrub Mottled foliage in shades of green and yellow, and purple flowers.
↕↔ 2.5m (8ft)

BUXUS SEMPERVIRENS 'ELEGANTISSIMA'
♡ **Evergreen shrub** page **95**

CAMELLIA x *WILLIAMSII* 'GOLDEN SPANGLES'
Evergreen shrub Lime-hating shrub, with bright pink flowers and gold-splashed leaves.
↕↔ 2.5m (8ft)

CEANOTHUS 'PERSHORE ZANZIBAR'
Evergreen shrub Fast-growing shrub with fluffy blue flowers in late spring, and lemon-yellow and bright green foliage.
↕↔ 2.5m (8ft)

CORNUS ALBA 'ELEGANTISSIMA'
♡ **Shrub** Dark red stems with grey-green leaves, edged with white, that turn pink in autumn.
↕↔ 3m (10ft)

CORNUS ALBA 'SPAETHII'
♡ **Shrub** page **138**

CORNUS ALTERNIFOLIA 'ARGENTEA'
♡ **Shrub** Tiers of horizontal branches, clothed with small leaves that are edged in white.
↕3m (10ft) ↔ 2.5m (8ft)

CORNUS MAS 'VARIEGATA'
♡ **Shrub** After yellow flowers in spring the plant is bright with white-edged leaves.
↕2.5m (8ft) ↔ 2m (6ft)

COTONEASTER ATROPURPUREUS 'VARIEGATUS'
♀ **Shrub** page **145**

ELAEAGNUS PUNGENS 'MACULATA'
♀ **Evergreen shrub** page **182**

EUONYMUS FORTUNEI 'SILVER QUEEN'
♀ **Evergreen shrub** page **200**

FUCHSIA MAGELLANICA VAR. *GRACILIS* 'VARIEGATA'
♀ **Hardy shrub** The leaves are coloured in smoky pinks and greys, with small, red and purple flowers.
↕ to 3m (10ft) ↔ 2–3m (6–10ft)

FUCHSIA MAGELLANICA VAR. *MOLINAE* 'SHARPITOR'
Shrub Slightly tender plant with pretty, white-edged leaves and pale pink flowers.
↕ to 3m (10ft) ↔ 2–3m (6–10ft)

HEDERA HELIX 'GLACIER'
♀ **Evergreen climber** page **246**

HIBISCUS SYRIACUS 'MEEHANII'
Shrub Sun-loving, with purple-blue flowers; leaves with broad white edges.
↕ 3m (10ft) ↔ 2m (6ft)

HYDRANGEA MACROPHYLLA 'TRICOLOR'
♀ **Shrub** Grey-green leaves marked with white, and pale pink flowers.
↕ 1.5m (5ft) ↔ 1.2m (4ft)

ILEX X *ALTACLERENSIS* 'LAWSONIANA'
♀ **Evergreen shrub** page **271**

LIGUSTRUM LUCIDUM 'EXCELSUM SUPERBUM'
♀ **Evergreen shrub** page **314**

OSMANTHUS HETEROPHYLLUS 'VARIEGATUS'
Evergreen shrub Holly-like leaves with broad yellow margins, and fragrant, tiny flowers in autumn.
↕↔ 2.5m (8ft)

PHILADELPHUS CORONARIUS 'VARIEGATUS'
♀ **Shrub** page **385**

PHILADELPHUS 'INNOCENCE'
Shrub Arching shrub with creamy-yellow leaves and semi-double white flowers.
↕ 3m (10ft) ↔ 2m (6ft)

PIERIS 'FLAMING SILVER'
♀ **Evergreen shrub** Acid-loving plant with narrow foliage, edged white, that is pink in spring.
↕↔ 2.5m (8ft)

PITTOSPORUM TENUIFOLIUM 'IRENE PATERSON'
♀ **Evergreen shrub** Slow-growing, slightly tender shrub with white-speckled leaves.
↕ 1.2m (4ft) ↔ 60cm (2ft)

PSEUDOPANAX LESSONII 'GOLD SPLASH'
♀ **Evergreen shrub, not hardy** page **426**

RHAMNUS ALATERNUS 'ARGENTEOVARIEGATA'
♀ **Evergreen shrub** page **434**

RHODODENDRON 'PRESIDENT ROOSEVELT'
Evergreen shrub Weakly-branching, acid-loving shrub with gold-splashed leaves and red flowers.
↕↔ 2m (6ft)

SAMBUCUS NIGRA 'PULVERULENTA'
Shrub Bright, slow-growing shrub for part-shade with young leaves heavily splashed with white.
↕↔ 2m (6ft)

WEIGELA FLORIDA 'VARIEGATA'
♀ **Shrub** page **541**

GOLD-LEAVED PLANTS

Gold-leaved plants bring a splash of sunshine to the garden. In contrast to plants with variegated leaves, most fully gold-leaved plants are rather prone to scorch in full sun, so are best in light shade. But avoid dark shade, or the leaves may become lime-green. Some plants are gold for only a part of their growth; their young, gold tips fade to green, but this contrast is still pleasing.

ACER CAPPADOCICUM 'AUREUM'
♀ **Tree** The leaves unfurl yellow in spring, turn green in summer and gold in autumn.
↕ 15m (50ft) ↔ 10m (30ft)

ACER SHIRASAWANUM 'AUREUM'
♀ **Tree** The bright yellow leaves turn red in autumn.
↕↔ 6m (20ft)

BEGONIA 'TIGER PAWS'
♀ **Perennial, not hardy** page **81**

CALLUNA VULGARIS 'BEOLEY GOLD'
♀ **Evergreen shrub** page **98**

CAREX ELATA 'AUREA'
♀ **Ornamental grass** page **109**

CAREX OSHIMENSIS 'EVERGOLD'
♀ **Ornamental grass** page **109**

CHAMAECYPARIS LAWSONIANA 'MINIMA AUREA'
♀ **Conifer** Small evergreen, conical shrub with lime-green and gold foliage.
↕ 1m (3ft)

CHAMAECYPARIS LAWSONIANA 'STARDUST'
♀ **Conifer** The foliage is yellow and fern-like.
↕ 15m (50ft) ↔ 8m (25ft)

CHAMAECYPARIS OBTUSA 'CRIPPSII'
♀ **Conifer** A slow-growing tree with gold foliage.
↕ 15m (50ft) ↔ 8m (25ft)

CHOISYA TERNATA 'SUNDANCE'
♀ **Evergreen shrub** page **121**

CORNUS ALBA 'AUREA'
Shrub Beautiful soft-gold foliage that is prone to sun-scorch on dry soils.
↕↔ 1m (3ft)

CORTADERIA SELLOANA 'AUREOLINEATA'
♀ **Ornamental grass** page **141**

ERICA ARBOREA 'ALBERT'S GOLD'
♀ **Evergreen shrub** Attractive, upright habit, with gold foliage but few flowers.
↕ 2m (6ft) ↔ 80cm (32in)

ERICA CARNEA 'FOXHOLLOW'
♀ **Evergreen shrub** page **187**

ERICA CARNEA 'WESTWOOD YELLOW'
♀ **Evergreen shrub** Upright habit, with yellow foliage, and pale pink flowers in winter.
↕ 20cm (8in) ↔ 30cm (12in)

ERICA x *STUARTII* 'IRISH LEMON'
♀ **Evergreen shrub** page **189**

FAGUS SYLVATICA 'DAWYCK GOLD'
♀ **Tree** An upright, narrow, compact tree with bright yellow leaves.
↕ 18m (60ft) ↔ 7m (22ft)

FRAXINUS EXCELSIOR 'JASPIDEA'
♀ **Tree** The winter shoots are yellow, as are the leaves in spring and autumn.
↕ 30m (100ft) ↔ 20m (70ft)

FUCHSIA 'GENII'
♀ **Shrub** page **212**

GLEDITSIA TRIACANTHOS 'SUNBURST'
♀ **Tree** page **232**

HAKONECHLOA MACRA 'AUREOLA'
♀ **Ornamental grass** page **235**

HEDERA HELIX 'BUTTERCUP'
♀ **Evergreen climber** page **246**

HOSTA 'SUM AND SUBSTANCE'
♀ **Perennial** page **261**

HUMULUS LUPULUS 'AUREUS'
♀ **Climbing perennial** page **262**

ILEX CRENATA 'GOLDEN GEM'
♀ **Evergreen shrub** Compact, small-leaved shrub with bright gold leaves and sparse, black berries.
↕ 1m (3ft) ↔ 1.2–1.5m (4–5ft)

IRIS PSEUDACORUS 'VARIEGATUS'
♀ **Perennial** page **278**

LAURUS NOBILIS 'AUREA'
♀ **Evergreen shrub** page **304**

LONICERA NITIDA
'BAGGESEN'S GOLD'
♀ **Evergreen shrub** page **323**

ORIGANUM VULGARE 'AUREUM'
♀ **Perennial herb** page **358**

PELARGONIUM CRISPUM
'VARIEGATUM'
♀ **Perennial, not hardy** page **372**

PHILADELPHUS CORONARIUS
'AUREUS'
♀ **Deciduous shrub** Gold young leaves turning greeny-yellow in summer, white fragrant flowers in early summer.
↕ 2.5m (8ft) ↔ 1.5m (5ft)

PHORMIUM 'YELLOW WAVE'
♀ **Perennial, not fully hardy** page **394**

PHYSOCARPUS OPULIFOLIUS
'DART'S GOLD'
♀ **Shrub** page **399**

RIBES SANGUINEUM
'BROCKLEBANKII'
♀ **Shrub** page **445**

ROBINIA PSEUDOACACIA 'FRISIA'
♀ **Tree** page **447**

SALVIA OFFICINALIS 'KEW GOLD'
♀ **Evergreen subshrub** Aromatic golden leaves sometimes flecked with green, mauve flower spikes in summer.
↕ 20–30cm (8–12in) ↔ 30cm (12in)

SAMBUCUS RACEMOSA
'SUTHERLAND GOLD'
♀ **Shrub** Finely-divided foliage of bright yellow that is best when plants are regularly pruned.
↕↔ 2m (6ft)

SPIRAEA JAPONICA 'GOLDFLAME'
♀ **Shrub** page **501**

DARK AND PURPLE FOLIAGE

The primary value of dark foliage in the garden is as a foil to other plants, though many are very beautiful in their own right. An adjacent purple-leaved plant makes gold, variegated and silver plants look even more brilliant, and is the perfect foil for white, pink, yellow and orange flowers. Most purple-leaved plants develop their best colour in full sun, looking dull and greenish in shade.

ACER PLATANOIDES 'CRIMSON KING'
♀ **Tree** page **36**

AEONIUM 'ZWARTKOP'
♀ **Shrub, not hardy** page **45**

AJUGA REPTANS 'ATROPURPUREA'
♀ **Perennial** page **48**

BERBERIS THUNBERGII F. *ATROPURPUREA*
♀ **Shrub** The deep purple foliage turns bright red in autumn before it falls, revealing the spiny stems.
↕2m (6ft) ↔ 2.5m (8ft)

BERBERIS THUNBERGII 'RED CHIEF'
♀ **Shrub** Deep, reddish-purple foliage on an upright shrub.
↕1.5m (5ft) ↔ 60cm (24in)

CERCIS CANADENSIS 'FOREST PANSY'
♀ **Tree** Pink flowers on bare twigs in early spring are followed by beautiful purple foliage.
↕↔ 10m (30ft)

CIMICIFUGA SIMPLEX 'BRUNETTE'
Perennial Coarsely divided purple foliage, and dark stems with white fluffy flowers in autumn.
↕1–1.2m (3–4ft) ↔ 60cm (24in)

CLEMATIS MONTANA VAR. *RUBENS* 'TETRAROSE'
♀ **Climber** page **128**

CORYLUS MAXIMA 'PURPUREA'
♀ **Shrub** page **143**

COTINUS COGGYGRIA 'ROYAL PURPLE'
♀ **Shrub** page **144**

COTINUS 'GRACE'
♀ **Shrub** page **145**

CRYPTOTAENIA JAPONICA F. *ATROPURPUREA*
Biennial Three-lobed leaves on upright stems, wholly coloured with purple, and tiny flowers.
↕↔ 60cm (24in)

ERICA CARNEA 'VIVELLII'
♀ **Evergreen shrub** page **187**

EUPHORBIA DULCIS 'CHAMELEON'
Perennial Purple foliage on bushy plants with lime-green bracts when in flower.
↕↔ 30cm (12in)

FAGUS SYLVATICA 'DAWYCK PURPLE'
♀ **Tree** This columnar tree has deep purple leaves.
↕20m (70ft) ↔ 5m (15ft)

FAGUS SYLVATICA 'PURPUREA PENDULA'
♀ **Tree** The deep purple leaves hang from pendent branches on this small, domed tree.
↕↔ 3m (10ft)

GERANIUM SESSILIFLORUM SUBSP. *NOVAE-ZELANDIAE* 'NIGRICANS'
Perennial Small, mat-forming plant with dull, bronze-purple foliage and small white flowers.
↕8cm (3in) ↔ 15cm (6in)

HEBE 'MRS. WINDER'
♀ **Evergreen shrub** A compact shrub with dark leaves that are purple when young, and violet-blue flowers.
↕1m (3ft) ↔ 1.2m (4ft)

HEDERA HELIX 'ATROPURPUREA'
♀ **Evergreen climber** page **246**

HEUCHERA MICRANTHA VAR. *DIVERSIFOLIA* 'PALACE PURPLE'
♀ **Perennial** page **255**

OPHIOPOGON PLANISCAPUS 'NIGRESCENS'
♀ **Perennial** page **357**

PENSTEMON DIGITALIS 'HUSKER RED'
Perennial Semi-evergreen perennial with red and purple foliage and pale pink flowers.
↕50–75cm (20–30in) ↔ 30cm (12in)

PHYSOCARPUS OPULIFOLIUS 'DIABLO'
Shrub Deep purple, almost brown leaves and clusters of small pink flowers.
↕2m (6ft) ↔ 1m (3ft)

RHEUM PALMATUM 'ATROSANGUINEUM'
Perennial Scarlet buds open to reveal large, architectural leaves that fade to deep green.
↕2m (6ft)

ROSA GLAUCA
♀ **Species rose** page **464**

SALVIA OFFICINALIS PURPURASCENS GROUP
♀ **Evergreen shrub** page **476**

SAMBUCUS NIGRA 'GUINCHO PURPLE'
♀ **Shrub** page **480**

SEDUM TELEPHIUM SUBSP. *MAXIMUM* 'ATROPURPUREUM'
♀ **Perennial** page **491**

TRADESCANTIA PALLIDA 'PURPUREA'
♀ **Perennial, not hardy** This sprawling plant has bright purple leaves and small pink flowers.
↕20cm (8in) ↔ 40cm (16in)

VIBURNUM SARGENTII 'ONONDAGA'
♀ **Shrub** Upright-growing shrub with purple leaves that turn red in autumn, and pale pink flowers.
↕2m (6ft)

VIOLA RIVINIANA 'PURPUREA'
Perennial Low-growing plant with small purple leaves and flowers, which seeds profusely.
↕10–20cm (4–8in) ↔ 20–40cm (8–16cm)

VITIS VINIFERA 'PURPUREA'
♀ **Climber** page **540**

WEIGELA FLORIDA 'FOLIIS PURPUREIS'
♀ **Shrub** page **541**

SILVER FOLIAGE

Most plants that have silver leaves have adapted to hot, sunny climates, and these are plants for full sun in the garden. They are also adapted to low rainfall in many cases, so they can form the basis of a drought garden or gravel garden. Their colour makes them ideal to associate with pink and white flowers, and with purple foliage. In addition, many have fragrant leaves.

ACACIA BAILEYANA
♀ **Shrub, not hardy** The foliage is less fine than *A. dealbata* but steely grey; the flowers are bright yellow.
↕5–8m (15–25ft) ↔ 3–6m (10–20ft)

ACCA SELLOWIANA
Evergreen shrub Green leaves with silver reverses, flowers with fleshy, red and white edible petals.
↕2m (6ft) ↔ 2.5m (8ft)

ACHILLEA TOMENTOSA
♀ **Perennial** page **41**

ANAPHALIS TRIPLINERVIS
♀ **Perennial** Clump-forming plant with silver-grey leaves and white flowers in late summer.
↕80–90cm (32–36in) ↔ 45–60cm (18–24in)

ANAPHALIS TRIPLINERVIS 'SOMMERSCHNEE'
♀ **Perennial** page **56**

ANTENNARIA MICROPHYLLA
♀ **Perennial** page **61**

ANTHEMIS PUNCTATA SUBSP. *CUPANIANA*
♀ **Perennial** page **61**

ARTEMISIA ABSINTHIUM 'LAMBROOK SILVER'
♀ **Perennial** page **69**

ARTEMISIA ALBA 'CANESCENS'
♀ **Perennial** Very finely divided grey leaves that form a feathery mass.
↕45cm (18in) ↔ 30cm (12in)

ARTEMISIA LUDOVICIANA 'SILVER QUEEN'
♀ **Perennial** page **69**

ARTEMISIA PONTICA
♀ **Evergreen perennial** A creeping perennial that has masses of upright stems with feathery leaves that form a mounded clump.
↕40–80cm (16–32in) ↔ indefinite

ATRIPLEX HALIMUS
Shrub Slightly tender, but wind-tolerant, fast-growing plant with small, shiny, silver leaves.
↕2m (6ft) ↔ 2.5m (8ft)

BRACHYGLOTTIS 'SUNSHINE'
♀ **Evergreen shrub** page **91**

CEDRUS ATLANTICA F. *GLAUCA*
♀ **Conifer** The blue Atlas cedar forms a large specimen tree with blue-grey foliage.
↕40m (130ft) ↔ 10m (30ft)

CONVOLVULUS CNEORUM
♀ **Evergreen shrub** page **136**

CYTISUS BATTANDIERI
♀ **Shrub** page **159**

DIANTHUS 'BECKY ROBINSON'
♀ **Perennial** page **172**

DIANTHUS 'HAYTOR WHITE'
♀ **Perennial** page **172**

ECHEVERIA AGAVOIDES
♀ **Succulent** page **181**

ELAEAGNUS 'QUICKSILVER'
♀ **Evergreen shrub** page **183**

ERICA TETRALIX 'ALBA MOLLIS'
♀ **Evergreen shrub** page **189**

EUCALYPTUS GUNNII
♀ **Evergreen tree** page **197**

HALIMIUM 'SUSAN'
♀ **Evergreen shrub** page **237**

HEBE PIMELEOIDES 'QUICKSILVER'
♀ **Evergreen shrub** Ground-hugging shrub with tiny silver leaves and pale lilac flowers.
↕30cm (12in) ↔ 60cm (24in)

HEBE PINGUIFOLIA 'PAGEI'
♀ **Evergreen shrub** page **243**

HEBE 'RED EDGE'
♀ **Evergreen shrub** Spreading, low shrub with grey leaves edged with red.
↕45cm (18in) ↔ 60cm (24in)

HELIANTHEMUM 'WISLEY PRIMROSE'
♀ **Evergreen shrub** page **249**

HELICHRYSUM SPLENDIDUM
♀ **Perennial** page **251**

HELICTOTRICHON SEMPERVIRENS
♀ **Ornamental grass** A tufted perennial forming a mound of grey-blue leaves, with taller flower stems in early summer.
↕1.4m (4½ft) ↔ 60cm (2ft)

LAVANDULA x *INTERMEDIA* DUTCH GROUP
♀ **Shrub** page **306**

PULMONARIA 'MARGERY FISH'
♀ **Perennial** page **427**

PYRUS SALICIFOLIA 'PENDULA'
♀ **Tree** page **432**

ROMNEYA COULTERI
♀ **Perennial** Vigorous, suckering plant with coarsely-toothed silver leaves and white, yellow-centred flowers.
↕2m (6ft) ↔ indefinite

SALIX 'BOYDII'
♀ **Shrub** page **469**

SALIX LANATA
♀ **Shrub** page **471**

SALVIA ARGENTEA
♀ **Perennial** page **472**

SALVIA DISCOLOR
♀ **Perennial, not hardy** page **473**

SANTOLINA CHAMAECYPARISSUS
♀ **Evergreen shrub** page **480**

SEDUM SPATHULIFOLIUM 'CAPE BLANCO'
♀ **Perennial** page **489**

SENECIO CINERARIA 'SILVER DUST'
♀ **Evergreen shrub** page **493**

SENECIO CINERARIA 'WHITE DIAMOND'
♀ **Evergreen shrub** The almost white leaves resemble oak leaves in shape.
↕30–40cm (12–16in) ↔ 30cm (12in)

SENECIO VIRAVIRA
♀ **Shrub, not fully hardy** The finely divided leaves are carried on sprawling stems that eventually produce creamy, pompon flowers.
↕60cm (24in) ↔ 1m (3ft)

VERBASCUM BOMBYCIFERUM
♀ **Perennial** page **529**

PLANTS FOR SPRING COLOUR

Spring is a frantic time in the garden, and at times it seems that every plant is trying to flower. The earliest flowers are demure and adapted to survive any snows, sleet and wind, but by April, flowers are bigger and bolder, and the yellows, white and blues of the spring bulbs are joined by masses of pink cherry blossom, showy magnolias, rhododendrons, clematis and wisteria.

ACER PSEUDOPLATANUS 'BRILLIANTISSIMUM'
♀ **Tree** page **37**

BERGENIA PURPURASCENS
♀ **Perennial** The bold, leathery leaves turn red in cold weather and clusters of bright, purplish flowers open in spring.
↕45cm (18in) ↔ 30cm (12in)

CALTHA PALUSTRIS
♀ **Perennial** page **99**

CAMELLIA X *WILLIAMSII* CULTIVARS
♀ **Evergreen shrubs** pages **104–105**

CHAENOMELES SPECIOSA 'MOERLOOSEI'
♀ **Shrub** page **115**

CLEMATIS ALPINA
♀ **Climber** The blue, bell-shaped flowers have white centres and are followed by fluffy seedheads.
↕2–3m (6–10ft)

CLEMATIS ALPINA 'FRANCES RIVIS'
♀ **Climber** page **128**

CLEMATIS MACROPETALA 'MARKHAM'S PINK'
♀ **Climber** page **128**

CLEMATIS MONTANA VAR. *RUBENS*
♀ **Climber** page **128**

CORYDALIS SOLIDA
♀ **Bulbous perennial** Tubular pink flowers are held above feathery, grey foliage.
↕25cm (10in) ↔ 20cm (8in)

CORYLOPSIS PAUCIFLORA
♀ **Shrub** page **142**

DAPHNE TANGUTICA
♀ **Evergreen shrub** The tips of shoots are studded with fragrant, pink and white flowers in late spring.
↕↔ 1m (3ft)

DICENTRA 'LUXURIANT'
♀ **Perennial** Deeply-lobed leaves and clusters of red flowers over a long season.
↕30cm (12in) ↔ 45cm (18in)

DODECATHEON MEADIA
♀ **Perennial** A clump-forming plant with clusters of magenta-pink flowers that resemble cyclamen.
↕40cm (16in) ↔ 25cm (10in)

EPIMEDIUM X *RUBRUM*
♀ **Perennial** page **184**

EPIMEDIUM X *VERSICOLOR* 'SULPHUREUM'
♀ **Perennial** Evergreen, clump-forming plant with divided leaves and pretty, pale yellow flowers.
↕↔ 30cm (12in)

EUPHORBIA X *MARTINII*
♀ **Evergreen subshrub** page **202**

EUPHORBIA POLYCHROMA
♀ **Perennial** page **203**

FORSYTHIA X *INTERMEDIA* 'LYNWOOD'
♀ **Shrub** page **208**

FOTHERGILLA MAJOR
♀ **Shrub** page **209**

HEPATICA NOBILIS
♀ **Perennial** page **255**

MAGNOLIA CAMPBELLII 'CHARLES RAFFILL'
♀ **Shrub** page **332**

MAGNOLIA 'ELIZABETH'
♀ **Shrub** page **332**

MAGNOLIA x *LOEBNERI* 'MERRILL'
♀ **Shrub** page **332**

MAGNOLIA x *SOULANGEANA* 'LENNEI'
♀ **Tree** Beautiful tree of spreading habit with large, deep purple flowers.
↕↔ 6m (20ft)

MAGNOLIA STELLATA
♀ **Tree** page **333**

MALUS FLORIBUNDA
♀ **Tree** page **336**

PIERIS JAPONICA 'MOUNTAIN FIRE'
♀ **Evergreen shrub** Acid-loving shrub with white flowers and red new growth that becomes bronze, then green.
↕ 4m (12ft) ↔ 3m (10ft)

PRIMULA VERIS
♀ **Perennial** page **419**

PRIMULA VIALII
♀ **Perennial** A short-lived perennial with dense heads of lilac flowers and red buds.
↕ 30–60cm (12–24in) ↔ 30cm (12in)

PRUNUS AVIUM 'PLENA'
♀ **Tree** page **424**

PRUNUS GLANDULOSA 'ALBA PLENA'
♀ **Shrub** page **421**

PRUNUS 'KANZAN'
♀ **Tree** page **424**

PRUNUS PADUS 'COLORATA'
♀ **Tree** page **424**

PRUNUS PADUS 'WATERERI'
♀ **Tree** page **424**

PRUNUS 'PANDORA'
♀ **Tree** page **424**

PRUNUS 'PINK PERFECTION'
♀ **Tree** page **425**

PRUNUS 'SHIROFUGEN'
♀ **Tree** page **425**

PRUNUS 'SHÔGETSU'
♀ **Tree** page **425**

PRUNUS 'SPIRE'
♀ **Tree** page **425**

PRUNUS x *SUBHIRTELLA* 'AUTUMNALIS ROSEA'
♀ **Tree** page **425**

PRUNUS 'UKON'
♀ **Tree** page **425**

PRUNUS x *YEDOENSIS*
♀ **Tree** page **425**

PULMONARIA RUBRA
♀ **Perennial** page **428**

RHODODENDRON SPECIES AND CULTIVARS
♀ **Shrubs** pages **438–444**

SAXIFRAGA 'TUMBLING WATERS'
♀ **Alpine** Rosettes of silvery-green leaves produce tall, arching stems bearing hundreds of tiny white flowers.
↕ 45cm (18in) ↔ 30cm (12in)

VIOLA 'MAGGIE MOTT'
♀ **Perennial** Dainty blue and white flowers on bushy plants.
↕ 15cm (6in) ↔ 25cm (10in)

WISTERIA FLORIBUNDA 'MULTIJUGA'
♀ **Climber** page **542**

WISTERIA FLORIBUNDA 'ROSEA'
♀ **Climber** Long racemes of pale pink flowers cascade from this vigorous plant.
↕ 9m (28ft)

PLANTS FOR SUMMER COLOUR

In gardens the summer months are often dominated by bedding, but there are many bright-flowered perennials, too, at their best. There are fewer shrubs in flower in mid-summer than in spring, but an important exception is the rose, without which gardens would be much the poorer. There are roses for every part of the garden, from climbers to ground cover.

ANCHUSA AZUREA 'LODDON ROYALIST'
♀ **Perennial** page **57**

AQUILEGIA VULGARIS 'NORA BARLOW'
♀ **Perennial** page **63**

BUDDLEJA DAVIDII 'DARTMOOR'
♀ **Shrub** The narrow leaves are deeply toothed and the flowers are a rich reddish purple.
↕3m (10ft) ↔ 5m (15ft)

BUDDLEJA GLOBOSA
♀ **Shrub** page **94**

CAMPANULA LACTIFLORA 'LODDON ANNA'
♀ **Perennial** page **107**

CLEMATIS 'BEES' JUBILEE'
♀ **Climber** page **129**

CLEMATIS X *DURANDII*
♀ **Perennial** Non-climbing hybrid that sprawls through other plants, with large, blue flowers.
↕1–2m (3–6ft)

CLEMATIS 'GIPSY QUEEN'
♀ **Climber** Velvety, purple flowers with red anthers throughout summer.
↕3m (10ft)

CLEMATIS 'PERLE D'AZUR'
♀ **Climber** page **133**

DEUTZIA X *HYBRIDA* 'MONT ROSE'
♀ **Shrub** page **170**

DICENTRA 'LANGTREES'
♀ **Perennial** Pale, pearly white flowers are held on glossy stems above the feathery grey foliage.
↕30cm (12in) ↔ 45cm (18in)

DIGITALIS LANATA
♀ **Perennial** Leafy stems with densely packed, small cream or fawn flowers.
↕60cm (24in) ↔ 30cm (12in)

ERICA CINEREA 'VELVET NIGHT'
♀ **Evergreen shrub** Exceptionally dark foliage is highlighted by deep purple flowers.
↕60cm (24in) ↔ 80cm (32in)

ERICA VAGANS 'LYONESSE'
♀ **Evergreen shrub** page **189**

ERICA VAGANS 'MRS D. F. MAXWELL'
♀ **Evergreen shrub** page **189**

EUPHORBIA SCHILLINGII
♀ **Perennial** page **203**

FUCHSIA HARDY TYPES
♀ **Shrubs** page **212**

FUCHSIA 'PHYLLIS'
♀ **Shrub** The semi-double cerise flowers are freely carried on an upright, hardy plant.
↕1–1.5m (3–5ft) ↔ 75–90cm (30–36in)

GENISTA TENERA 'GOLDEN SHOWER'
♀ **Shrub** page **224**

GERANIUMS, HARDY, LARGE
♀ **Perennials** pages **228–229**

HEBE 'GREAT ORME'
♀ **Evergreen shrub** page **241**

HEMEROCALLIS CULTIVARS
♀ **Perennials** page **254**

HEMEROCALLIS 'PINK DAMASK'
♀ **Perennial** Large, dark salmon-pink flowers held above arching leaves.
↕1m (3ft)

HYDRANGEA SERRATA 'BLUEBIRD'
♀ **Shrub** page **268**

HYPERICUM 'HIDCOTE'
♀ **Shrub** page **269**

BEARDED IRIS
♀ **Perennials** pages **284–287**

IRIS 'ARCTIC FANCY'
♀ **Perennial** White petals heavily edged with deep violet.
↕50cm (20in)

IRIS 'BLUE-EYED BRUNETTE'
♀ **Perennial** Large flowers are pale brown with a lilac-blue mark by the base of the beard.
↕90cm (3ft)

IRIS LAEVIGATA
♀ **Perennial** page **278**

JASMINUM HUMILE 'REVOLUTUM'
♀ **Shrub** Fragrant, bright yellow flowers are set against divided foliage for most of summer.
↕2.5m (8ft) ↔ 3m (10ft)

LATHYRUS LATIFOLIUS 'WHITE PEARL'
♀ **Climber** This herbaceous perennial has pure white, scentless flowers.
↕2m (6ft)

OENOTHERA FRUTICOSA 'FYRVERKERI'
♀ **Perennial** page **354**

PENSTEMON CULTIVARS
♀ **Perennials** pages **378–379**

PHOTINIA X *FRASERI* 'RED ROBIN'
♀ **Evergreen shrub** page **396**

PHYGELIUS AEQUALIS 'YELLOW TRUMPET'
♀ **Evergreen shrub** page **396**

PLATYCODON GRANDIFLORUS
♀ **Perennial** page **406**

POTENTILLA 'GIBSON'S SCARLET'
♀ **Perennial** page **411**

ROSA 'JUST JOEY'
♀ **Large-flowered bush rose** page **449**

ROSA 'MANY HAPPY RETURNS'
♀ **Cluster-flowered bush rose** page **451**

ROSA 'MOUNTBATTEN'
♀ **Cluster-flowered bush rose** page **451**

ROSA 'SWEET DREAM'
♀ **Patio rose** page **457**

ROSA 'TEQUILA SUNRISE'
♀ **Large-flowered bush rose** Bright yellow flowers heavily edged with scarlet.
↕75cm (30in) ↔ 60cm (24in)

ROSES, OLD GARDEN
♀ **Deciduous shrubs** pages **460–1**

ROSES, RAMBLER
♀ **Deciduous climbers** pages **454–5**

SALVIA X *SYLVESTRIS* 'MAINACHT'
♀ **Perennial** page **479**

THALICTRUM DELAVAYI 'HEWITT'S DOUBLE'
♀ **Perennial** page **510**

TRADESCANTIA X *ANDERSONIANA* 'J. C. WEUGELIN'
♀ **Perennial** page **517**

TRADESCANTIA X *ANDERSONIANA* 'OSPREY'
♀ **Perennial** page **518**

PLANTS FOR AUTUMN COLOUR

Autumn is a season of great change in the garden, and although many annuals and perennials continue their summer display, it is the colours of leaves and fruits (see also pp.502–3) that most capture the imagination. This should be the most spectacular time of all in the garden, as borders erupt in fiery orange and red shades before the more sombre displays of winter.

ACER GROSSERI VAR. *HERSII*
♀ **Tree** page **32**

ACER PALMATUM 'BLOODGOOD'
♀ **Small tree** page **34**

ACER PALMATUM 'GARNET'
♀ **Shrub** page **34**

ACER PALMATUM 'OSAKAZUKI'
♀ **Shrub or small tree** page **35**

ACER RUBRUM 'OCTOBER GLORY'
♀ **Tree** page **38**

AMELANCHIER LAMARCKII
♀ **Large shrub** page **56**

ANEMONE HUPEHENSIS 'HADSPEN ABUNDANCE'
♀ **Perennial** page **59**

ANEMONE HUPEHENSIS 'SEPTEMBER CHARM'
♀ **Perennial** Pale pink flowers on neat growth.
↕ 60–90cm (24–36in) ↔ 40cm (16in)

ASTER AMELLUS 'KING GEORGE'
♀ **Perennial** page **72**

ASTER ERICOIDES 'PINK CLOUD'
♀ **Perennial** The bushy plants are covered with small, pink flowers in autumn.
↕ 1m (3ft) ↔ 30cm (12in)

ASTER LATERIFOLIUS 'HORIZONTALIS'
♀ **Perennial** page **73**

BERBERIS WILSONIAE
♀ **Shrub** page **87**

CEANOTHUS X *DELILEANUS* 'GLOIRE DE VERSAILLES'
♀ **Shrub** page **113**

CLEMATIS 'ALBA LUXURIANS'
♀ **Climber** page **132**

CLEMATIS 'DUCHESS OF ALBANY'
♀ **Climber** page **132**

CLEMATIS X *TRITERNATA* 'RUBROMARGINATA'
♀ **Climber** Strong shoots bear masses of small, cross-shaped deep pink and white flowers.
↕ 5m (15ft)

CORNUS KOUSA VAR. *CHINENSIS*
♀ **Tree** page **139**

CORTADERIA SELLOANA 'SUNNINGDALE SILVER'
♀ **Ornamental grass** page **141**

COTONEASTER MICROPHYLLUS
♀ **Evergreen shrub** An arching shrub with pinkish-red berries in autumn.
↕ 1m (3ft) ↔ 1.5m (5ft)

DAHLIA CULTIVARS
Perennials pages **162–163**

EUCRYPHIA X *NYMANSENSIS* 'NYMANSAY'
♀ **Evergreen tree** page **198**

EUONYMUS ALATUS
♀ **Shrub** page **198**

EUONYMUS EUROPAEUS 'RED CASCADE'
♀ **Shrub** page **199**

FUCHSIA 'MRS POPPLE'
♀ **Shrub** page **212**

GENTIANA SEPTEMFIDA
♀ **Perennial** page **225**

GINKGO BILOBA
♀ **Hardy tree** This deciduous conifer has broad leaves that turn butter-yellow before they drop in autumn.
↕ to 30m (100ft) ↔ 8m (25ft)

HELIANTHUS 'LODDON GOLD'
♀ **Perennial** page **249**

HIBISCUS SYRIACUS 'OISEAU BLEU'
♀ **Shrub** page **256**

INDIGOFERA AMBLYANTHA
♀ **Shrub** Slender stems bear feathery leaves and upright stems of tiny pink flowers.
↕ 2m (6ft) ↔ 2.5m (8ft)

LIRIODENDRON TULIPIFERA
♀ **Tree** page **320**

LIRIOPE MUSCARI
♀ **Perennial** page **321**

MALUS 'JOHN DOWNIE'
♀ **Tree** page **336**

NANDINA DOMESTICA
♀ **Shrub** page **345**

NYSSA SINENSIS
♀ **Tree** page **353**

NYSSA SYLVATICA
♀ **Tree** page **353**

PARTHENOCISSUS TRICUSPIDATA
♀ **Climber** page **369**

PHYGELIUS CAPENSIS
♀ **Shrub, not fully hardy** page **396**

PRUNUS SARGENTII
♀ **Tree** page **425**

PRUNUS X *SUBHIRTELLA* 'AUTUMNALIS ROSEA'
♀ **Tree** page **425**

PSEUDOLARIX AMABILIS
♀ **Tree** A deciduous conifer that is grown for its attractive conical shape and golden autumn colour.
↕ 15–20m (50–70ft) ↔ 6–12m (20–40ft)

RHUS TYPHINA 'DISSECTA'
♀ **Shrub** page **445**

RUDBECKIA FULGIDA VAR. *DEAMII*
♀ **Perennial** Daisy-like flowers of orange, with black centres.
↕ 60cm (24in) ↔ 45cm (18in)

RUDBECKIA 'GOLDQUELLE'
♀ **Perennial** page **468**

SALVIA ULIGINOSA
♀ **Perennial** page **479**

SCHIZOSTYLIS COCCINEA 'MAJOR'
♀ **Perennial** page **486**

SEDUM 'RUBY GLOW'
♀ **Perennial** page **488**

SEDUM SPECTABILE 'ICEBERG'
Perennial Fleshy, pale-leaved plant with pure white flowers.
↕ 30–45cm (12–18in) ↔ 35cm (14in)

SORBUS REDUCTA
♀ **Shrub** page **499**

SORBUS VILMORINII
♀ **Shrub or small tree** page **500**

TRICYRTIS FORMOSANA
♀ **Perennial** Erect stems with glossy leaves, and pale pink, starry flowers with darker spots.
↕ 80cm (32in) ↔ 45cm (18in)

VIBURNUM PLICATUM 'MARIESII'
♀ **Shrub** page **536**

VITIS COIGNETIAE
♀ **Climber** page **540**

PLANTS FOR WINTER INTEREST

Gardeners who do not think that anything happens in gardens in winter miss out on some of the most exciting plants of all. Delicate scents and brightly coloured flowers continue to appear in all but the coldest climates. Many grow happily in shady positions and this is the ideal place to plant a winter garden, preferably by a door or where it can be seen from a window.

ACER GRISEUM
♀ **Tree** page **32**

ACER PENSYLVANICUM 'ERYTHROCLADUM'
♀ **Deciduous tree** page **36**

ASPLENIUM SCOLOPENDRIUM CRISTATUM GROUP
Evergreen fern Erect, leathery fronds with broadened, irregular tips, which look bold in winter.
↕45cm (18in) ↔ 60cm (24in)

AUCUBA JAPONICA 'CROTONIFOLIA'
♀ **Evergreen shrub** page **78**

BERGENIA PURPURASCENS 'BALLAWLEY'
♀ **Perennial** page **88**

CALLUNA VULGARIS 'ROBERT CHAPMAN'
♀ **Heather** page **98**

CHAMAECYPARIS OBTUSA 'NANA AUREA'
♀ **Conifer** This dwarf conifer is rounded with a flat top and yellow foliage.
↕2m (6ft)

CHIMONANTHUS PRAECOX 'GRANDIFLORUS'
♀ **Shrub** page **120**

CLEMATIS CIRRHOSA VAR. *BALEARICA*
♀ **Climber** This evergreen climber has pale cream, bell-shaped flowers that are fragrant.
↕2.5–3m (8–10ft)

CORNUS ALBA 'SIBIRICA'
♀ **Shrub** page **137**

CORNUS MAS
♀ **Shrub or small tree** page **139**

CORNUS MAS 'AUREA'
Shrub Masses of tiny yellow flowers in winter followed by yellow spring foliage that fades to green in summer.
↕↔ 5m (15ft)

CYCLAMEN CILICIUM
♀ **Tuberous perennial** page **157**

CYCLAMEN COUM PEWTER GROUP
♀ **Bulb** page **158**

DAPHNE BHOLUA 'GURKHA'
♀ **Shrub** page **164**

ERICA CARNEA 'ANN SPARKES'
♀ **Evergreen shrub** page **187**

ERICA CARNEA 'PINK SPANGLES'
♀ **Evergreen shrub** This bright heather has pink flowers that are pink and white as they first open.
↕15cm (6in) ↔ 45cm (18in)

ERICA CARNEA 'VIVELLII'
♀ **Evergreen shrub** page **187**

ERICA X *DARLEYENSIS* 'FURZEY'
♀ **Evergreen shrub** This small shrub has dark foliage and deep pink flowers.
↕30cm (12in) ↔ 60cm (24in)

ERICA X *DARLEYENSIS* 'J. W. PORTER'
♀ **Evergreen shrub** Deep green foliage tipped with cream and red in spring, and deep pink flowers.
↕30cm (12in) ↔ 60cm (24in)

HAMAMELIS X *INTERMEDIA* CULTIVARS
♀ **Shrubs** pages **238–239**

HELLEBORUS ARGUTIFOLIUS
♀ **Perennial** page **252**

HELLEBORUS FOETIDUS
♀ **Perennial** page **253**

HELLEBORUS NIGER
♀ **Perennial** page **253**

HELLEBORUS X *NIGERCORS*
♀ **Perennial** Clump-forming plant with short, branched stems of white flowers, flushed green and pink.
↕30cm (12in) ↔ 1m (3ft)

ILEX AQUIFOLIUM 'GOLDEN MILKBOY'
♀ **Holly** page **273**

ILEX X *MESERVEAE* 'BLUE PRINCESS'
♀ **Holly** page **274**

IRIS UNGUICULARIS
♀ **Perennial** page **289**

JUNIPERUS COMMUNIS 'REPANDA'
♀ **Conifer** The foliage of this ground-hugging conifer is often bronze in winter.
↕20cm (8in) ↔ 1m (3ft)

LONICERA X *PURPUSII* 'WINTER BEAUTY'
♀ **Shrub** page **325**

MAHONIA X *MEDIA* 'LIONEL FORTESCUE'
♀ **Evergreen shrub** Divided leaves with leaflets like holly leaves and upright spikes of bright yellow flowers.
↕5m (15ft) ↔ 4m (12ft)

MAHONIA X *MEDIA* 'WINTER SUN'
♀ **Evergreen shrub** The appeal of this prickly, upright shrub is its scented yellow winter flowers.
↕5m (15ft) ↔ 4m (12ft)

PULMONARIA RUBRA
♀ **Evergreen perennial** page **428**

RUBUS THIBETANUS
♀ **Shrub** page **467**

SALIX BABYLONICA VAR. *PEKINENSIS* 'TORTUOSA'
♀ **Tree** page **469**

SALIX HASTATA 'WEHRHAHNII'
♀ **Shrub** page **470**

SKIMMIA JAPONICA 'NYMANS'
♀ **Evergreen shrub** This spreading shrub is female and bears good clusters of red berries.
↕1m (3ft) ↔ 2m (6ft)

STACHYURUS PRAECOX
♀ **Shrub** page **503**

SYMPHORICARPUS X *DOORENBOSII* 'WHITE HEDGE'
Shrub Suckering shrub of upright habit with white berries.
↕2m (6ft) ↔ indefinite

VIBURNUM X *BODNANTENSE* 'DAWN'
♀ **Shrub** page **534**

VIBURNUM FARRERI
♀ **Shrub** page **535**

VIBURNUM TINUS 'EVE PRICE'
♀ **Evergreen shrub** page **537**

BERRYING PLANTS

If birds do not enjoy the feast as soon as they ripen, berries can enhance the garden for many months. Red berries are most common, but there are black, white, yellow, pink, blue and even purple berries, to be included in almost any garden. Pale berries look best against a dark background such as an evergreen hedge, and red berries are attractive against a wintery, sunny sky.

ACTAEA ALBA
♀ **Perennial** Clump-forming plant with divided leaves and fluffy flowers followed by pearly, white berries with black eyes.
↕ 90cm (36in) ↔ 45–60cm (18–24in)

ARBUTUS UNEDO 'RUBRA'
♀ **Evergreen tree** The pink flowers and red, globular fruits are both at their best in autumn.
↕↔ 8m (25ft)

ARUM ITALICUM 'MARMORATUM'
♀ **Perennial** Evergreen, marbled foliage and spikes of red berries in autumn when the leaves die down.
↕ 30cm (12in) ↔ 15cm (6in)

BERBERIS DICTYOPHYLLA
♀ **Shrub** This deciduous shrub is at its best in winter when the white shoots are studded with red berries.
↕ 2m (6ft) ↔ 1.5m (5ft)

BERBERIS x *STENOPHYLLA* 'CORALLINA COMPACTA'
♀ **Evergreen shrub** page **85**

BERBERIS VERRUCULOSA
♀ **Evergreen shrub** page **87**

CALLICARPA BODINIERI VAR. *GIRALDII* 'PROFUSION'
♀ **Shrub** page **96**

CELASTRUS ORBICULATUS
Climber Strong-growing climber with yellow autumn colour and yellow fruits opening to reveal red seeds.
↕ 14m (46ft)

CLERODENDRON TRICHOTOMUM VAR. *FARGESII*
♀ **Shrub** Fast-growing plant with fragrant white flowers and turquoise berries set against red calyces.
↕↔ 5m (15ft)

CORIARIA TERMINALIS VAR. *XANTHOCARPA*
Shrub, not fully hardy Slightly tender, arching subshrub with small leaves and clusters of translucent yellow berries.
↕ 1m (3ft) ↔ 2m (6ft)

CORNUS 'NORMAN HADDEN'
♀ **Evergreen tree** Some leaves turn yellow and drop each autumn, when the cream and pink flowers are followed by large red fruits.
↕↔ 8m (25ft)

COTONEASTER CONSPICUUS 'DECORUS'
♀ **Evergreen shrub** page **146**

COTONEASTER 'ROTHSCHILDIANUS'
♀ **Evergreen shrub** An arching shrub with golden yellow berries in autumn after white flowers in summer.
↕↔ 5m (15ft)

EUONYMUS PLANIPES
♀ **Shrub** The foliage is bright red in autumn and falls to reveal red capsules containing orange seeds.
↕↔ 3m (10ft)

GAULTHERIA MUCRONATA 'WINTERTIME'
♀ **Evergreen shrub** page **220**

HIPPOPHAE RHAMNOIDES
♀ **Shrub** page **257**

HYPERICUM KOUYTCHENSE
♀ **Shrub** page **270**

ILEX AQUIFOLIUM & CULTIVARS
♀ **Evergreen shrubs** pages **272–273**

ILEX × MESERVEAE 'BLUE ANGEL'
♀ **Evergreen shrub** Compact, slow-growing shrub with glossy, dark, bluish-green leaves and red berries.
↕4m (12ft) ↔ 2m (6ft)

ILEX VERTICILLATA 'WINTER RED'
Shrub Deciduous shrub with white flowers in spring and masses of small red berries in winter.
↕2.5–3m (8–10ft) ↔ 3m (10ft)

LEYCESTERIA FORMOSA
Shrub Tall, arching stems tipped with white flowers within maroon bracts, followed by purple berries.
↕↔ 2m (6ft)

LONICERA NITIDA 'BAGGESEN'S GOLD'
♀ **Evergreen shrub** page **323**

LONICERA PERICLYMENUM 'GRAHAM THOMAS'
♀ **Climber** page **324**

PHYSALIS ALKEKENGI
♀ **Perennial** Creeping plant with upright stems: orange lanterns containing orange berries follow white flowers.
↕60–75cm (24–30in) ↔ 90cm (36in)

PYRACANTHA 'CADROU'
Evergreen shrub Spiny shrub with white flowers and red berries, resistant to scab.
↕↔ 2m (6ft)

ROSA 'FRU DAGMAR HASTRUP'
♀ **Shrub rose** page **464**

ROSA MOYESII 'GERANIUM'
♀ **Shrub rose** Arching, prickly stems with neat red flowers and large, long red hips.
↕2.5m (8ft) ↔ 1.5m (5ft)

ROSA 'SCHARLACHGLUT'
♀ **Shrub rose** Vigorous long-stemmed rose that can be trained as a climber, with showy scarlet flowers and bright scarlet hips.
↕3m (10ft) ↔ 2m (6ft)

SAMBUCUS RACEMOSA 'PLUMOSA AUREA'
Shrub Divided yellow leaves in summer and clusters of small, red berries.
↕↔ 3m (10ft)

SKIMMIA JAPONICA 'FRUCTU ALBO'
Evergreen shrub Neat evergreen with white flowers and bright, white fruits.
↕60cm (24in) ↔ 1m (3ft)

SORBUS ARIA 'LUTESCENS'
♀ **Tree** page **498**

SORBUS HUPEHENSIS VAR. *OBTUSA*
♀ **Tree** page **498**

TROPAEOLUM SPECIOSUM
♀ **Perennial climber** page **521**

VIBURNUM DAVIDII
♀ **Evergreen shrub** page **535**

VIBURNUM OPULUS 'XANTHOCARPUM'
♀ **Shrub** page **536**

CONIFERS FOR SMALL GARDENS

Conifers provide an amazing range of shapes, sizes, colours and textures. They can be used to give upright accents in borders, as dense screens, and for evergreen ground cover. They tolerate most soils but most, except yew, need sun. Many change colour with the seasons, and are especially brilliant in early summer when new growth contrasts with older foliage.

ABIES BALSAMEA F. *HUDSONIA*
♀ **Conifer** This very dwarf form grows into a rounded shrub, but does not bear cones.
↕60cm (24in) ↔ 1m (3ft)

ABIES KOREANA 'SILBERLOCKE'
♀ **Conifer** Attractive twisted foliage that reveals the silver reverse to the needles, and attractive cones.
↕10m (30ft) ↔ 6m (20ft)

ABIES LASIOCARPA 'COMPACTA'
♀ **Conifer** Slow-growing conical tree with blue-grey leaves.
↕3–5m (10–15ft) ↔ 2–3m (6–10ft)

ABIES NORDMANNIANA 'GOLDEN SPREADER'
♀ **Conifer** Dwarf, slow-growing plant with bright gold foliage.
↕1m (3ft) ↔ 1.5m (5ft)

CHAMAECYPARIS LAWSONIANA 'CHILWORTH SILVER'
♀ **Conifer** Slow-growing conical shrub with silver-grey foliage.
↕1.5m (5ft)

CHAMAECYPARIS LAWSONIANA 'ELLWOOD'S GOLD'
♀ **Conifer** page **117**

CHAMAECYPARIS OBTUSA 'NANA GRACILIS'
♀ **Conifer** page **118**

CHAMAECYPARIS OBTUSA 'TETRAGONA AUREA'
♀ **Conifer** page **119**

CRYPTOMERIA JAPONICA 'ELEGANS COMPACTA'
♀ **Conifer** page **155**

CRYPTOMERIA JAPONICA 'VILMORINIANA'
♀ **Conifer** Forms a tight ball of foliage that is green in summer but bronze in winter.
↕↔ 45cm (18in)

JUNIPERUS CHINENSIS 'BLAAUW'
♀ **Conifer** Forms a dense, upright shrub with blue-grey leaves.
↕1.2m (4ft) ↔ 1m (3ft)

JUNIPERUS CHINENSIS 'OBELISK'
♀ **Conifer** Grows slowly into an interesting, upright shape with bluish-green leaves.
↕2.5m (8ft) ↔ 60cm (24in)

JUNIPERUS COMMUNIS 'COMPRESSA'
♀ **Conifer** page **293**

JUNIPERUS × *PFITZERIANA* 'PFITZERIANA'
♀ **Conifer** page **293**

JUNIPERUS PROCUMBENS 'NANA'
♀ **Conifer** page **294**

JUNIPERUS SCOPULORUM 'BLUE HEAVEN'
♀ **Conifer** Neat in habit, with blue leaves and a conical shape.
↕2m (6ft) ↔ 60cm (24in)

JUNIPERUS SQUAMATA 'BLUE STAR'
♀ **Conifer** page **294**

JUNIPERUS SQUAMATA 'HOLGER'
♀ **Conifer** Spreading evergreen with bluish foliage which contrasts with the yellowish new growth.
↕↔ 2m (6ft)

MICROBIOTA DECUSSATA
♀ **Conifer** Spreading conifer with fine foliage that turns bronze in winter.
↕ 1m (3ft) ↔ indefinite

PICEA ABIES 'NIDIFORMIS'
♀ **Conifer** This slow-growing plant grows outwards to form a "nest" in the centre of the plant.
↕ 1.5m (5ft) ↔ 3–4m (10–12ft)

PICEA GLAUCA VAR. *ALBERTIANA* 'CONICA'
♀ **Conifer** page **400**

PICEA MARIANA 'NANA'
♀ **Conifer** page **401**

PICEA PUNGENS 'KOSTER'
♀ **Conifer** page **401**

PINUS MUGO 'MOPS'
♀ **Conifer** page **404**

PINUS PARVIFLORA 'ADCOCK'S DWARF'
♀ **Conifer** A dwarf cultivar of the Japanese white pine, with greyish leaves.
↕ 2m (6ft)

PINUS SYLVESTRIS 'BEUVRONENSIS'
♀ **Conifer** A rounded, dwarf cultivar of the Scots pine.
↕ 1m (3ft)

TAXUS BACCATA 'DOVASTONII AUREA'
♀ **Conifer** page **509**

TAXUS BACCATA 'FASTIGIATA AUREOMARGINATA'
♀ **Conifer** Upright accent plant with leaves margined in yellow, and red-fleshed (poisonous) berries.
↕ 3–5m (10–15ft) ↔ 1–2.5m (3–8ft)

TAXUS BACCATA 'FASTIGIATA'
♀ **Conifer** page **510**

TAXUS BACCATA 'REPENS AUREA'
♀ **Conifer** This spreading form of yew has golden leaves.
↕↔ 1–1.5m (3–5ft)

THUJA OCCIDENTALIS 'HOLMSTRUP'
♀ **Conifer** page **511**

THUJA OCCIDENTALIS 'RHEINGOLD'
♀ **Conifer** page **512**

THUJA OCCIDENTALIS 'SMARAGD'
♀ **Conifer** A dwarf, conical bush with bright green leaves.
↕ 1m (3ft) ↔ 80cm (32in)

THUJA ORIENTALIS 'AUREA NANA'
♀ **Conifer** page **512**

THUJA PLICATA 'STONEHAM GOLD'
♀ **Conifer** page **513**

TSUGA CANADENSIS 'JEDDELOH'
♀ **Conifer** page **521**

TREES FOR SMALL GARDENS

Trees add shade and character to gardens, but large forest trees should never be planted in small gardens or too near homes. Beech, oaks and ash are often cheap to buy, but it is best too look for small trees that will give interest over a long period during the year. Consider the shade they will cast: dense evergreens can create areas that are dry and dark where little will grow.

ACER DAVIDII 'ERNEST WILSON'
Deciduous tree Unlobed leaves turn orange in autumn before falling to show the green, white-streaked branches.
↕8m (25ft) ↔ 10m (30ft)

ACER PALMATUM CULTIVARS
♀ **Deciduous trees or large shrubs** pages **34–35**

AMELANCHIER X *GRANDIFLORA* 'BALLERINA'
♀ **Deciduous tree** page **55**

BETULA UTILIS VAR. *JACQUEMONTII*
♀ **Deciduous tree** page **90**

CERCIS SILIQUASTRUM
♀ **Deciduous tree** page **115**

CORNUS 'EDDIE'S WHITE WONDER'
♀ **Deciduous tree** Multi-stemmed tree that bears deep purple, small flowers surrounded by large white bracts.
↕6m (20ft) ↔ 5m (15ft)

CRATAEGUS LAEVIGATA 'PAUL'S SCARLET'
♀ **Deciduous tree** page **149**

GENISTA AETNENSIS
♀ **Deciduous tree or large shrub** page **223**

GLEDITSIA TRIACANTHOS 'RUBYLACE'
Deciduous tree Elegant divided foliage that is bright wine-red when young, fading to bronzed green.
↕12m (40ft) ↔ 10m (30ft)

LABURNUM X *WATERERI* 'VOSSII'
♀ **Deciduous tree** page **300**

LIGUSTRUM LUCIDUM
♀ **Evergreen tree or large shrub** page **313**

MAGNOLIA 'HEAVEN SCENT'
♀ **Deciduous tree** Goblet-shaped pink flowers, with white interiors, in spring and early summer.
↕↔ 10m (30ft)

MAGNOLIA X *LOEBNERI* 'LEONARD MESSEL'
♀ **Deciduous tree** page **333**

MALUS CORONARIA 'CHARLOTTAE'
Deciduous tree Spreading, with fragrant, semi-double, pale pink flowers in spring.
↕↔ 9m (28ft)

MALUS TSCHONOSKII
♀ **Deciduous tree** page **337**

PRUNUS SERRULA
♀ **Deciduous tree** page **423**

PYRUS CALLERYANA 'CHANTICLEER'
♀ **Deciduous tree** page **431**

SALIX CAPREA 'KILMARNOCK'
♀ **Deciduous tree** page **470**

SALIX 'ERYTHROFLEXUOSA'
Deciduous tree Semi-weeping tree with twisted, orange-yellow shoots.
↕↔ 5m (15ft)

SORBUS 'JOSEPH ROCK'
♀ **Deciduous tree** page **499**

STYRAX JAPONICUS
♀ **Deciduous tree** page **504**

STYRAX OBASSIA
♀ **Deciduous tree** page **504**

HEDGE PLANTS WITH ATTRACTIVE FOLIAGE

Formal hedging plants must be tolerant of regular clipping. Those that have one flush of growth each year, such as yew, only need clipping once a season, unlike privet that may require trimming twice or three times. Evergreens are most popular, but deciduous plants still reduce wind speed, and are often cheap to buy. (N.B Dimensions below are ultimate sizes for unclipped plants.)

BUXUS SEMPERVIRENS
'ELEGANTISSIMA'
♀ **Evergreen shrub** page **95**

BUXUS SEMPERVIRENS
'SUFFRUTICOSA'
♀ **Evergreen shrub** page **95**

CHAMAECYPARIS LAWSONIANA
'FLETCHERI'
♀ **Conifer** Dense, grey foliage on an erect, compact shrub.
↕12m (40ft)

CHAMAECYPARIS LAWSONIANA
'LANE'
♀ **Conifer** page **117**

CHAMAECYPARIS LAWSONIANA
'PEMBURY BLUE'
♀ **Conifer** page **117**

x *CUPRESSOCYPARIS LEYLANDII*
Conifer Very vigorous plant that can be managed if trimmed at an early stage, and then regularly.
↕35m (120ft) ↔ 5m (15ft)

x *CUPRESSOCYPARIS LEYLANDII*
'HAGGERSTON GREY'
♀ **Conifer** page **156**

x *CUPRESSOCYPARIS LEYLANDII*
'ROBINSON'S GOLD'
♀ **Conifer** The best gold form, with foliage that is bronze when young.
↕35m (120ft) ↔ 5m (15ft)

FAGUS SYLVATICA
♀ **Deciduous tree** Common beech and its purple form retain their dead leaves in winter if trimmed to 2m (6ft).
↕25m (80ft) ↔ 15m (50ft)

LIGUSTRUM OVALIFOLIUM 'AUREUM'
♀ **Evergreen shrub** Good choice where a bright yellow hedge is required, and regular clipping is possible.
↕↔ 4m (12ft)

PRUNUS x *CISTENA*
♀ **Deciduous shrub** page **421**

PRUNUS LAUROCERASUS
♀ **Evergreen shrub** Common laurel requires careful pruning but can be attractive, and withstands hard pruning well.
↕8m (25ft) ↔ 10m (30ft)

PRUNUS LAUROCERASUS
'OTTO LUYKEN'
♀ **Evergreen shrub** page **422**

PRUNUS LUSITANICA
♀ **Evergreen shrub** Pleasant evergreen with dark green leaves on red stalks, and white flowers if not clipped.
↕↔ 20m (70ft)

TAXUS BACCATA
♀ **Evergreen tree** page **509**

FLOWERING HEDGES

Flowering hedges add much more than structural elements and security to the garden: they can become a focus of attention. Many flowering shrubs that tolerate pruning can be used but, because of the pruning required to maintain flowering at its best, they may not be suitable for dense, formal hedges, or boundary hedges where year-round screening is required.

ESCALLONIA 'APPLE BLOSSOM'
♀ **Evergreen shrub** page **194**

FORSYTHIA X *INTERMEDIA* 'LYNWOOD'
♀ **Shrub** page **208**

FUCHSIA 'RICCARTONII'
♀ **Shrub** page **212**

HEBE 'MIDSUMMER BEAUTY'
♀ **Evergreen shrub** Bright green leaves and purple flowers, fading to white, on short spikes in mid- to late summer.
↕1m (3ft) ↔ 1.2m (4ft)

HYPERICUM 'ROWALLANE'
♀ **Shrub** Semi-evergreen, bearing clusters of yellow, cupped flowers in summer on arching stems.
↕2m (6ft) ↔ 1m (3ft)

LAVANDULA ANGUSTIFOLIA 'HIDCOTE'
♀ **Evergreen shrub** page **305**

OSMANTHUS X *BURKWOODII*
♀ **Evergreen shrub** page **359**

PHILADELPHUS CORONARIUS 'AUREUS'
♀ **Shrub** Golden yellow foliage that may scorch in full sun on poor soil, and fragrant white flowers.
↕2.5m (8ft) ↔ 1.5m (5ft)

POTENTILLA FRUTICOSA 'PRIMROSE BEAUTY'
♀ **Shrub** page **410**

PRUNUS X *CISTENA*
♀ **Shrub** page **421**

PRUNUS LUSITANICA SUBSP. *AZORICA*
♀ **Shrub, borderline hardy** page **423**

RHODODENDRON 'HINO-MAYO'
♀ **Evergreen shrub** page **436**

RIBES SANGUINEUM 'PULBOROUGH SCARLET'
♀ **Shrub** page **446**

ROSA 'BUFF BEAUTY'
♀ **Modern shrub rose** page **462**

ROSA 'CHINATOWN'
♀ **Cluster-flowered bush rose** page **450**

ROSA 'FELICIA'
♀ **Modern shrub rose** page **462**

SPIRAEA X *VANHOUTTEI*
♀ **Shrub** page **502**

SYRINGA PUBESCENS SUBSP. *MICROPHYLLA* 'SUPERBA'
♀ **Shrub** page **506**

VIBURNUM TINUS 'GWENLLIAN'
♀ **Evergreen shrub** Dense shrub with pinkish flowers that open from deep pink buds.
↕↔ 3m (10ft)

SPINY HEDGES

There are places in the garden, usually around the edges, where the physical barrier of a hedge is not enough, and plants with spines are needed to ensure privacy and prevent the access of animals. Though these plants have many advantages, pruning and clipping must be carefully done, and dropped twigs can cause more discomfort in future when weeding at the base.

BERBERIS DARWINII
♀ **Evergreen shrub** page **84**

BERBERIS X *OTTAWENSIS* 'SUPERBA'
♀ **Shrub** page **85**

BERBERIS X *STENOPHYLLA*
♀ **Evergreen shrub** Long, arching shoots are covered with small orange flowers in spring, and spines all year.
↕3m (10ft) ↔ 5m (15ft)

BERBERIS THUNBERGII
♀ **Shrub** Spiny stems have purple leaves that turn red before they fall in autumn.
↕1m (3ft) ↔ 2.5m (8ft)

CRATAEGUS MONOGYNA
Tree The common hawthorn forms a quick-growing spiny hedge, but is not as attractive as some.
↕10m (30ft) ↔ 8m (25ft)

ILEX AQUIFOLIUM & CULTIVARS
♀ **Evergreen shrubs** pages **272–273**

MAHONIA JAPONICA
♀ **Evergreen shrub** page **334**

MAHONIA X *MEDIA* 'BUCKLAND'
♀ **Evergreen shrub** page **335**

PONCIRUS TRIFOLIATA
Shrub Angular green shoots with vicious spines, fragrant white flowers and orange-like fruits in autumn.
↕↔ 5m (15ft)

PRUNUS SPINOSA
Tree The blackthorn is a dense shrub, with white spring flowers and sloes in autumn.
↕5m (15ft) ↔ 4m (12ft)

PYRACANTHA 'ORANGE GLOW'
♀ **Evergreen shrub** page **430**

PYRACANTHA 'WATERERI'
♀ **Evergreen shrub** page **431**

ROSA GLAUCA
♀ **Shrub rose** page **464**

ROSA RUGOSA 'ALBA'
♀ **Shrub rose** Thickets of prickly stems with white flowers followed by large red hips.
↕↔ 1–2.5m (3–8ft)

ROSA RUGOSA 'RUBRA'
♀ **Shrub rose** page **465**

GROUND COVER FOR SUN

Many plants with creeping, trailing or clump-forming habits can be planted to form ground cover in sunny gardens. However, most will only suppress new weeds, and very few will actively smother existing weeds, so clear the soil of all perennial weeds before you plant. When planting, mix different plants to create interest and add a few taller plants to prevent a flat effect.

ALCHEMILLA MOLLIS
♀ **Perennial** page **49**

ARTEMISIA STELLERIANA 'BOUGHTON SILVER'
Perennial Divided, evergreen, silver foliage that forms dense mats.
↕15cm (6in) ↔ 30–45cm (12–18in)

CAMPANULA GLOMERATA 'SUPERBA'
♀ **Perennial** page **106**

CEANOTHUS THYRSIFLORUS VAR. *REPENS*
♀ **Evergreen shrub, not fully hardy** page **113**

CORNUS CANADENSIS
♀ **Perennial** page **138**

DICENTRA 'STUART BOOTHMAN'
♀ **Perennial** page **176**

ERICA x *DARLEYENSIS* 'JENNY PORTER'
♀ **Evergreen shrub** page **187**

GERANIUM 'JOHNSON'S BLUE'
♀ **Perennial** page **227**

GERANIUM x *OXONIANUM* 'WARGRAVE PINK'
♀ **Perennial** page **229**

HOSTA FORTUNEI VAR. *AUREOMARGINATA*
♀ **Perennial** page **259**

JUNIPERUS SQUAMATA 'BLUE CARPET'
♀ **Conifer** Low-growing plant with shoots that lift from the ground at a gentle angle.
↕30–45cm (12–18in) ↔ 1.5–1.8m (5–6ft)

LAMIUM MACULATUM 'WHITE NANCY'
♀ **Perennial** page **301**

OSTEOSPERMUM JUCUNDUM
♀ **Perennial, borderline hardy** page **360**

PERSICARIA VACCINIFOLIA
♀ **Perennial** page **382**

PHALARIS ARUNDINACEA 'PICTA'
♀ **Perennial grass** page **383**

PHLOMIS RUSSELIANA
♀ **Perennial** page **386**

PHLOX SUBULATA 'MCDANIEL'S CUSHION'
♀ **Alpine** The mossy foliage is covered with starry, pink flowers in late spring.
↕5–15cm (2–6in) ↔ 50cm (20in)

POTENTILLA MEGALANTHA
♀ **Perennial** page **411**

ROSA GROUND COVER TYPES
♀ **Shrubs** pages **458–459**

ROSMARINUS OFFICINALIS 'SEVERN SEA'
♀ **Evergreen shrub** A mound-forming plant with arching branches and bright blue flowers.
↕1m (3ft) ↔ 1.5m (5ft)

SEMPERVIVUM CILIOSUM
♀ **Alpine** page **492**

VERONICA GENTIANOIDES
♀ **Perennial** page **532**

VIOLA 'NELLIE BRITTON'
♀ **Perennial** page **539**

GROUND COVER FOR SHADE

Shade is often considered to be a problem but there are lots of plants to use as ground cover that do not need full sun. However, the more dense the shade, the less choice there is. Luckily, those that tolerate the worst conditions are evergreen, though slow growing. In less hostile conditions many of these plants will quickly spread: an excellent alternative to grass under trees.

ADIANTUM VENUSTUM
♀ **Fern** page **44**

AJUGA REPTANS 'CATLIN'S GIANT'
♀ **Perennial** Very large, purple leaves and tall, blue flower spikes.
↕20cm (8in) ↔ 60–90cm (24–36in)

BERGENIA 'SILBERLICHT'
♀ **Perennial** page **88**

CONVALLARIA MAJALIS
♀ **Perennial** page **135**

COTONEASTER DAMMERI
♀ **Evergreen shrub** Spreading shrub bearing white flowers and red berries.
↕20cm (8in) ↔ 2m (6ft)

EPIMEDIUM × *PERRALCHICUM*
♀ **Perennial** page **184**

EUONYMUS FORTUNEI 'EMERALD GAIETY'
♀ **Evergreen shrub** Bushy, with white-edged leaves tinted pink in winter.
↕1m (3ft) ↔ 1.5m (5ft)

EUPHORBIA AMYGDALOIDES VAR. *ROBBIAE*
♀ **Perennial** page **200**

GAULTHERIA PROCUMBENS
♀ **Evergreen shrub** page **221**

GERANIUM MACRORRHIZUM 'CZAKOR'
Perennial Mats of scented foliage tinted with purple in autumn, and magenta flowers in summer.
↕50cm (20in) ↔ 60cm (24in)

GERANIUM SYLVATICUM 'ALBUM'
♀ **Perennial** For moist soil, with deeply-lobed leaves and small, white flowers.
↕75cm (30in) ↔ 60cm (24in)

HEDERA HIBERNICA
♀ **Evergreen climber** page **247**

HEUCHERA 'RED SPANGLES'
♀ **Perennial** page **256**

HOSTA 'FRANCES WILLIAMS'
♀ **Perennial** page **259**

OMPHALODES CAPPADOCICA
♀ **Perennial** page **355**

PACHYSANDRA TERMINALIS
♀ **Perennial** page **362**

POLYSTICHUM SETIFERUM
♀ **Fern** page **409**

SANGUINARIA CANADENSIS 'PLENA'
♀ **Perennial** For moist soil, with large glaucous leaves and white, double flowers.
↕15cm (6in) ↔ 30cm (12in)

TIARELLA CORDIFOLIA
♀ **Perennial** page **516**

TOLMIEA 'TAFF'S GOLD'
♀ **Perennial** page **516**

TRACHYSTEMON ORIENTALIS
Perennial Large, rough, heart-shaped leaves and borage-like flowers.
↕30cm (12in) ↔ 1m (3ft)

VINCA MINOR 'ARGENTEOVARIEGATA'
♀ **Perennial** Pale blue flowers among grey-green and cream leaves.
↕15cm (6in) ↔ 1m (36in)

PLANTS WITH SCENTED FOLIAGE

While flowers tend to have sweet, fruity perfumes, leaf scents tend to be more spicy or resinous, though some, especially scented pelargoniums, mimic other plants. Some plants waft their perfume onto the air, and others need gentle stroking. It is likely that these plants evolved their scents to make themselves less appealing to insect pests; gardeners find them irresistible.

ALOYSIA TRIPHYLLA
♀ **Shrub** page **54**

AMICIA ZYGOMERIS
Perennial Unusual plant, related to beans, with grey-green foliage that smells of cucumber when crushed.
↕2.2m (7ft) ↔ 1.2m (4ft)

CALOCEDRUS DECURRENS
♀ **Conifer** A columnar tree with foliage that is sweetly scented when crushed.
↕20–40m (70–130ft) ↔ 2–9m (6–28ft)

CALYCANTHUS OCCIDENTALIS
Shrub Large leaves with a spicy scent and brick-red flowers that smell of vinegar.
↕3m (10ft) ↔ 4m (12ft)

CERCIDIPHYLLUM JAPONICUM
♀ **Tree** In autumn the leaves of this graceful tree turn orange and red, and smell of caramel.
↕20m (70ft) ↔ 15m (50ft)

CHAMAEMELUM NOBILE 'TRENEAGUE'
Perennial This non-flowering form of chamomile hugs the soil, and its foliage smells fruity when gently crushed.
↕10cm (4in) ↔ 45cm (18in)

CISTUS X *HYBRIDUS*
♀ **Evergreen shrub** page **127**

CISTUS LADANIFER
♀ **Evergreen shrub** The dark green leaves are sticky and fragrant; the flowers are white with a yellow eye.
↕2m (6ft) ↔ 1.5m (5ft)

HELICHRYSUM ITALICUM
♀ **Evergreen shrub** Narrow, silver foliage on a small shrub that smells of curry.
↕60cm (24in) ↔ 1m (3ft)

HOUTTUYNIA CORDATA 'FLORE PLENO'
Perennial A rather invasive, creeping plant with purplish leaves that have a strong citrus smell when crushed.
↕15–30cm (6–12in) ↔ indefinite

LAVANDULA ANGUSTIFOLIA 'TWICKEL PURPLE'
♀ **Evergreen shrub** page **305**

LAVANDULA STOECHAS
♀ **Shrub** The purple flowerheads are topped with purple bracts.
↕↔ 60cm (24in)

MELISSA OFFICINALIS 'AUREA'
Perennial Form of lemon balm with yellow splashes on the leaves.
↕1m (3ft) ↔ 45cm (18in)

MENTHA SUAVEOLENS 'VARIEGATA'
Perennial Variegated apple mint has a pleasant fragrance and showy leaves.
↕1m (3ft) ↔ indefinite

MONARDA 'CAMBRIDGE SCARLET'
♀ **Perennial** page **342**

MONARDA 'SCORPION'
Perennial Whorls of bracts and violet flowers on tall, leafy stems.
↕1.5m (5ft) ↔ 1m (3ft)

ORIGANUM LAEVIGATUM
♀ **Perennial** page **357**

ORIGANUM LAEVIGATUM 'HERRENHAUSEN'
♀ **Perennial** page **358**

PELARGONIUMS, SCENTED-LEAVED
♀ **Perennials, not hardy** pages **372–373**

PERILLA FRUTESCENS VAR. *CRISPA*
♀ **Annual** page **380**

PEROVSKIA ATRIPLICIFOLIA
♀ **Shrub** The upright stems carry tiny, blue flowers in autumn but the greyish leaves are fragrant all summer.
↕ 1.2m (4ft) ↔ 1m (3ft)

PEROVSKIA 'BLUE SPIRE'
♀ **Subshrub** page **380**

PROSTANTHERA CUNEATA
♀ **Evergreen shrub** A slightly tender plant with small, mint-scented leaves and pretty white flowers in summer.
↕↔ 30–90cm (12–36in)

PROSTANTHERA ROTUNDIFOLIA
♀ **Shrub, not hardy** In late spring the mint-scented leaves are smothered in lilac flowers.
↕ 2–4m (6–12ft) ↔ 1–3m (3–10ft)

PSEUDOTSUGA MENZIESII
♀ **Conifer** A large tree with resinous foliage.
↕ 25–50m (80–160ft) ↔ 6–10m (20–30ft)

PTELEA TRIFOLIATA 'AUREA'
♀ **Tree** The gold foliage and bark is strongly scented.
↕ 5m (15ft)

ROSA EGLANTERIA
♀ **Species rose** Very thorny, arching shoots with small pink flowers, and foliage that smells of apples when wet.
↕↔ 2.5m (8ft)

ROSMARINUS OFFICINALIS 'SILVER SPIRES'
Evergreen shrub The culinary rosemary, but with silver-variegated foliage on an upright plant.
↕ 1m (3ft) ↔ 60cm (24in)

SALVIA DISCOLOR
♀ **Shrub, not hardy** page **473**

SALVIA OFFICINALIS 'ICTERINA'
♀ **Subshrub** page **476**

SKIMMIA X *CONFUSA* 'KEW GREEN'
♀ **Evergreen shrub** page **494**

PLANTS WITH SCENTED FLOWERS

Fragrance is too often forgotten when planting a garden. Yet there are as many shades of fragrance as there are of colours: they can affect mood, take you back to your childhood, or whisk you off to a far-off land with a single sniff. The most strongly scented flowers are often white or insignificant in appearance, but have evolved to make their presence felt in other ways.

ABELIA CHINENSIS
Shrub Spreading, with heads of small pale pink flowers in late summer.
↕1.5m (5ft) ↔ 2.5m (8ft)

BUDDLEJA ALTERNIFOLIA
♀ **Shrub** page **92**

CAMELLIA 'INSPIRATION'
♀ **Evergreen shrub** page **99**

CAMELLIA JAPONICA 'ADOLPHE AUDUSSON'
♀ **Evergreen shrub** page **100**

CAMELLIA JAPONICA 'ELEGANS'
♀ **Evergreen shrub** page **100**

CHIMONANTHUS PRAECOX 'GRANDIFLORUS'
♀ **Shrub** page **120**

CHIMONANTHUS PRAECOX 'LUTEUS'
♀ **Shrub** After the large leaves fall, pale yellow flowers scent the winter air.
↕4m (12ft) ↔ 3m (10ft)

CHOISYA TERNATA
♀ **Evergreen shrub** page **121**

CHOISYA 'AZTEC PEARL'
♀ **Evergreen shrub** Narrowly divided, deep green leaves and white, pink-tinged flowers in spring and autumn.
↕↔ 2.5m (8ft)

CLEMATIS MONTANA F. *GRANDIFLORA*
♀ **Climber** page **128**

DAPHNE BHOLUA 'GURKHA'
♀ **Shrub** page **164**

DAPHNE TANGUTICA RETUSA GROUP
♀ **Evergreen shrubs** page **165**

DIANTHUS 'DORIS'
♀ **Perennial** page **172**

ERICA ERIGENA 'GOLDEN LADY'
♀ **Evergreen shrub** page **187**

HAMAMELIS (WITCH HAZELS)
♀ **Shrubs** pages **238–239**

HOSTA 'HONEYBELLS'
♀ **Perennial** page **260**

JASMINUM OFFICINALE
♀ **Climber** A strong, twining climber with white flowers in summer which have an intense, sweet scent.
↕12m (40ft)

JASMINUM OFFICINALE 'ARGENTEOVARIEGATUM'
♀ **Climber** page **292**

LILIUM PINK PERFECTION GROUP
♀ **Bulb** page **317**

LONICERA CAPRIFOLIUM
♀ **Climber** The Italian honeysuckle has pink and cream, fragrant flowers in summer.
↕6m (20ft)

LONICERA PERICLYMENUM 'BELGICA'
♀ **Climber** In early summer this honeysuckle produces creamy-yellow flowers streaked with maroon.
↕7m (22ft)

LONICERA PERICLYMENUM 'GRAHAM THOMAS'
♀ **Climber** page **324**

MAGNOLIA GRANDIFLORA 'GOLIATH'
♀ **Evergreen tree** page **331**

MAHONIA X *MEDIA* 'CHARITY'
♀ **Evergreen shrub** page **335**

OSMANTHUS DELAVAYI
♀ **Evergreen shrub** page **359**

PAEONIA LACTIFLORA 'DUCHESSE DE NEMOURS'
♀ **Perennial** page **364**

PHILADELPHUS 'BEAUCLERK'
♀ **Shrub** page **384**

PHILADELPHUS 'BELLE ETOILE'
♀ **Shrub** page **384**

PHLOX PANICULATA 'WHITE ADMIRAL'
♀ **Perennial** Large heads of pure white flowers with a sweet, peppery scent, borne in summer.
↕1m (3ft)

PITTOSPORUM TENUIFOLIUM
♀ **Evergreen shrub** page **404**

PITTOSPORUM TOBIRA
♀ **Evergreen shrub or small tree** page **405**

PRIMULA FLORINDAE
♀ **Perennial** page **414**

ROSA 'ALBERTINE'
♀ **Rambler rose** page **454**

ROSA 'ARTHUR BELL'
♀ **Cluster-flowered bush rose** page **450**

ROSA 'BLESSINGS'
♀ **Large-flowered bush rose** page **448**

ROSA 'COMPASSION'
♀ **Climbing rose** page **452**

ROSA 'GRAHAM THOMAS'
♀ **Modern shrub rose** page **463**

ROSA 'ICEBERG CLIMBING'
♀ **Climbing rose** This fine rose has many pure white flowers all summer.
↕2.5m (8ft)

ROSA 'JUST JOEY'
♀ **Large-flowered bush rose** page **449**

ROSA 'MARGARET MERRIL'
♀ **Cluster-flowered bush rose** page **451**

ROSA 'PEACE'
♀ **Large-flowered bush rose** page **449**

ROSA 'PENELOPE'
♀ **Modern shrub rose** page **463**

ROSA 'REMEMBER ME'
♀ **Large-flowered bush rose** page **449**

SARCOCOCCA HOOKERIANA VAR. *DIGYNA*
♀ **Evergreen shrub** page **483**

SKIMMIA JAPONICA 'RUBELLA'
♀ **Evergreen shrub** page **495**

SMILACINA RACEMOSA
♀ **Perennial** page **495**

SYRINGA MEYERI 'PALIBIN'
♀ **Shrub** page **505**

SYRINGA VULGARIS 'MME LEMOINE'
♀ **Shrub** page **507**

ULEX EUROPAEUS 'FLORE PLENO'
♀ **Evergreen shrub** Spiny bush that has a few of its double flowers, scented of coconut, open almost all year.
↕2.5m (8ft) ↔2m (6ft)

VIBURNUM X *BURKWOODII* 'PARK FARM HYBRID'
♀ **Evergreen shrub** Upright shrub with bronze new leaves and deep pink, scented flowers in late spring.
↕3m (10ft) ↔2m (6ft)

VIBURNUM CARLESII 'AURORA'
♀ **Shrub** Bushy shrub with pink flowers in late spring, opening from red buds.
↕↔2m (6ft)

PLANTS FOR PAVING

Plants help to break up large expanses of paving or gravel. The clean surface also helps prevent the delicate flowers of small plants becoming splashed with soil, and reflects heat back up to sun-loving plants. Few plants tolerate much treading; use only the toughest, such as thymes and chamomile, where there is heavy foot traffic. Less busy areas can be home to dwarf shrubs and alpines.

ACAENA 'BLUE HAZE'
Perennial A vigorous, spreading perennial with divided, grey-blue leaves and round, white flowerheads that are followed by red burrs.
↕ 10–15cm (4–6in) ↔ 1m (3ft)

ACAENA MICROPHYLLA
♀ **Perennial** page **31**

ACHILLEA AGERATIFOLIA
♀ **Perennial** page **39**

ACHILLEA x *LEWISII* 'KING EDWARD'
♀ **Perennial** page **40**

AETHIONEMA 'WARLEY ROSE'
♀ **Shrub** page **46**

AJUGA REPTANS 'ATROPURPUREA'
♀ **Perennial** page **48**

ANTHEMIS PUNCTATA SUBSP. *CUPANIANA*
♀ **Perennial** page **61**

ARENARIA MONTANA
♀ **Perennial** page **66**

ARMERIA JUNIPERIFOLIA
♀ **Subshrub** page **68**

CAMPANULA COCHLEARIFOLIA
♀ **Perennial** page **106**

CHAMAEMELUM NOBILE 'TRENEAGUE'
Perennial This non-flowering form of chamomile hugs the soil, and its foliage smells fruity when gently crushed.
↕ 10cm (4in) ↔ 45cm (18in)

DIANTHUS 'PIKE'S PINK'
♀ **Perennial** page **173**

DIASCIA 'JOYCE'S CHOICE'
♀ **Perennial** Early-flowering, with long spikes of pale, apricot-pink flowers in summer.
↕ 30cm (12in) ↔ 45cm (18in)

ERIGERON KARVINSKIANUS
♀ **Perennial** page **190**

ERINUS ALPINUS
♀ **Perennial** page **190**

HELIANTHEMUM 'RHODANTHE CARNEUM'
♀ **Evergreen shrub** page **248**

LYSIMACHIA NUMMULARIA 'AUREA'
♀ **Perennial** page **329**

PENSTEMON RUPICOLA
♀ **Dwarf shrub** Tiny compared with border penstemons: evergreen, with leathery leaves and small, tubular deep pink flowers in early summer.
↕ 10cm (4in) ↔ 45cm (18in)

PHLOX DOUGLASII 'CRACKERJACK'
♀ **Perennial** page **388**

PHLOX DOUGLASII 'RED ADMIRAL'
♀ **Perennial** page **388**

PRATIA PEDUNCULATA
Perennial Mildly invasive creeping plant with tiny leaves and star-shaped, pale blue flowers in summer.
↕ 1.5cm (½in) ↔ indefinite

SAPONARIA OCYMOIDES
♀ **Perennial** page **481**

SEDUM ACRE 'AUREUM'
Perennial Rather invasive succulent with tiny shoots and leaves that are yellow when young, and yellow flowers.
↕5cm (2in) ↔ 60cm (24in)

SEDUM KAMTSCHATICUM 'VARIEGATUM'
♀ **Perennial** page **488**

SEDUM SPATHULIFOLIUM 'PURPUREUM'
♀ **Perennial** page **489**

SEDUM SPURIUM 'SCHORBUSER BLUT'
♀ **Perennial** page **490**

SOLEIROLIA SOLEIROLII 'AUREA'
Perennial Slightly tender but very invasive plant with bright lime-green, tiny leaves that form mounds and cushions.
↕5cm (2in) ↔ 1m (3ft)

THYMUS x *CITRIODORUS* 'BERTRAM ANDERSON'
♀ **Evergreen shrub** page **514**

THYMUS x *CITRIODORUS* 'SILVER QUEEN'
♀ **Evergreen shrub** The leaves of this cultivar are variegated, and look good with the lavender-pink flowers.
↕30cm (12in) ↔ 25cm (10in)

THYMUS 'PINK CHINTZ'
♀ **Perennial** Trailing stems that root as they grow, with greyish leaves and pink flowers loved by bees.
↕25cm (10in) ↔ 45cm (18in)

THYMUS POLYTRICHUS SUBSP. *BRITANNICUS* 'ALBUS'
♀ **Subshrub** A mat-forming woody plant with hairy leaves and white flowers.
↕5cm (2in) ↔ 60cm (24in)

THYMUS SERPYLLUM VAR. *COCCINEUM*
♀ **Subshrub** page **515**

VIOLA 'JACKANAPES'
♀ **Perennial** page **539**

ARCHITECTURAL PLANTS

Every garden needs plants that are larger than life, with the sort of shape or texture that cannot be ignored. Too many of these plants will give a sub-tropical effect, with bold leaves and spiky shapes, but if carefully placed among less extraordinary plants, they become the focus of a view, or a "full stop" in the border. Make use of light and shade to emphasise bold silhouettes.

ACANTHUS SPINOSUS
♀ **Perennial** page **31**

AESCULUS PARVIFLORA
♀ **Shrub** page **46**

AGAVE AMERICANA
♀ **Succulent** Viciously spiny plant with steel-grey leaves that curve to make a magnificent rosette.
↕2m (6ft) ↔ 3m (10ft)

AILANTHUS ALTISSIMA
♀ **Tree** This can be a tall tree but will produce divided leaves 1.2m (4ft) long if cut back hard every year.
↕25m (80ft) ↔ 15m (50ft)

ARAUCARIA HETEROPHYLLA
♀ **Conifer** page **64**

BETULA NIGRA
♀ **Tree** page **89**

BETULA PENDULA 'YOUNGII'
♀ **Tree** page **89**

CATALPA BIGNONIOIDES 'AUREA'
♀ **Tree** Spreading tree that can be pruned hard annually for large, gold leaves.
↕↔ 10m (30ft)

CHAMAECYPARIS NOOTKATENSIS 'PENDULA'
♀ **Conifer** page **118**

CHAMAECYPARIS PISIFERA 'FILIFERA AUREA'
♀ **Conifer** A broad, arching shrub with whip-like, golden shoots.
↕12m (40ft) ↔ 5m (15ft)

CORYLUS AVELLANA 'CONTORTA'
♀ **Shrub** page **143**

CROCOSMIA MASONIORUM
♀ **Perennial** page **152**

CYCAS REVOLUTA
♀ **Palm-like tree** page **157**

ERYNGIUM GIGANTEUM
♀ **Biennial** Rosettes of deep green leaves, and spiny, white stems and flowerheads in the second year.
↕90cm (36in) ↔ 30cm (12in)

EUCALYPTUS PAUCIFLORA SUBSP. *NIPHOPHILA*
♀ **Evergreen tree** page **197**

EUPHORBIA CHARACIAS
♀ **Perennial** page **201**

FAGUS SYLVATICA 'PENDULA'
♀ **Tree** A tree of huge proportions with horizontal and arching branches cascading to the ground.
↕15m (50ft) ↔ 20m (70ft)

FARGESIA NITIDA
♀ **Ornamental grass** page **205**

GUNNERA MANICATA
♀ **Perennial** page **233**

HELIANTHUS 'MONARCH'
♀ **Perennial** page **250**

KNIPHOFIA CAULESCENS
♀ **Perennial** page **297**

MACLEAYA × *KEWENSIS* 'KELWAY'S CORAL PLUME'
♀ **Perennial** page **330**

MELIANTHUS MAJOR
♀ **Perennial, not hardy** page **340**

PAEONIA DELAVAYI
♀ **Shrub** page **363**

PAULOWNIA TOMENTOSA
♀ **Tree** page **371**

PHORMIUM COOKIANUM SUBSP. *HOOKERI* 'CREAM DELIGHT'
♀ **Perennial** page **392**

PHORMIUM TENAX
♀ **Perennial** page **393**

PHORMIUM TENAX PURPUREUM GROUP
♀ **Perennials** page **394**

PHYLLOSTACHYS AUREA
♀ **Bamboo** The golden bamboo, with yellow-brown canes and yellow-green leaves.
↕2–10m (6–30ft) ↔ indefinite

PHYLLOSTACHYS AUREOSULCATA 'AUREOCAULIS'
Bamboo Large bamboo with bright, golden canes and narrow leaves.
↕3–6m (10–20ft) ↔ indefinite

PHYLLOSTACHYS NIGRA
♀ **Bamboo** page **398**

PHYLLOSTACHYS NIGRA VAR. *HENONIS*
♀ **Bamboo** page **399**

PLEIOBLASTUS VARIEGATUS
♀ **Bamboo** page **407**

PRUNUS 'AMANOGAWA'
♀ **Tree** Very slender, upright growth, resembling a Lombardy poplar, with semi-double pink flowers in spring.
↕8m (25ft) ↔ 4m (12ft)

PRUNUS 'KIKU-SHIDARE-ZAKURA'
♀ **Tree** page **422**

RODGERSIA AESCULIFOLIA
♀ **Perennial** Creeping rhizomes produce clumps of large leaves like those of horse chestnuts, and pink flowers.
↕2m (6ft) ↔ 1m (3ft)

SORBARIA TOMENTOSA VAR. *ANGUSTIFOLIA*
♀ **Shrub** A spreading shrub with feathery leaves, red stems and fluffy, white flowerheads.
↕ ↔ 3m (10ft)

STIPA GIGANTEA
♀ **Ornamental grass** page **503**

TRACHYCARPUS FORTUNEI
♀ **Hardy palm** Slow-growing but cold-tolerant palm with fan-shaped leaves and a furry trunk with age.
↕20m (70ft) ↔ 2.5m (8ft)

VIBURNUM PLICATUM 'PINK BEAUTY'
♀ **Shrub** A spreading shrub with horizontal tiers of branches, covered with white flowers turning to pink.
↕3m (10ft) ↔ 4m (12ft)

WOODWARDIA RADICANS
♀ **Hardy fern** A large, evergreen fern with huge, arching fronds.
↕2m (6ft) ↔ 3m (10ft)

YUCCA FILAMENTOSA
♀ **Evergreen shrub** A clump-forming, stemless plant with soft leaves and spires of creamy flowers.
↕75cm (30in) ↔ 1.5m (5ft)

YUCCA FILAMENTOSA 'BRIGHT EDGE'
♀ **Shrub** page **543**

YUCCA FLACCIDA 'IVORY'
♀ **Shrub** page **544**

YUCCA GLORIOSA
♀ **Evergreen shrub** Erect trunks with grey-green narrow, sharp-tipped leaves, and large clusters of white flowers.
↕↔ 2m (6ft)

SHRUBS AND CLIMBERS FOR COLD WALLS

The coldest walls or fences are sunless nearly all year, but the even temperatures and often moist soil suits ivies, climbing hygrangeas and some shrubs. Walls that receive only morning sun can be a problem in frost-prone climates: rapid thawing can damage shoots and flowers, as with camellias. However, this is the perfect site for some clematis, climbing roses, chaenomeles and honeysuckles.

AKEBIA QUINATA
Semi-evergreen climber Twining stems with dark green, divided leaves and scented, purple flowers in spring.
↕ 10m (30ft)

CAMELLIA × WILLIAMSII 'FRANCIS HANGER'
Evergreen shrub Glossy leaves form a good foil to the white, golden-centred flowers.
↕ ↔ 1.5m (5ft)

CHAENOMELES SPECIOSA 'GEISHA GIRL'
Shrub Bushy plant with semi-double flowers of pale apricot pink.
↕ ↔ 1.5m (5ft)

CLEMATIS 'CARNABY'
Mid-season clematis Compact climber with large pink flowers, with a deeper centre to each petal.
↕ 2.5m (8ft) ↔ 1m (3ft)

CLEMATIS 'HELSINGBORG'
♀ **Early clematis** Masses of dainty, deep purple-blue flowers are followed by fluffy seedheads.
↕ 2–3m (6–10ft) ↔ 1.5m (5ft)

CLEMATIS 'HENRYI'
♀ **Mid-season clematis** page **130**

CLEMATIS 'MINUET'
♀ **Late-season clematis** page **133**

CLEMATIS 'NELLY MOSER'
♀ **Mid-season clematis** page **130**

CLEMATIS 'NIOBE'
♀ **Mid-season clematis** page **131**

CLEMATIS 'VENOSA VIOLACEA'
♀ **Late-season clematis** page **133**

CODONOPSIS CONVOLVULACEA
♀ **Perennial climber** page **134**

CORYLOPSIS PAUCIFLORA
♀ **Shrub** page **142**

COTONEASTER HORIZONTALIS
♀ **Shrub** page **146**

DAPHNE ODORA 'AUREOMARGINATA'
Evergreen shrub Low-growing, mounded shrub with leaves edged with gold and pale pink, fragrant flowers.
↕ ↔ 1.5m (5ft)

EUONYMUS FORTUNEI 'EMERALD 'N' GOLD'
♀ **Evergreen shrub** page **199**

FORSYTHIA SUSPENSA
♀ **Shrub** page **209**

GARRYA ELLIPTICA 'JAMES ROOF'
♀ **Evergreen shrub** page **219**

HEDERA CANARIENSIS 'GLOIRE DE MARENGO'
♀ **Evergreen climber** Silvery-green leaves variegated with white, and tinged pink in winter.
↕ 4m (12ft)

HEDERA COLCHICA 'DENTATA'
♀ **Evergreen climber** page **245**

HEDERA COLCHICA 'SULPHUR HEART'
♀ **Evergreen climber** page **245**

HEDERA HELIX 'GOLDHEART'
Evergreen climber Red stems with deep green leaves marked with a central gold splash.
↕ 8m (25ft)

HYDRANGEA ANOMALA SUBSP. *PETIOLARIS*
♀ **Climber** page **264**

JASMINUM HUMILE
Evergreen shrub Sparsely branched, arching shrub with bright yellow flowers in summer.
↕ ↔ 2.5m (8ft)

JASMINUM NUDIFLORUM
♀ **Shrub** page **291**

KERRIA JAPONICA 'PLENIFLORA'
♀ **Shrub** Vigorous, upright plant with slender, green stems and double orange/gold flowers.
↕ ↔ 3m (10ft)

LONICERA JAPONICA 'HALLIANA'
♀ **Evergreen climber** Strong-growing climber with scented white flowers that age to yellow.
↕ 10m (30ft)

MUEHLENBECKIA COMPLEXA
Climber Masses of thread-like, dark, twisting stems with tiny violin-shaped leaves.
↕ 3m (10ft)

PARTHENOCISSUS HENRYANA
♀ **Climber** page **369**

PARTHENOCISSUS QUINQUEFOLIA
♀ **Climber** Vigorous, deciduous climber with leaves divided into five leaflets that take on vivid red shades in autumn.
↕ 15m (50ft)

PILEOSTEGIA VIBURNOIDES
♀ **Evergreen climber** Oblong, dark green leaves and clusters of fluffy, white flowers.
↕ 6m (20ft)

PYRACANTHA 'HARLEQUIN'
Evergreen shrub Prickly shrub with white-variegated leaves, white flowers and red berries.
↕ 1.5m (5ft) ↔ 2m (6ft)

ROSA 'ALBÉRIC BARBIER'
♀ **Rambler rose** page **454**

ROSA 'DUBLIN BAY'
♀ **Climbing rose** page **452**

ROSA 'HANDEL'
♀ **Climbing rose** page **453**

ROSA 'MERMAID'
♀ **Climbing rose** Strong, thorny climber with dark, shiny leaves and single, primrose yellow flowers.
↕ 6m (20ft)

SCHIZOPHRAGMA INTEGRIFOLIUM
♀ **Climber** Large climber with toothed, dark green leaves and showy white flowers.
↕ 12m (40ft)

PLANTS FOR WARM WALLS

Reserve the warmest, sunniest garden walls to grow plants that are slightly tender in your climate. However, these sites can also be very dry, especially if the border is narrow and in the shadow of a roof, and it may be difficult to establish plants. Walls that receive only the afternoon sun are more gentle to plants, which may grow more quickly because there may be more moisture.

ALONSOA WARCSEWICZII
♀ **Perennial, not fully hardy** page **54**

ABELIA 'EDWARD GOUCHER'
♀ **Semi-evergreen shrub** page **28**

ABELIA FLORIBUNDA
♀ **Evergreen shrub, not hardy** page **28**

ABELIA X *GRANDIFLORA*
♀ **Semi-evergreen shrub** page **29**

ABELIA X *GRANDIFLORA* 'FRANCIS MASON'
♀ **Evergreen shrub** Yellow and dark green leaves; pale pink flowers in late summer.
↕ ↔ 1.5m (6ft)

ABUTILON 'KENTISH BELLE'
♀ **Shrub, not hardy** page **29**

ACACIA DEALBATA
♀ **Evergreen tree** Not very hardy, but beautiful, with feathery grey-green foliage and fragrant yellow spring flowers.
↕ 15–30m (50–100ft) ↔ 6–10m (20–30ft)

ACTINIDIA KOLOMIKTA
♀ **Climber** page **43**

CALLISTEMON CITRINUS 'SPLENDENS'
♀ **Evergreen shrub, not hardy** page **97**

CAMPSIS X *TAGLIABUANA* 'MADAME GALEN'
♀ **Climber** page **108**

CARPENTERIA CALIFORNICA
♀ **Evergreen shrub, not hardy** page **110**

CEANOTHUS 'AUTUMNAL BLUE'
♀ **Evergreen shrub, not hardy** page **112**

CEANOTHUS 'CONCHA'
Evergreen shrub Dark blue flowers.
↕ ↔ 3m (10ft)

CESTRUM PARQUI
♀ **Shrub** Weak-stemmed bush with narrow leaves and clusters of lime-green, tubular, night-scented flowers in summer.
↕ ↔ 2m (6ft)

CLEMATIS CIRRHOSA 'FRECKLES'
♀ **Early clematis** Purple-tinted, cream-spotted leaves; bell-shaped winter flowers.
↕ 2.5–3m (8–10ft) ↔ 1.5m (5ft)

CLEMATIS 'ETOILE VIOLETTE'
♀ **Late-season clematis** page **132**

CLEMATIS 'JACKMANNII'
♀ **Late-season clematis** page **133**

CLEMATIS 'LASURSTERN'
♀ **Mid-season clematis** page **130**

CLEMATIS REHDERIANA
♀ **Late-season clematis** page **133**

CLEMATIS 'THE PRESIDENT'
♀ **Mid-season clematis** page **131**

CLEMATIS VITICELLA 'PURPUREA PLENA ELEGANS'
♀ **Late-season clematis** page **133**

CLIANTHUS PUNICEUS
♀ **Evergreen shrub, not hardy** page **134**

CYTISUS BATTANDIERI
♀ **Semi-evergreen shrub** page **159**

ECCREMOCARPUS SCABER
♀ **Climber, not hardy** page **180**

ESCALLONIA 'LANGLEYENSIS'
♀ **Evergreen shrub** page **195**

FREMONTODENDRON 'CALIFORNIA GLORY'
♀ **Shrub, not hardy** page **210**

HEDYCHIUM GARDNERIANUM
♀ **Perennial, not hardy** Large heads of spidery, showy, sweetly scented cream flowers.
↕ ↔ 2–2.2m (6–7ft)

IPOMOEA INDICA
♀ **Climber, not hardy** page **277**

JASMINUM X *STEPHANENSE*
♀ **Climber** Fast-growing, with clusters of pink, fragrant flowers in summer.
↕ 5m (15ft)

JOVELLANA VIOLACEA
♀ **Semi-evergreen shrub** Rather tender, weak shrub with fine foliage and bell-like, pale violet flowers in summer.
↕ 60cm (24in) ↔ 1m (3ft)

LAPAGERIA ROSEA
♀ **Climber, not hardy** page **301**

LONICERA X *ITALICA*
♀ **Climber** page **323**

LONICERA X *TELLMANNIANA*
♀ **Climber** page **325**

MAGNOLIA GRANDIFLORA 'EXMOUTH'
♀ **Evergreen shrub** page **330**

PASSIFLORA CAERULEA
♀ **Climber, borderline hardy** page **370**

PHYGELIUS
♀ **Shrubs, borderline hardy** pages **396–8**

PITTOSPORUM TENUIFOLIUM 'SILVER QUEEN'
♀ **Evergreen shrub** Wiry black twigs support grey-green, white-edged leaves, and purple, scented flowers in autumn.
↕ 1–4m (3–12ft) ↔ 2m (6ft)

RHODANTHEMUM HOSMARIENSE
♀ **Subshrub** page **435**

RIBES SPECIOSUM
♀ **Shrub** Spiny shrub with bristly stems, small glossy leaves and pendulous red flowers, resembling fuchsias.
↕ ↔ 2m (6ft)

ROSES, CLIMBING
♀ **Deciduous climbers** pages **452–453**

SOLANUM CRISPUM 'GLASNEVIN'
♀ **Climber, not hardy** page **496**

SOLANUM JASMINOIDES 'ALBUM'
♀ **Climber, not hardy** page **496**

THUNBERGIA GRANDIFLORA
♀ **Climber, not hardy** page **513**

TRACHELOSPERMUM JASMINOIDES
♀ **Evergreen climber, not hardy** page **517**

VESTIA FOETIDA
♀ **Shrub** A short-lived, rather tender plant with unpleasantly-scented leaves and prolific, pendulous yellow flowers.
↕ 2m (6ft) ↔ 1.5m (5ft)

VITIS 'BRANT'
♀ **Climber** An ornamental vine with green leaves that turn red in autumn, and black, edible grapes.
↕ 7m (22ft)

WISTERIA FLORIBUNDA 'ALBA'
♀ **Climber** page **542**

WISTERIA SINENSIS
♀ **Climber** page **543**

WISTERIA SINENSIS 'SIERRA MADRE'
♀ **Evergreen shrub** Attractive cultivar with bicoloured, fragrant flowers.
↕ 9m (28ft)

ZAUSCHNERIA CALIFORNICA 'DUBLIN'
♀ **Perennial, not hardy** page **545**

PLANTS FOR BEES AND BUTTERFLIES

Plants that will attract these fascinating and useful garden visitors usually have simple, tubular or daisy-like flowers, especially in pinks and purples; avoid double-flowered varieties. Butterflies also like fruity scents. Remember that their caterpillar stage needs different food plants. Nettles are well known as the food of tortoiseshells, but long grass and other weeds support many species.

AJUGA REPTANS 'BRAUNHERZ'
♀ **Perennial** Creeping plant with glossy, purplish leaves and blue flowers in spring.
↕ 15cm (6in) ↔ 90cm (36in)

ALLIUM SCHOENOPRASUM 'FORESCATE'
Perennial An ornamental form of chives with abundant heads of pink flowers.
↕ 60cm (24in)

ASCLEPIAS INCARNATA
Perennial Thick, upright stems support small heads of curious pale pink flowers followed by interesting seed heads.
↕ 1.2m (4ft) ↔ 60cm (24in)

ASTER AMELLUS 'SONIA'
Perennial Hairy, leafy plant with pale pink, yellow-centred flowers.
↕ 60cm (24in) ↔ 45cm (18in)

ASTER 'ANDENKEN AN ALMA PÖTSCHKE'
♀ **Perennial** page 72

ASTER x *FRIKARTII* 'MÖNCH'
♀ **Perennial** page 73

ASTER TURBINELLUS
♀ **Perennial** Wiry, almost black stems with small leaves and diffuse heads of lilac flowers in late summer.
↕ 1.2m (4ft) ↔ 60cm (2ft)

BUDDLEJA AURICULATA
Evergreen shrub Small clusters of white and orange, scented flowers in autumn.
↕ ↔ 3m (10ft)

BUDDLEJA DAVIDII CULTIVARS
♀ **Shrubs** page 93

BUDDLEJA DAVIDII 'BLACK KNIGHT'
♀ **Shrub** Large spikes of deep purple flowers, best if pruned hard as growth begins.
↕ 3m (10ft) ↔ 5m (15ft)

BUDDLEJA 'LOCHINCH'
♀ **Shrub, borderline hardy** page 94

CALLUNA VULGARIS 'WICKWAR FLAME'
♀ **Evergreen shrub** Mats of gold leaves turn red in winter; pink flowers in summer.
↕ 50cm (20in) ↔ 65cm (26in)

CARYOPTERIS x *CLANDONENSIS* 'HEAVENLY BLUE'
♀ **Shrub** page 110

CEANOTHUS 'PUGET BLUE'
♀ **Evergreen shrub** Billowing mass of fine foliage covered with mid-blue flowers.
↕ ↔ 2.2m (7ft)

CYTISUS x *BEANII*
♀ **Shrub** page 160

DAHLIA MERCKII
Perennial, not hardy Tuberous-rooted plant with slender, translucent stems and pale mauve, often nodding flowers.
↕ 2m (6ft) ↔ 1m (3ft)

DIGITALIS PURPUREA 'SUTTON'S APRICOT'
♀ **Biennial** Tall-stemmed foxglove with pale, apricot-pink flowers in summer.
↕ 1–2m (3–6ft)

ECHINACEA PURPUREA
Perennial Stiff stems carry large, daisy-like flowers with pink petals and dark centres.
↕ 1m (3ft) ↔ 45cm (18in)

ECHIUM VULGARE 'BLUE BEDDER'
Annual Bushy plant with bristly, greyish-green leaves and soft blue flowers.
↕ ↔ 45cm (18in)

ERICA VAGANS 'BIRCH GLOW'
♀ **Evergreen shrub** page **189**

ERICA VAGANS 'VALERIE PROUDLEY'
♀ **Evergreen shrub** Gold foliage, and white flowers in summer.
↕ 15cm (6in) ↔ 30cm (12in)

ERICA X *VEITCHII* 'EXETER'
♀ **Evergreen shrub, not hardy** page **186**

ERYNGIUM PLANUM
Perennial Branched stems of steely blue, with small, light blue, spiky flower heads, emerge from evergreen leaf rosettes.
↕ 90cm (36in) ↔ 45cm (18in)

HELIOTROPIUM 'PRINCESS MARINA'
♀ **Shrub, not hardy** page **252**

HOHERIA GLABRATA
♀ **Tree** page **258**

HYSSOPUS OFFICINALIS
Evergreen shrub Herb with green leaves and spikes of blue flowers in summer.
↕ 60cm (24in) ↔ 1m (3ft)

LAMIUM ORVALA
Perennial A choice, clump-forming plant that does not creep, and has large leaves and purplish flowers in spring.
↕ 60cm (24in) ↔ 30cm (12in)

LAVANDULA ANGUSTIFOLIA 'LODDON PINK'
Evergreen shrub Compact shrub with grey leaves and spikes of pale pink flowers.
↕ 45cm (18in) ↔ 60cm (24in)

LUNARIA REDIVIVA
Perennial Pale lilac, fragrant flowers followed by translucent seedheads.
↕ 60–90cm (24–36in) ↔ 30cm (12in)

MENTHA LONGIFOLIA BUDDLEJA MINT GROUP
Perennial Tall stems of greyish leaves and tall heads of tightly packed pink flowers.
↕ ↔ 1m (3ft)

MONARDA 'CROFTWAY PINK'
♀ **Perennial** page **343**

ORIGANUM LAEVIGATUM 'HERRENHAUSEN'
♀ **Perennial** page **358**

PAPAVER ORIENTALE 'CEDRIC MORRIS'
♀ **Perennial** page **367**

PAPAVER RHOEAS 'MOTHER OF PEARL'
Annual Easy to grow from seed, with delicate flowers in pastel shades.
↕ 90cm (36in) ↔ 30cm (12in)

PENSTEMON 'SOUR GRAPES'
Perennial Intriguing flowers in shades of greyish-blue, pink and mauve on spikes above large, green leaves.
↕ 60cm (24in) ↔ 45cm (18in)

PRUNELLA GRANDIFLORA 'LOVELINESS'
♀ **Perennial** page **420**

ROSMARINUS OFFICINALIS 'MISS JESSOP'S UPRIGHT'
♀ **Evergreen shrub** page **466**

SEDUM 'HERBSTFREUDE'
♀ **Perennial** Upright, unbranched stems with pale green, fleshy leaves, and deep pink flowers in flat heads in late summer. Also sold as 'Autumn Joy'.
↕ ↔ 60cm (24in)

TAGETES 'NAUGHTY MARIETTA'
Annual Bushy plants with single, yellow flowers marked with red.
↕ 30–40cm (12–16in)

TRACHELIUM CAERULEUM
♀ **Perennial** Usually grown as an annual, this wiry, upright plant produces flat heads of small purple flowers in summer.
↕ 1m (3ft) ↔ 30cm (12in)

PLANTS TO ATTRACT GARDEN BIRDS

Native and visiting birds will visit gardens to feed on a wide variety of plants, especially those bearing berries and seeds (see also pages 586 and 615). Unfortunately, their feeding necessarily means that the food source – and the attractive autumn display – does not last long, so it is worth offering them a variety of plants, and providing extra food on a regular basis.

ARBUTUS MENZIESII
♀ **Tree** page **65**

ATRIPLEX HORTENSIS VAR. *RUBRA*
Annual Vigorous plant with deep red leaves that contrast well with other plants; its seeds are loved by birds.
↕1.2m (4ft) ↔ 30cm (12in)

BERBERIS THUNBERGII
♀ **Shrub** Green leaves and small, yellow flowers in summer become red leaves and berries in autumn.
↕2m (6ft) ↔ 2.5m (8ft)

CORTADERIA SELLOANA 'PUMILA'
♀ **Ornamental grass** A compact pampas grass with short flower spikes.
↕1.5m (5ft) ↔ 1.2m (4ft)

COTONEASTER LACTEUS
♀ **Evergreen shrub** page **147**

COTONEASTER SIMONSII
♀ **Shrub** page **147**

CRATAEGUS X *LAVALLEI* 'CARRIEREI'
♀ **Tree** page **150**

CYNARA CARDUNCULUS
♀ **Perennial** page **159**

DAPHNE MEZEREUM
Shrub Upright branches, fragrant pink flowers in spring, red berries in autumn.
↕1.2m (4ft) ↔ 1m (3ft)

HEDERA HELIX & CULTIVARS
Evergreen climbers pages **246–247**
When ivy reaches its flowering stage the black berries are attractive to many birds; ivy also provides valuable nesting sites.

HELIANTHUS ANNUUS 'MUSIC BOX'
Annual Multi-coloured sunflowers that produce heads of seeds that may be harvested to feed birds in later months.
↕70cm (28in) ↔ 60cm (24in)

ILEX AQUIFOLIUM 'HANDSWORTH NEW SILVER'
♀ **Evergreen shrub** page **273**

LONICERA PERICLYMENUM 'SEROTINA'
♀ **Climber** page **324**

MAHONIA AQUIFOLIUM
Evergreen shrub A suckering shrub with gently spiny leaves, and yellow flowers followed by black berries.
↕1m (3ft) ↔ 1.5m (5ft)

MALUS ZUMI 'GOLDEN HORNET'
♀ **Tree** page **337**

MISCANTHUS SINENSIS
♀ **Ornamental grass** This grass forms clumps of long, arching leaves and silver or pink flowerheads in late summer.
↕2.5m (8ft) ↔ 1.2m (4ft)

ONOPORDUM NERVOSUM
♀ **Biennial** A large, silvery, prickly plant with thistle-like purple flowers.
↕2.5m (8ft) ↔ 1m (3ft)

PAPAVER SOMNIFERUM 'WHITE CLOUD'
Annual White, double flowers and pretty seedheads attractive to birds.
↕1m (3ft) ↔ 30cm (12in)

PRUNUS PADUS
Tree A spreading tree with pendent spikes of small white flowers followed by black berries.
↕15m (50ft) ↔ 10m (30ft)

PYRACANTHA 'MOHAVE'
Evergreen shrub Dense, spiny growth with dark green leaves and bright red berries.
↕4m (12ft) ↔ 5m (15ft)

RIBES ODORATUM
Shrub Weakly-branched shrub with fresh green leaves, yellow scented flowers in spring and black berries in late summer.
↕↔ 2m (6ft)

ROSA FILIPES 'KIFTSGATE'
♀ **Climbing rose** page **452**

ROSA PIMPINELLIFOLIA
Species shrub rose This very spiny bush has single white flowers followed by purplish-black hips.
↕1m (3ft) ↔ 1.2m (4ft)

ROSA 'SCABROSA'
♀ **Rugosa shrub rose** Deep pink flowers and bright red hips on a mounded bush with deeply veined foliage.
↕↔ 1.7m (5½ft)

SAMBUCUS NIGRA 'AUREOMARGINATA'
Shrub Fast-growing plant for any soil, with yellow-edged leaves, white flowers and heads of black elderberries.
↕↔ 6m (20ft)

SILYBUM MARIANUM
Biennial Rosettes of spiny leaves, veined with white; prickly mauve seedheads and thistle seeds.
↕1.5m (5ft) ↔ 60–90cm (24–36in)

SORBUS AUCUPARIA 'FASTIGIATA'
Tree Upright, narrow tree with red berries in late summer after white spring flowers.
↕8m (25ft) ↔ 5m (15ft)

VIBURNUM BETULIFOLIUM
Shrub Spectacular displays of red berries follow white flowers in summer when several plants are grown together.
↕↔ 3m (10ft)

VIBURNUM OPULUS
Shrub Strong-growing shrub with white flowers in summer, bright autumn colour and red berries.
↕5m (15ft) ↔ 4m (12ft)

VITIS VINIFERA 'PURPUREA'
♀ **Climber** page **540**

FLOWERS FOR CUTTING

It is useful to be able to cut flowers from the garden, either to use on their own or to add to bought flowers. Many annuals are grown especially for cutting, but other garden plants can supply flowers for the house without spoiling the display. To produce many smaller stems for cutting, pinch out the shoots of free-branching plants such as asters and delphiniums in early summer.

ACHILLEA 'CORONATION GOLD'
♀ **Perennial** page **39**

ACONITUM 'BRESSINGHAM SPIRE'
♀ **Perennial** page **42**

ASTER 'LITTLE CARLOW'
♀ **Perennial** page **74**

ASTER PRINGLEI 'MONTE CASSINO'
♀ **Perennial** Thin stems of narrow, upright habit, forming a dense bush with needle-like leaves and small white flowers.
↕1m (3ft) ↔ 30cm (12in)

ASTILBE X *ARENDSII* 'FANAL'
♀ **Perennial** page **74**

ASTRANTIA MAJOR 'SHAGGY'
♀ **Perennial** The bracts around the clusters of flowers are longer than usual.
↕30–90cm (12–36in) ↔ 45cm (18in)

BAPTISIA AUSTRALIS
♀ **Perennial** page **79**

CALLISTEPHUS MILADY SUPER MIXED
♀ **Annuals** page **97**

CAMPANULA LACTIFLORA 'PRICHARD'S VARIETY'
♀ **Perennial** Compact cultivar with heads of violet-blue flowers in mid-summer.
↕75cm (30in) ↔ 45cm (18in)

CAMPANULA PERSICIFOLIA 'CHETTLE CHARM'
Perennial Mats of deep green foliage and thin stems with white, blue-tinted flowers.
↕1m (3ft) ↔ 30cm (12in)

CHRYSANTHEMUMS
♀ **Perennials, not hardy** pages **122–125**

CLEMATIS 'VYVYAN PENNELL'
♀ **Climber** page **109**

CROCOSMIA X *CROCOSMIIFLORA* 'SOLFATERRE'
♀ **Perennial** page **151**

DELPHINIUM 'BELLAMOSUM'
Perennial Well-branched stems with thin spikes of deep blue flowers for a long period.
↕1–1.2m (3–4ft) ↔ 45cm (18in)

DELPHINIUM 'BRUCE'
♀ **Perennial** page **166**

DELPHINIUM 'SUNGLEAM'
♀ **Perennial** page **169**

BORDER CARNATIONS (*DIANTHUS*)
♀ **Perennials** page **171**

DIANTHUS 'CORONATION RUBY'
♀ **Perennial** A laced pink with pink and ruby-red flowers with a clove scent.
↕38cm (15in) ↔ 30cm (12in)

ERYNGIUM X *TRIPARTITUM*
♀ **Perennial** page **192**

GEUM 'MRS J. BRADSHAW'
♀ **Perennial** Hairy basal leaves, wiry branched stems and double scarlet flowers.
↕40–60cm (16–24in) ↔ 60cm (24in)

KNIPHOFIA 'ROYAL STANDARD'
♀ **Perennial** page **298**

LATHYRUS ODORATUS (SWEET PEAS)
♀ **Annual climbers** page **303**

LEUCANTHEMUM X *SUPERBUM* 'WIRRAL SUPREME'
♀ **Perennial** page **310**

NARCISSUS 'MERLIN'
♀ **Bulb** page **349**

NARCISSUS 'WHITE LION'
♀ **Bulb** Double, white flowers too heavy to stand up in the garden; best when cut.
↕40cm (16in)

OSTEOSPERMUM 'WHIRLIGIG'
♀ **Subshrub, not hardy** page **361**

PAEONIA LACTIFLORA 'SARAH BERNHARDT'
♀ **Perennial** page **364**

PHLOX MACULATA 'ALPHA'
♀ **Perennial** page **390**

PHLOX MACULATA 'OMEGA'
♀ **Perennial** Conical heads of fragrant, small white flowers with a deep pink eye.
↕90cm (3ft) ↔ 45cm (18in)

PHYSOSTEGIA VIRGINIANA 'VIVID'
♀ **Perennial** page **400**

ROSA 'ALEXANDER'
♀ **Large-flowered bush rose** page **448**

ROSA 'ICEBERG'
♀ **Cluster-flowered bush rose** page **451**

ROSA 'ROYAL WILLIAM'
♀ **Large-flowered bush rose** page **449**

ROSA 'SILVER JUBILEE'
♀ **Large-flowered bush rose** page **449**

ROSES, MODERN SHRUB
♀ **Deciduous shrubs** pages **462–463**

RUDBECKIA FULGIDA VAR. *SULLIVANTII* 'GOLDSTURM'
♀ **Perennial** page **468**

RUDBECKIA 'GOLDQUELLE'
♀ **Perennial** page **468**

SCABIOSA CAUCASICA 'CLIVE GREAVES'
♀ **Perennial** page **485**

SCHIZOSTYLIS COCCINEA 'SUNRISE'
♀ **Perennial** page **486**

SOLIDAGO 'GOLDENMOSA'
♀ **Perennial** page **497**

X *SOLIDASTER LUTEUS* 'LEMORE'
♀ **Perennial** This generic hybrid produces heads of yellow daisy-like flowers.
↕90cm (3ft) ↔ 30cm (12in)

TANACETUM COCCINEUM 'BRENDA'
♀ **Perennial** page **508**

TULIPA 'SORBET'
♀ **Bulb** Late blooms are pale pink with carmine streaks and flashes.
↕60cm (24in)

VERONICA SPICATA SUBSP. *INCANA*
♀ **Perennial** page **533**

FLOWERS FOR DRYING

Dried flowers prolong the beauty of summer throughout the year. Many are easy to grow and dry, by simply hanging them upside-down in a shady, airy position. If dried in warm sand or silica gel, then kept in a dry atmosphere, almost any flowers can be used. Select young, unblemished flowers that are not fully open and remove most of the leaves before tying them into bunches.

ACHILLEA FILIPENDULINA 'GOLD PLATE'
♀ **Perennial** page **40**

ACHILLEA 'MOONSHINE'
♀ **Perennial** page **41**

AMARANTHUS HYPOCHONDRIACUS 'GREEN THUMB'
♀ **Annual** Leafy plants produce upright, branched spikes of pale green flowers.
↕60cm (24in) ↔ 30cm (12in)

ASTRANTIA MAXIMA
♀ **Perennial** page **76**

BRACTEANTHA BRIGHT BIKINI SERIES
♀ **Annuals** page **91**

CATANANCHE CAERULEA 'MAJOR'
♀ **Perennial** Cornflower-like flowers of papery texture on wiry stems above narrow, greyish leaves.
↕50–90cm (20–36in) ↔ 30cm (12in)

CENTAUREA CYANUS 'FLORENCE PINK'
Annual Upright-growing annual with a bushy habit and pink flowers.
↕35cm (14in) ↔ 45cm (18in)

CONSOLIDA AJACIS GIANT IMPERIAL SERIES
Annual Larkspur producing elegant spires of flowers: like annual delphiniums.
↕60–90cm (2–3ft) ↔ 35cm (14in)

CORTADERIA SELLOANA 'SUNNINGDALE SILVER'
♀ **Ornamental grass** page **141**

ECHINOPS RITRO
♀ **Perennial** page **181**

GOMPHRENA HAAGEANA 'STRAWBERRY FIELDS'
♀ **Annual** page **233**

HYDRANGEA MACROPHYLLA CULTIVARS
♀ **Shrubs** page **266**

HYDRANGEA SERRATA 'BLUEBIRD'
♀ **Shrub** page **268**

LAGURUS OVATUS
♀ **Ornamental grass** page **300**

LIMONIUM SINUATUM 'ART SHADES'
Perennial Usually grown as annuals, with crispy flowers in shades of pink, salmon, orange, pink and blue.
↕60cm (24in) ↔ 30cm (12in)

LIMONIUM SINUATUM 'FOREVER GOLD'
♀ **Perennial, not hardy** page **319**

NIGELLA DAMASCENA 'MULBERRY ROSE'
♀ **Annual** Feathery foliage and flowers in purplish-pink, and inflated seed pods.
↕45cm (18in) ↔ 23cm (9in)

RHODANTHE MANGLESII 'SUTTON'S ROSE'
♀ **Annual** Wiry plants with greyish leaves and white or pink flowers with a strawlike texture.
↕60cm (24in) ↔ 15cm (6in)

SEDUM SPECTABILE 'BRILLIANT'
♀ **Perennial** page **490**

PLANTS WITH ORNAMENTAL SEEDHEADS

Although they may lack the bright colours of the flowers, there is much beauty to be enjoyed in the seedheads of plants. Some may be cut and preserved to decorate the home, while others can be left in the garden to bring straw or bronze tones to the winter scene, and look especially good when covered with frost, until birds pull them apart in their hunt for food.

ALLIUM CRISTOPHII
♀ **Bulb** page **51**

ASTILBE CHINENSIS VAR. *PUMILA*
♀ **Perennial** Dumpy, pink flower spikes become rust-brown as they age.
↕25cm (10in) ↔ 20cm (8in)

CLEMATIS 'BILL MACKENZIE'
♀ **Climber** page **132**

CLEMATIS MACROPETALA 'WHITE SWAN'
Climber A very compact, early-flowering clematis with white blooms and silver seedheads.
↕1m (3ft)

COTINUS COGGYGRIA
♀ **Shrub** Green leaves in summer that turn scarlet in autumn, with feathery, smoke-like seedheads.
↕↔ 5m (15ft)

HYDRANGEA PANICULATA 'GRANDIFLORA'
♀ **Shrub** page **267**

HYOSCYAMUS NIGER
Annual Extremely poisonous plant with sinister, veined flowers and beautiful seedheads resembling shuttlecocks.
↕60–120cm (2–4ft) ↔ 1m (3ft)

IRIS FOETIDISSIMA
♀ **Perennial** Pale blue and brown flowers develop into green pods that split in autumn to reveal orange seeds.
↕30–90cm (12–36in) ↔ 30cm (12in)

IRIS FOETIDISSIMA 'VARIEGATA'
♀ **Perennial** page **282**

IRIS 'SHELFORD GIANT'
♀ **Perennial** Sheaves of long, green leaves, tall stems of yellow and white flowers and distinctive ribbed seedheads.
↕1.8m (6ft) ↔ 60cm (24in)

NIGELLA ORIENTALIS 'TRANSFORMER'
Annual Bushy annual with finely-divided leaves and small yellow flowers that produce umbrella-like seed pods.
↕45cm (18in) ↔ 22–30cm (9–12in)

PAEONIA LUTEA VAR. *LUDLOWII*
♀ **Shrub** page **365**

PAPAVER SOMNIFERUM 'HEN AND CHICKENS'
Annual The single flowers are followed by curious pods that are surrounded by a ring of tiny pods.
↕1.2m (4ft) ↔ 30cm (12in)

PHYSALIS ALKEKENGI
♀ **Perennial** Vigorous, suckering perennial with bright orange, "Chinese lantern" fruits in autumn.
↕60–75cm (24–30in) ↔ 90cm (3ft)

SCABIOSA STELLATA 'DRUMSTICK'
♀ **Annual** Wiry-stemmed, hairy annual with pale lilac flowers and round, crispy seedheads.
↕30cm (12in) ↔ 23cm (9in)

COTTAGE GARDEN-STYLE PLANTS

The idealised image of a cottage garden is in summer with bees lazily buzzing around roses, lilies, hollyhocks and peonies, but in fact the authentic cottage garden was a glorious mixture because it contained old-fashioned plants discarded by wealthier gardeners. Cottage garden flowers are often scented, usually herbaceous, and always evocative of a gentler age.

ACONITUM 'SPARK'S VARIETY'
♈ **Perennial** page **42**

ALCEA ROSEA 'NIGRA'
Biennial "Black-flowered" hollyhock.
↕2m (6ft) ↔ 60cm (24in)

CALENDULA 'FIESTA GITANA'
♈ **Annual** page **96**

CAMPANULA 'BURGHALTII'
♈ **Perennial** A hybrid that forms mounds of mid-green leaves and large, tubular flowers of greyish blue.
↕60cm (2ft) ↔ 30cm (12in)

CAMPANULA PERSICIFOLIA 'WHITE CUP AND SAUCER'
Perennial The pure white flowers are bell-shaped with a white, circular disc, like a saucer, below the bloom.
↕90cm (36in) ↔ 30cm (12in)

CAMPANULA PORTENSCHLAGIANA
♈ **Perennial** page **107**

DELPHINIUMS
♈ **Perennials** pages **166–169**

DIANTHUS 'GRAN'S FAVOURITE'
♈ **Perennial** page **172**

DICENTRA SPECTABILIS
♈ **Perennial** page **175**

ERYNGIUM ALPINUM
♈ **Perennial** page **191**

ESCHSCHOLZIA CALIFORNICA
♈ **Annual** page **196**

GERANIUM MACRORRHIZUM 'ALBUM'
♈ **Perennial** Scented, evergreen leaves and white flowers in summer.
↕50cm (20in) ↔ 60cm (24in)

GERANIUM x *OXONIANUM* 'A.T. JOHNSON'
♈ **Perennial** This clump-forming plant has silvery-pink flowers.
↕↔ 30cm (12in)

GERANIUM SANGUINEUM 'ALBUM'
♈ **Perennial** A compact plant with divided leaves and white flowers.
↕20cm (8in) ↔ 30cm (12in)

GERANIUM SYLVATICUM 'MAYFLOWER'
♈ **Perennial** page **229**

GEUM 'LADY STRATHEDEN'
♈ **Perennial** page **230**

KNIPHOFIA TRIANGULARIS
♈ **Perennial, borderline hardy** page **299**

LAVATERA, ANNUALS
♈ **Annuals** page **308**

LILIUM CANDIDUM
♈ **Bulb** page **314**

LUPINUS POLYPHYLLUS 'BAND OF NOBLES'
♈ **Perennial** A good seed mixture with tall spikes of bicolored flowers.
↕1.5m (5ft) ↔ 75cm (30in)

LUPINUS 'THE CHATELAINE'
Perennial Tall spires of bicoloured flowers in deep pink and white.
↕90cm (36in) ↔ 75cm (30in)

LYCHNIS CHALCEDONICA
♀ **Perennial** page **327**

PAEONIA LACTIFLORA 'BOWL OF BEAUTY'
♀ **Perennial** page **363**

PAEONIA OFFICINALIS 'RUBRA PLENA'
♀ **Perennial** page **365**

PAPAVER ORIENTALE 'BEAUTY OF LIVERMERE'
♀ **Perennial** page **366**

PAPAVER ORIENTALE 'MRS PERRY'
♀ **Perennial** Salmon-pink flowers with petals like satin above coarse foliage.
↕ 75cm (30in) ↔ 60–90cm (24–36in)

PAPAVER RHOEAS SHIRLEY MIXED
♀ **Annuals** page **367**

PHLOX 'KELLY'S EYE'
♀ **Perennial** page **390**

PHLOX PANICULATA CULTIVARS
♀ **Perennials** page **391**

PRIMULA 'WANDA'
♀ **Perennial** page **407**

ROSA 'BALLERINA'
♀ **Patio rose** page **456**

ROSA 'FANTIN-LATOUR'
♀ **Old garden rose** page **461**

ROSA 'GERTRUDE JEKYLL'
♀ **English shrub rose** page **463**

ROSA XANTHINA 'CANARY BIRD'
♀ **Shrub rose** page **465**

SAXIFRAGA X *URBIUM*
♀ **Perennial** Mats of evergreen leaf rosettes and sprays of dainty, white flowers in late spring.
↕ 30cm (12in) ↔ indefinite

SCABIOSA CAUCASICA 'MISS WILLMOTT'
♀ **Perennial** page **485**

SCHIZOSTYLIS COCCINEA 'SUNRISE'
♀ **Perennial** page **486**

VERBASCUM 'COTSWOLD BEAUTY'
♀ **Perennial** page **529**

VERBASCUM 'PINK DOMINO'
♀ **Perennial** Dark green leaves give rise to unbranched stems of rounded, deep pink flowers.
↕ 1.2m (4ft) ↔ 30cm (12in)

VIOLA CORNUTA
♀ **Perennial** page **538**

PLANTS FOR A ROCK GARDEN

Rock gardens and rockeries are good places to grow small plants, raising them closer to observers' eyes. But the raised beds also allow the soil to be tailored to suit these plants, which often require perfect drainage, or specific soil mixes. Most of the plants below are easily grown and require no special treatment, making them yet more attractive: rock gardening can be an addictive hobby.

ADIANTUM PEDATUM
♈ **Hardy fern** page **43**

ALLIUM MOLY
♈ **Bulbous perennial** page **52**

ANCHUSA CESPITOSA
♈ **Alpine** White-eyed blue flowers appear between narrow leaves in spring.
↕5–10cm (2–4in) ↔ 15–20cm (6–8in)

ANDROSACE CARNEA SUBSP. *LAGGERI*
♈ **Perennial** page **57**

ANDROSACE SEMPERVIVOIDES
♈ **Alpine** Leaf rosettes form loose mats, and pink flowers open in late spring.
↕2.5–5cm (1–2in) ↔ 15–20cm (6–8in)

ARENARIA MONTANA
♈ **Perennial** page **66**

ARMERIA MARITIMA 'VINDICTIVE'
♈ **Perennial** Hummocks of narrow leaves; slender stems of deep pink flowers.
↕15cm (6in) ↔ 20cm (8in)

AUBRIETA X *CULTORUM*
'BRESSSINGHAM PINK'
Perennial Cushions of green leaves covered with double, pink flowers in spring.
↕5cm (2in) ↔ 60cm (24in)

AURINIA SAXATILIS
♈ **Perennial** page **78**

BERBERIS THUNBERGII 'BAGATELLE'
♈ **Shrub** page **86**

CAMPANULA 'BIRCH HYBRID'
♈ **Perennial** A strong-growing, spreading plant with deep blue bell-flowers.
↕10cm (4in) ↔ 50cm (40in)

CAMPANULA CHAMISSONIS 'SUPERBA'
♈ **Alpine** The rosettes of pale green leaves disappear under pale blue flowers in early summer.
↕5cm (2in) ↔ 20cm (8in)

CROCUS CORSICUS
♈ **Spring bulb** page **153**

DAPHNE PETRAEA 'GRANDIFLORA'
♈ **Evergreen shrub** page **164**

DIANTHUS 'LA BOURBOULE'
♈ **Perennial** page **173**

DIASCIA BARBERAE 'RUBY FIELD'
♈ **Perennial** Mats of heart-shaped leaves are covered with deep salmon-pink flowers in summer.
↕25cm (10in) ↔ 60cm (24in)

DRYAS OCTOPETALA
♈ **Evergreen shrub** Mats of deep green leaves, with white flowers in spring, then fluffy seedheads.
↕10cm (4in) ↔ 1m (36in)

GENTIANA ACAULIS
♈ **Perennial** page **224**

GERANIUM CINEREUM 'BALLERINA'
♈ **Perennial** page **226**

GEUM MONTANUM
♈ **Perennial** page **230**

GYPSOPHILA 'ROSENSCHLEIER'
♈ **Perennial** page **235**

HELIANTHEMUM 'FIRE DRAGON'
♈ **Shrub** page **247**

HELIANTHEMUM 'HENFIELD BRILLIANT'
♀ **Shrub** page **248**

IRIS CRISTATA
♀ **Perennial** In moist soil the creeping stems produce dainty fans of foliage, and lilac-blue and white flowers in late spring.
↕10cm (4in)

IRIS 'KATHARINE HODGKIN'
♀ **Bulbous perennial** page **283**

IRIS LACUSTRIS
♀ **Perennial** page **288**

LATHYRUS VERNUS
♀ **Perennial** page **302**

LEPTOSPERMUM SCOPARIUM 'KIWI'
♀ **Shrub** page **310**

LEWISIA COTYLEDON
♀ **Perennial** page **312**

LEWISIA TWEEDYI
♀ **Perennial** page **312**

LINUM 'GEMMELL'S HYBRID'
♀ **Perennial** page **320**

MUSCARI AUCHERI
♀ **Bulbous perennial** page **344**

NARCISSUS 'HAWERA'
♀ **Bulb** page **347**

OMPHALODES CAPPADOCICA 'CHERRY INGRAM'
♀ **Perennial** page **356**

OXALIS ADENOPHYLLA
♀ **Bulbous perennial** page **362**

OXALIS ENNEAPHYLLA 'ROSEA'
Bulb Clumps of frilly, grey-green leaves; pink flowers in early summer.
↕8cm (3in) ↔ 15cm (6in)

PHLOX DIVARITICA 'CHATTAHOOCHEE'
♀ **Perennial** page **387**

PHLOX DOUGLASII 'BOOTHMAN'S VARIETY'
♀ **Perennial** page **387**

PHLOX NANA 'MARY MASLIN'
Perennial Spreading stems with bright scarlet flowers in summer.
↕20cm (8in) ↔ 30cm (12in)

PITTOSPORUM TENUIFOLIUM 'TOM THUMB'
♀ **Evergreen shrub** page **405**

PULSATILLA HALLERI
♀ **Perennial** page **429**

RAMONDA MYCONI
♀ **Perennial** page **432**

RANUNCULUS CALANDRINOIDES
♀ **Perennial** page **433**

RANUNCULUS GRAMINEUS
♀ **Perennial** page **434**

ROSMARINUS OFFICINALIS PROSTRATUS GROUP
♀ **Shrub** page **466**

SAPONARIA x *OLIVANA*
♀ **Perennial** page **482**

SAXIFRAGA 'JENKINSIAE'
♀ **Alpine** page **484**

SAXIFRAGA 'SOUTHSIDE SEEDLING'
♀ **Perennial** page **484**

SEMPERVIVUM TECTORUM
♀ **Perennial** page **493**

SILENE SCHAFTA
♀ **Perennial** page **494**

TULIPA LINIFOLIA BATALINII GROUP
♀ **Bulb** page **523**

TULIPA TURKESTANICA
♀ **Bulb** page **523**

VERBASCUM 'LETITIA'
♀ **Subshrub** page **531**

VERONICA PROSTRATA
♀ **Perennial** page **533**

VERONICA SPICATA SUBSP. *INCANA*
♀ **Perennial** page **533**

HOT COLOUR SCHEMES

A grouping of plants in warm, vivid colours looks best in full sunlight; if your garden can offer a really sunny spot, take care when choosing an assortment of plants that they will all enjoy the heat. Mixing in some plants with dark bronze and purple foliage (see p.574) will heighten the fiery effect. Remember that hot colours jump towards the eye and can make spaces seem smaller.

ALONSOA WARSCEWICZII
♀ **Perennial, not hardy** page **54**

ALSTROEMERIA LIGTU HYBRIDS
♀ **Perennial** page **55**

ARCTOTIS X *HYBRIDA* 'FLAME'
♀ **Perennial, not hardy** Reddish-gold daisy-flowers with silver petal backs and silver-grey leaves. 'Mahogany' and 'Red Magic' are also recommended.
↕45–50cm (18–20in) ↔ 30cm (12in)

ASTILBE X *ARENDSII* 'FANAL'
♀ **Perennial** page **74**

BASSIA SCOPARIA F. *TRICHOPHYLLA*
♀ **Annual** Feathery leaves that start off green, then turn fiery red ageing to purple over the summer.
↕0.3–1.5m (1–5ft) ↔ 30–45cm (12–18in)

BEGONIA 'ILLUMINATION ORANGE'
♀ **Perennial, not hardy** page **82**

CALENDULA 'FIESTA GITANA'
♀ **Annual** page **96**

CALLISTEMON CITRINUS 'SPLENDENS'
♀ **Evergreen shrub, not hardy** page **97**

CAMPSIS X *TAGLIABUANA* 'MADAME GALEN'
♀ **Climber** page **108**

CARTHAMUS TINCTORIA 'ORANGE GOLD'
♀ **Annual** Clusters of bright, thistle-like tufted flowers, good for drying.
↕30–60cm (12–24in) ↔ 30cm (12in)

CHRYSANTHEMUM 'AMBER YVONNE ARNAUD'
♀ **Perennial, not fully hardy** page **122**

CHRYSANTHEMUM 'WENDY'
♀ **Perennial, not fully hardy** page **124**

COREOPSIS TINCTORIA 'SUNRISE'
♀ **Annual** Upright stems among clumps of mid-green leaves bearing solitary, daisy-like flowers, attractive to bees.
↕60cm (24in) ↔ 30cm (12in)

COREOPSIS VERTICILLATA 'GRANDIFLORA'
♀ **Perennial** Loose clusters of dark yellow flowers in early summer.
↕60–80cm (24–32in) ↔ 45cm (18in)

CROCOSMIA 'LUCIFER'
♀ **Perennial** page **152**

DAHLIA 'HAMARI GOLD'
♀ **Perennial, not fully hardy** page **163**

DAHLIA 'ZORRO'
♀ **Perennial, not fully hardy** page **163**

DIASCIA 'RUPERT LAMBERT'
♀ **Perennial, not fully hardy** Mat-forming, with spikes of spurred, deep warm pink flowers from summer to autumn.
↕20cm (8in) ↔ to 50cm (20in)

EMBOTHRIUM COCCINEUM 'NORQUINCO'
♀ **Shrub, borderline hardy** Upright, with spidery scarlet flowers in early summer.
↕20cm (8in) ↔ 30cm (12in)

ESCHSCHOLZIA CALIFORNICA
♀ **Annual** page **196**

FUCHSIA 'CORALLE'
♀ **Shrub, not hardy** page **214**

GAZANIAS
Perennials, not hardy page **222**

HELIANTHEMUM 'FIRE DRAGON'
♀ **Evergreen shrub** page **247**

HELIOPSIS 'GOLDGEFIEDER'
♀ **Perennial** Double daisy-like golden-yellow flowerheads with a green centre, on stiff stems with coarse leaves.
↕ 90cm (36in) ↔ 60cm (24in)

IRIS 'APRICORANGE'
♀ **Perennial** page **284**

IRIS 'SUN MIRACLE'
♀ **Perennial** page **287**

KNIPHOFIA 'BEE'S SUNSET'
♀ **Perennial** page **297**

LIGULARIA 'GREGYNOG GOLD'
♀ **Evergreen shrub** page **313**

LONICERA ETRUSCA 'DONALD WATERER'
♀ **Climber** Dark, red-bloomed stems and buds, bright orange berries.
↕ 4m (12ft)

LOTUS BERTHELOTII
♀ **Subshrub, not hardy** page **326**

LYCHNIS CHALCEDONICA
♀ **Perennial** page **327**

MIMULUS CUPREUS 'WHITECROFT SCARLET'
♀ **Perennial, not fully hardy** Spreading stems with tubular, scarlet summer flowers.
↕ 10cm (4in) ↔ 15cm (6in)

PAPAVER ORIENTALE 'AGLAIA'
♀ **Perennial** Salmon flowers with cherry-pink shading at the petal bases.
↕ 45–90cm (18–36in) ↔ 60–90cm (24–36in)

PAPAVER ORIENTALE 'LEUCHTFEUER'
♀ **Perennial** Orange flowers with black blotches at the petal bases, silvery leaves.
↕ 45–90cm (18–36in) ↔ 60–90cm (24–36in)

PELARGONIUM 'VOODOO'
♀ **Perennial, not hardy** page **376**

PENSTEMON 'CHESTER SCARLET'
♀ **Perennial, not fully hardy** page **379**

PENSTEMON 'SCHOENHOLZERI'
♀ **Perennial** page **379**

PERSICARIA AMPLEXICAULIS 'FIRETAIL'
♀ **Perennial** Robust clump-forming plant with tall, bright red "bottlebrush" flower heads on upright stems.
↕ ↔ to 1.2m (4ft)

PHORMIUM 'SUNDOWNER'
♀ **Perennial, not fully hardy** page **393**

PHYGELIUS x *RECTUS* 'DEVIL'S TEARS'
♀ **Shrub** page **397**

POTENTILLA 'GIBSON'S SCARLET'
♀ **Perennial** page **411**

RHODODENDRON 'SPEK'S ORANGE'
♀ **Deciduous shrub** page **439**

ROSA ALEXANDER ('HARLEX')
♀ **Large-flowered bush rose** page **448**

RUDBECKIA FULGIDA VAR. SULLIVANTII 'GOLDSTURM'
♀ **Perennial** page **468**

SALVIA COCCINEA 'PSEUDOCOCCINEA'
♀ **Perennial, not fully hardy** page **473**

STREPTOSOLEN JAMESII
♀ **Climber, not hardy** Sprawling stems with small dark green leaves and rounded clusters of saucer-shaped orange flowers.
↕ 2–3m (6–10ft) ↔ 1–2.5m (3–8ft)

TROPAEOLUM MAJUS 'HERMINE GRASHOFF'
♀ **Annual climber** page **520**

VELTHEIMIA BRACTEATA
♀ **Bulbous perennial, not hardy** page **528**

COOL COLOUR SCHEMES

Cool blues, purples and creamy-whites always look elegant, and can be used to create an illusion of distance. As dusk falls, cool colours appear to glow in the fading light, a bonus in gardens used for evening entertaining. Mix in plenty of lush foliage to damp down summer heat still further, or for a lighter, brighter look, choose some plants with grey and silvery leaves (see p.576).

AEONIUM HAWORTHII
♀ **Succulent** page **44**

AGERATUM 'BLUE DANUBE'
♀ **Annual** Low, bushy plants covered in fluffy lavender-blue flowers. Deeper blue 'Blue Horizon' is also recommended.
↕20cm (8in) ↔ 30cm (12in)

ALLIUM CAERULEUM
♀ **Bulb** page **50**

ANCHUSA AZUREA 'LODDON ROYALIST'
♀ **Perennial** page **57**

ARUNCUS DIOICUS
♀ **Perennial** In moist ground, produces clumps of ferny leaves and feathery cream flower plumes.
↕20cm (8in) ↔ 30cm (12in)

BAPTISIA AUSTRALIS
♀ **Perennial** page **79**

CAMASSIA LEICHTLINII
♀ **Bulb** In late spring, bears spires of star-shaped, greeny-white flowers.
↕20cm (8in) ↔ 30cm (12in)

CAMPANULA GLOMERATA 'SUPERBA'
♀ **Perennial** page **106**

CENTAUREA CYANUS 'BLUE DIADEM'
♀ **Annual** Deep blue cornflower.
↕20–80cm (8–32in) ↔ 15cm (6in)

CONVOLVULUS SABATIUS
♀ **Trailing perennial** page **136**

CYANANTHUS LOBATUS
♀ **Perennial** page **156**

CYNARA CARDUNCULUS
♀ **Perennial** page **159**

DELPHINIUM 'GIOTTO'
♀ **Perennial** page **167**

DELPHINIUM 'LORD BUTLER'
♀ **Perennial** page **168**

DELPHINIUM 'THAMESMEAD'
♀ **Perennial** page **169**

ECHINOPS RITRO SUBSP. *RUTHENICUS*
♀ **Perennial** Metallic-blue globe thistle with silvery, cobwebby leaves.
↕60–90cm (24–36in) ↔ 45cm (18in)

ERYNGIUM BOURGATII 'OXFORD BLUE'
♀ **Perennial** Silver-veined spiny leaves and branching stems of blue thistle-flowers with silver bracts.
↕15–45cm (6–18in) ↔ 30cm (12in)

ERYNGIUM X *TRIPARTITUM*
♀ **Perennial** page **192**

GALTONIA VIRIDIFLORA
♀ **Bulb** In late summer, bears spires of snowdrop-like greeny-white flowers among grey-green straplike leaves.
↕to 1m (3ft) ↔ 10cm (4in)

HEBE ALBICANS
♀ **Shrub** page **240**

HEBE HULKEANA 'LILAC HINT'
Shrub Glossy green leaves and, in late spring and early summer, long sprays of lavender-blue flowers.
↕↔ 60cm (24in)

HELIOTROPIUM 'CHATSWORTH'
♀ **Perennial, not hardy** Popular summer bedding with deep mauve, very fragrant flowers.
↕60–100cm (2–3ft) ↔ 30–45cm (12–18in)

HOSTA 'LOVE PAT'
♀ **Perennial** page **260**

IPHEION 'ROLF FIEDLER'
♀ **Bulb, not hardy** Spring-flowering, with star-shaped blue flowers and blue-green leaves.
↕10–12cm (4–5in)

IRIS 'ORINOCO FLOW'
♀ **Perennial** page **285**

IRIS SIBIRICA 'SMUDGER'S GIFT'
♀ **Perennial** page **279**

IRIS SIBIRICA 'UBER DEN WOLKEN'
Perennial page **279**

LAVANDULA X *INTERMEDIA* DUTCH GROUP
♀ **Shrub** page **306**

LINUM NARBONENSE 'HEAVENLY BLUE'
♀ **Perennial, borderline hardy** Forms clumps covered with saucer-shaped pale blue flowers, each lasting a single day.
↕30–60cm (12–24in) ↔ 45cm (18in)

MOLUCELLA LAEVIS 'PIXIE BELLS'
♀ **Annual** Pale green leaves and flower spires conspicuous for the pale green, shell-like calyces surrounding each tiny flower.
↕60–90cm (24–36in) ↔ 23cm (9in)

MYOSOTIS 'BOUQUET'
♀ **Annual** Floriferous, compact blue forget-me-not. The dwarf 'Ultramarine' is also recommended.
↕12–20cm (4–8in) ↔ 15cm (6in)

PHLOX PANICULATA 'LE MAHDI'
♀ **Perennial** page **391**

PULMONARIA SACCHARATA ARGENTEA GROUP
♀ **Evergreen perennial** page **428**

RUTA GRAVEOLENS 'JACKMAN'S BLUE'
♀ **Shrub** Intensely grey-blue feathery leaves; the dull yellow flowerheads can be trimmed off to the benefit of the foliage. Wear gloves: contact with foliage may cause an allergic reaction.
↕ ↔ 60cm (24in)

SALVIA CACALIIFOLIA
♀ **Perennial, not hardy** page **472**

SALVIA GUARANITICA 'BLUE ENIGMA'
♀ **Perennial, not fully hardy** page **474**

SCABIOSA CAUCASICA 'CLIVE GREAVES'
♀ **Perennial** page **485**

TEUCRIUM FRUTICANS 'AZUREUM'
♀ **Shrub, not fully hardy** White-woolly stem, grey-blue leaves and, in summer, short spires of whorled, deep blue flowers.
↕60–100cm (2–3ft) ↔ 2m (6ft)

THALICTRUM DELAVAYI 'HEWITT'S DOUBLE'
♀ **Perennial** page **510**

VERONICA 'SHIRLEY BLUE'
♀ **Perennial** Grey-green, hairy leaves and spires of saucer-shaped blue flowers from late spring to mid-summer.
↕↔ 30cm (12in)

PLANTS FOR WHITE GARDENS

Single-colour gardens are popular with many gardeners, and the most planted are white gardens, perhaps because so many white flowers are also scented. Consider also leaf colour and foliage, including variegated, silver and grey-leaved plants. To relieve the sameness of the scheme, it is often helpful to add a few cream or pale blue flowers – these will actually enhance the effect.

ANEMONE BLANDA 'WHITE SPLENDOUR'
♀ **Bulb** page **58**

ANEMONE x *HYBRIDA* 'HONORINE JOBERT'
♀ **Perennial** page **59**

ASTILBE 'IRRLICHT'
Perennial The coarsely cut dark foliage is a good contrast to the upright, white fluffy flowers.
↕↔ 50cm (18in)

CLEMATIS 'MARIE BOISSELOT'
♀ **Climber** page **130**

COSMOS BIPINNATUS 'SONATA WHITE'
♀ **Annual** page **144**

CRAMBE CORDIFOLIA
♀ **Perennial** page **149**

CROCUS SIEBERI 'ALBUS'
♀ **Bulb** page **153**

DAHLIA 'HAMARI BRIDE'
♀ **Perennial, not hardy** This semi-cactus dahlia has pure white flowers and is popular for exhibition.
↕ 1.2m (4ft) ↔ 60cm (2ft)

DELPHINIUM 'SANDPIPER'
♀ **Perennial** page **169**

DEUTZIA SETCHUENENSIS VAR. *CORYMBIFLORA*
♀ **Shrub** Masses of white flowers produced on a bush with brown, peeling bark.
↕ 2m (6ft) ↔ 1.5m (5ft)

DICENTRA SPECTABILIS 'ALBA'
♀ **Perennial** page **175**

DICTAMNUS ALBUS
♀ **Perennial** page **176**

DIGITALIS PURPUREA F. *ALBIFLORA*
♀ **Biennial** page **178**

ECHINACEA PURPUREA 'WHITE LUSTRE'
Perennial The stiff stems have creamy-white flowers with golden cones.
↕ 80cm (32in) ↔ 45cm (18in)

ERICA CARNEA 'SPRINGWOOD WHITE'
♀ **Evergreen shrub** page **187**

ERICA TETRALIX 'ALBA MOLLIS'
♀ **Evergreen shrub** page **189**

GILLENIA TRIFOLIATA
♀ **Perennial** page **231**

GYPSOPHILA PANICULATA 'BRISTOL FAIRY'
♀ **Perennial** page **234**

HOHERIA SEXSTYLOSA
♀ **Evergreen tree** page **258**

HYDRANGEA ARBORESCENS 'ANNABELLE'
♀ **Shrub** page **265**

HYDRANGEA PANICULATA 'FLORIBUNDA'
♀ **Shrub** page **267**

HYDRANGEA QUERCIFOLIA
♀ **Shrub** page **268**

IRIS 'BEWICK SWAN'
Perennial The ruffled, white flowers have bright orange beards.
↕ 1m (36in) ↔ 30cm (12in)

IRIS CONFUSA
♀ **Bulbous perennial** page **280**

IRIS SIBIRICA 'CREME CHANTILLY'
♀ **Perennial** page **279**

IRIS SIBIRICA 'HARPSWELL HAPPINESS'
♀ **Perennial** page **279**

IRIS SIBIRICA 'MIKIKO'
♀ **Perennial** page **279**

JASMINUM POLYANTHUM
♀ **Climber** page **292**

LUNARIA ANNUA VAR. *ALBIFLORA*
♀ **Biennial** The white-flowered form of common honesty.
↕ 90cm (3ft) ↔ 30cm (12in)

MALVA MOSCHATA F. *ALBA*
♀ **Perennial** page **338**

NARCISSUS 'EMPRESS OF IRELAND'
♀ **Bulb** page **348**

PAEONIA OBOVATA VAR. *ALBA*
♀ **Perennial** A choice plant with rounded, greyish leaflets and pure white flowers, most attractive in bud.
↕↔ 60–70cm (24–28in)

PAPAVER ORIENTALE 'BLACK AND WHITE'
♀ **Perennial** page **366**

PARAHEBE CATARRACTAE
♀ **Subshrub** page **368**

PHYSOSTEGIA VIRGINIANA 'SUMMER SNOW'
♀ **Perennial** Pure white flowers in spikes on upright stems above mid-green foliage.
↕ 1.2m (4ft) ↔ 60cm (24in)

PULMONARIA OFFICINALIS 'SISSINGHURST WHITE'
♀ **Perennial** page **427**

PULSATILLA VULGARIS 'ALBA'
♀ **Perennial** page **430**

RANUNCULUS ACONITIFOLIUS 'FLORE PLENO'
♀ **Perennial** page **433**

ROSA 'ICEBERG'
♀ **Cluster-flowered bush rose** page **451**

ROSA 'MADAME HARDY'
♀ **Shrub rose** page **461**

ROSA 'MARGARET MERRIL'
♀ **Cluster-flowered bush rose** page **451**

ROSA 'RAMBLING RECTOR'
♀ **Rambler rose** page **455**

RUBUS 'BENENDEN'
♀ **Shrub** page **467**

TULIPA 'PURISSIMA'
Bulb The compact stems carry very large, pure white flowers in mid-spring.
↕ 35cm (14in)

ZANTEDESCHIA AETHIOPICA
♀ **Perennial** page **544**

PLANTS FOR CLAY SOIL

Clay soil is difficult to dig, either wet or dry. It is prone to harbour slugs, and it is slow to warm up in spring; it is often described as a cold soil. However, it is usually rich in nurients and if plenty of organic matter is added it can be very fertile. Plants to avoid are those from upland areas, such as alpines, or those on the borderline of hardiness in your climate.

CAMPANULA LATILOBA 'HIDCOTE AMETHYST'
♀ **Perennial** Stocky stems of mauve-purple, cup-shaped flowers in mid-summer.
↕ 90cm (36in) ↔ 45cm (18in)

CLEMATIS 'POLISH SPIRIT'
♀ **Climber** Late-flowering, with small, single, purple flowers with red anthers.
↕ 5m (15ft) ↔ 2m (6ft)

CORNUS STOLONIFERA 'FLAVIRAMEA'
♀ **Shrub** page **140**

COTONEASTER X *WATERERI* 'JOHN WATERER'
♀ **Evergreen shrub** page **148**

DEUTZIA X *ELEGANTISSIMA* 'ROSEALIND'
♀ **Shrub** page **170**

DIGITALIS GRANDIFLORA
♀ **Perennial** page **177**

FILIPENDULA PURPUREA
♀ **Perennial** page **207**

GERANIUM PSILOSTEMON
♀ **Perennial** page **229**

HEMEROCALLIS 'STELLA DE ORO'
♀ **Perennial** page **254**

HYDRANGEA PANICULATA 'KYUSHU'
♀ **Shrub** Erect-growing cultivar with glossy leaves and large creamy-white flowers.
↕ 3–7m (10–22ft) ↔ 2.5m (8ft)

IRIS SIBIRICA CULTIVARS
♀ **Perennials** page **279**

LONICERA NITIDA 'SILVER LINING'
Evergreen shrub Each small leaf of this mound-shaped shrub has a thin, white margin.
↕↔ 1.5m (5ft)

MAHONIA AQUIFOLIUM 'APOLLO'
♀ **Evergreen shrub** page **334**

NARCISSUS 'JUMBLIE'
♀ **Bulb** page **347**

PERSICARIA CAMPANULATA
Perennial Small clusters of pink, bell-shaped flowers on spreading stems.
↕↔ 90cm (36in)

PHILADELPHUS 'MANTEAU D'HERMINE'
♀ **Shrub** page **385**

POTENTILLA FRUTICOSA 'TANGERINE'
♀ **Shrub** Twiggy, with yellow flowers flushed red throughout summer.
↕ 1m (3ft) ↔ 1.5m (5ft)

ROSA 'AMBER QUEEN'
♀ **Cluster-flowered bush rose** page **450**

SPIRAEA JAPONICA 'ANTHONY WATERER'
♀ **Shrub** page **501**

SYMPHYTUM 'GOLDSMITH'
Perennial This spreading plant has heart-shaped leaves edged in gold, with pale blue, cream and pink flowers.
↕↔ 30cm (12in)

SYRINGA VULGARIS 'KATHERINE HAVEMEYER'
♀ **Shrub** page **507**

PLANTS FOR SANDY SOIL

The advantages of sandy soils include the ability to dig, hoe and prepare them at almost any time of the year because they drain quickly after rain. However, water drains quickly through, taking nutrients with it, so it is important to dig in organic matter to improve the soil structure, retain moisture and improve fertility. Silver-leaved and slightly tender plants are very suitable.

BUDDLEJA DAVIDII 'WHITE PROFUSION'
♀ **Shrub** page **93**

CALLUNA VULGARIS 'DARKNESS'
♀ **Evergreen shrub** page **98**

CERATOSTIGMA WILLMOTTIANUM
♀ **Shrub** page **114**

ECHINOPS BANNATICUS 'TAPLOW BLUE'
♀ **Perennial** A robust, thistle-like plant with globular blue flowerheads.
↕ 1.2m (4ft) ↔ 60cm (2ft)

ERICA CINEREA 'EDEN VALLEY'
♀ **Evergreen shrub** page **188**

ERYNGIUM × *OLIVERIANUM*
♀ **Perennial** page **191**

ESCALLONIA 'PEACH BLOSSOM'
♀ **Evergreen shrub** Bushy, arching, with peach-pink and white flowers in late spring.
↕↔ 2.5m (8ft)

EUCALYPTUS DALRYMPLEANA
♀ **Evergreen tree** A fast-growing tree with green adult leaves and white bark.
↕ 20m (70ft) ↔ 8m (25ft)

GENISTA HISPANICA
Shrub Small, spiny, covered with small yellow flowers in early summer.
↕ 75cm (30in) ↔ 1.5m (5ft)

GLADIOLUS COMMUNIS SUBSP. *BYZANTINUS*
♀ **Bulb** page **232**

KNIPHOFIA 'LITTLE MAID'
♀ **Perennial** page **298**

LAVATERA ARBOREA 'VARIEGATA'
Biennial An evergreen, slightly tender, intensely variegated plant with purple flowers in the second year.
↕ 3m (10ft) ↔ 1.5m (5ft)

LAVATERA 'ROSEA'
♀ **Shrub** page **307**

LIMNANTHES DOUGLASII
♀ **Annual** page **319**

OENOTHERA MACROCARPA
♀ **Perennial** page **354**

PENSTEMON 'HIDCOTE PINK'
♀ **Perennial** A good bedding penstemon: spikes of small, tubular, pink flowers.
↕ 60–75cm (24–30in) ↔ 45cm (18in)

PENSTEMON PINIFOLIUS 'MERSEA YELLOW'
Evergreen shrub Dwarf, spreading, with narrow yellow flowers in summer.
↕ 40cm (16in) ↔ 25cm (10in)

PERSICARIA AFFINIS 'SUPERBA'
♀ **Perennial** page **381**

PERSICARIA BISTORTA 'SUPERBA'
♀ **Perennial** page **381**

POTENTILLA FRUTICOSA 'DAYDAWN'
♀ **Shrub** page **410**

POTENTILLA NEPALENSIS 'MISS WILLMOTT'
♀ **Perennial** page **412**

SANTOLINA ROSMARINIFOLIA 'PRIMROSE GEM'
♀ **Evergreen shrub** page **481**

PLANTS FOR CHALK SOILS

The majority of plants will grow in soil that is neutral or slightly alkaline, but some positively prefer limy or chalky soil. These include plants that are just as important as the rhododendrons and camellias of acid soils, and include delphiniums, clematis and dianthus. But quality of soil is important: it must be improved with organic matter, and poor, thin chalky soils are difficult to plant.

AQUILEGIA VULGARIS 'NIVEA'
♀ **Perennial** page 62

BUDDLEJA DAVIDII 'PINK DELIGHT'
♀ **Shrub** Bright pink flowers are produced on thick, conical spikes.
↕3m (10ft) ↔ 5m (15ft)

BUDDLEJA DAVIDII 'ROYAL RED'
Shrub page 93

BUDDLEJA X *WEYERIANA* 'SUNGOLD'
♀ **Shrub** An attractive hybrid with spikes of golden yellow flowers in summer.
↕4m (12ft) ↔ 3m (10ft)

CERCIS SILIQUASTRUM
♀ **Tree** page 115

CHAENOMELES SPECIOSA 'NIVALIS'
♀ **Shrub** A variety of Japanese quince with pure white flowers in spring.
↕2.5m (8ft) ↔ 5m (15ft)

CLEMATIS 'COMTESSE DE BOUCHAUD'
♀ **Climber** page 133

CLEMATIS X *JOUINIANA* 'PRAECOX'
♀ **Climber** This unusual hybrid is a scrambler, with a frothy mass of tiny white and blue flowers.
↕2–3m (6–10ft)

CLEMATIS 'MISS BATEMAN'
♀ **Climber** page 130

CONVALLARIA MAJALIS VAR. *ROSEA*
Perennial This lily of the valley has dusky pink flowers with the familiar sweet scent.
↕23cm (9in) ↔ 30cm (12in)

COTONEASTER STERNIANUS
♀ **Evergreen shrub** page 148

DELPHINIUM 'BLUE NILE'
♀ **Perennial** page 166

DEUTZIA SCABRA 'PRIDE OF ROCHESTER'
Shrub The stems have attractive, peeling, brown bark but the chief merit is the scented, double, pale pink flowers.
↕3m (10ft) ↔ 2m (6ft)

DIANTHUS ALPINUS
♀ **Perennial** Cushions of grey foliage produce scented flowers in shades of pink.
↕8cm (3in) ↔ 10cm (4in)

DIANTHUS DELTOIDES
♀ **Perennial** Mats of deep green leaves are covered with small pink flowers for several weeks in summer.
↕20cm (8in) ↔ 30cm (12in)

DIANTHUS 'MONICA WYATT'
♀ **Perennial** page 173

EUONYMUS EUROPAEUS 'RED CASCADE'
♀ **Shrub** page 199

FRAXINUS ORNUS
♀ **Tree** Bushy, round-headed tree with showy white flowers and bright, purple autumn colour.
↕↔ 15m (50ft)

FUCHSIA 'HEIDI ANN'
♀ **Shrub, not hardy** An upright, bushy plant with double lilac and cerise flowers.
↕↔ 45cm (18in)

FUCHSIA 'PROSPERITY'
♀ **Shrub** A vigorous, hardy plant with crimson and pink double flowers.
↕↔ 45cm (18in)

GALANTHUS 'MAGNET'
♀ **Bulb** page **218**

HELIANTHEMUM 'JUBILEE'
♀ **Shrub** This small shrub has bright green leaves and double yellow flowers.
↕ 20cm (8in) ↔ 30cm (12in)

HELLEBORUS ORIENTALIS
Perennial Nodding flowers in shades of pink, white and green are produced in early spring.
↕↔ 45cm (18in)

ILEX AQUIFOLIUM 'SILVER QUEEN'
♀ **Evergreen tree** page **273**

MAGNOLIA x *KEWENSIS* 'WADA'S MEMORY'
♀ **Shrub** page **333**

MAGNOLIA 'RICKI'
♀ **Shrub** page **333**

MAGNOLIA SALICIFOLIA
♀ **Shrub** page **333**

MAGNOLIA x *SOULANGEANA* 'RUSTICA RUBRA'
♀ **Shrub** page **333**

MAGNOLIA WILSONII
♀ **Shrub** page **333**

MORUS NIGRA
♀ **Tree** The black mulberry is a hardy, long-lived, picturesque tree with tasty fruit.
↕ 12m (40ft) ↔ 15m (50ft)

PHILADELPHUS 'SYBILLE'
♀ **Deciduous shrub** Arching shrub with cup-shaped, intensely fragrant white flowers in early summer.
↕ 1.2m (4ft) ↔ 2m (6ft)

PRUNUS 'OKAME'
♀ **Tree** page **424**

PRUNUS 'TAIHAKU'
♀ **Tree** page **425**

PRUNUS TENELLA 'FIRE HILL'
♀ **Shrub** Deep pink, single flowers cover the upright branches in spring.
↕↔ 1.5m (5ft)

PULSATILLA VULGARIS
♀ **Perennial** page **429**

SYRINGA VULGARIS 'CHARLES JOLY'
♀ **Shrub** page **506**

VERBASCUM 'COTSWOLD BEAUTY'
♀ **Perennial** page **529**

PLANTS FOR ACID SOIL

Acid soils contain low quantities of calcium, a plant nutrient found in limestone and chalk, but some of the most beautiful plants, including rhododendrons, heathers and pieris, have adapted to grow well only in soils where it is deficient. Most also benefit from light shade and rich soil. Where soil is not ideal, grow plants in large pots or tubs of ericaceous compost.

ACER JAPONICUM 'ACONITIFOLIUM'
♀ **Shrub** page **33**

ACER PALMATUM VAR. *DISSECTUM*
♀ **Shrub** The finely cut leaves turn yellow in autumn, on a mounded shrub.
↕2m (6ft) ↔ 3m (10ft)

ACER PALMATUM 'SEIRYU'
♀ **Shrub** An upright shrub with divided leaves that turn orange in autumn.
↕2m (6ft) ↔ 1.2m (4ft)

CAMELLIA 'LEONARD MESSEL'
♀ **Evergreen shrub** page **102**

CAMELLIA SASANQUA 'NARUMIGATA'
♀ **Evergreen shrub** page **103**

CAMELLIA X *WILLIAMSII* 'DEBBIE'
♀ **Evergreen shrub** The semi-double flowers are deep pink.
↕2–5m (6–15ft) ↔ 1–3m (3–10ft)

CASSIOPE 'EDINBURGH'
♀ **Evergreen shrub** page **111**

CRINODENDRON HOOKERIANUM
♀ **Evergreen shrub** page **150**

DABOECIA CANTABRICA 'BICOLOR'
♀ **Evergreen shrub** page **161**

DABOECIA CANTABRICA 'WILLIAM BUCHANAN'
♀ **Evergreen shrub** page **161**

ENKIANTHUS CAMPANULATUS
♀ **Shrub** page **183**

ERICA ARBOREA VAR. *ALPINA*
♀ **Evergreen shrub** page **186**

ERICA ARBOREA 'ESTRELLA GOLD'
♀ **Evergreen shrub** Fragrant white flowers appear among the lime-green foliage and yellow shoot tips.
↕1.2m (4ft) ↔ 75cm (30in)

ERICA CILIARIS 'CORFE CASTLE'
♀ **Shrub** page **188**

ERICA CILIARIS 'DAVID MCLINTOCK'
♀ **Shrub** page **188**

ERICA CINEREA 'C. D. EASON'
♀ **Shrub** page **188**

ERICA CINEREA 'FIDDLER'S GOLD'
♀ **Shrub** page **188**

ERICA CINEREA 'WINDLEBROOKE'
♀ **Shrub** page **188**

ERICA ERIGENA 'W. T. RATCLIFF'
♀ **Evergreen shrub** A compact plant with green foliage and white flowers.
↕75cm (30in) ↔ 55cm (22in)

EUCRYPHIA NYMANSENSIS 'NYMANSAY'
♀ **Evergreen shrub** page **198**

HAMAMELIS X *INTERMEDIA* 'PALLIDA'
♀ **Shrub** page **239**

HYDRANGEA MACROPHYLLA 'BLUE WAVE'
♀ **Shrub** This lacecap provides a delicate but showy display, with large and small blue flowers in each head.
↕1.5m (5ft) ↔ 2m (6ft)

IRIS DOUGLASIANA
♀ **Perennial** page **281**

KALMIA LATIFOLIA
♀ **Shrub** page **295**

LEIOPHYLLUM BUXIFOLIUM
♀ **Perennial** page **309**

LITHODORA DIFFUSA 'HEAVENLY BLUE'
♀ **Evergreen shrub** page **321**

MAGNOLIA LILIIFLORA 'NIGRA'
♀ **Shrub** page **331**

MECONOPSIS BETONICIFOLIA
♀ **Perennial** page **339**

MECONOPSIS GRANDIS
♀ **Perennial** page **340**

PIERIS 'FOREST FLAME'
♀ **Evergreen shrub** page **402**

PIERIS JAPONICA 'BLUSH'
♀ **Evergreen shrub** page **403**

PRIMULA PULVERULENTA
♀ **Perennial** page **418**

RHODODENDRON 'BEETHOVEN'
♀ **Evergreen azalea** page **436**

RHODODENDRON 'DORA AMATEIS'
♀ **Dwarf rhododendron** page **444**

RHODODENDRON 'GINNY GEE'
♀ **Dwarf rhododendron** A compact plant bearing pink flowers that fade almost to white with age.
↕↔ 60–90cm (24–36in)

RHODODENDRON IMPEDITUM
♀ **Dwarf rhododendron** Grey-green leaves are almost hidden by lavender-blue flowers in spring.
↕↔ 60cm (24in)

RHODODENDRON 'PALESTRINA'
♀ **Evergreen azalea** page **437**

RHODODENDRON 'ROSE BUD'
♀ **Evergreen azalea** page **437**

RHODODENDRON 'VUYK'S ROSYRED'
♀ **Evergreen azalea** page **437**

SARRACENIA FLAVA
♀ **Perennial** page **483**

SKIMMIA JAPONICA 'RUBELLA'
♀ **Evergreen shrub** page **495**

SOLLYA HETEROPHYLLA
♀ **Cimber, not hardy** page **497**

TRILLIUM GRANDIFLORUM
♀ **Perennial** page **519**

VACCINIUM CORYMBOSUM
♀ **Shrub** page **526**

VACCINIUM GLAUCOALBUM
♀ **Shrub** page **527**

VACCINIUM VITIS-IDAEA KORALLE GROUP
♀ **Shrubs** page **527**

PLANTS FOR POOR SOIL

While it is true that most plants grow better in well-prepared soil, there are some that grow well in poor soil that has not had much preparation or cultivation. Many annuals evolved to take advantage of open sites, disappearing as the soil improves and larger plants invade the area, so they are a good choice, but there are shrubs and perennials that will also survive.

ACHILLEA 'MOONSHINE'
♀ **Perennial** page **41**

ARTEMISIA ABROTANUM
♀ **Shrub** This small shrub has green, finely-divided, pleasantly fragrant leaves.
↕↔ 1m (3ft)

BUDDLEJA DAVIDII 'EMPIRE BLUE'
♀ **Shrub** page **93**

CEANOTHUS 'BLUE MOUND'
♀ **Evergreen shrub** page **112**

CYTISUS PRAECOX 'ALLGOLD'
♀ **Shrub** page **160**

CYTISUS PRAECOX 'WARMINSTER'
♀ **Shrub** Arching shoots, thickly set with creamy yellow flowers in spring.
↕ 1.2m (4ft) ↔ 1.5m (5ft)

CISTUS X *PURPUREUS*
♀ **Evergreen shrub** page **127**

ERYSIMUM CHEIRI 'HARPUR CREWE'
♀ **Perennial** page **193**

ESCHSCHOLZIA CAESPITOSA 'SUNDEW'
Annual Neat, low-growing, with divided grey leaves and pale yellow flowers.
↕↔ 15cm (6in)

FALLOPIA BALDSCHUANICA
♀ **Climber** page **204**

FESTUCA GLAUCA 'BLAUFUCHS'
♀ **Ornamental grass** page **207**

GAILLARDIA 'DAZZLER'
♀ **Perennial** page **217**

GENISTA LYDIA
♀ **Shrub** page **223**

HEBE OCHRACEA 'JAMES STIRLING'
♀ **Evergreen shrub** page **242**

IBERIS SEMPERVIRENS
♀ **Evergreen shrub** page **270**

KOLKWITZIA AMABILIS 'PINK CLOUD'
♀ **Shrub** page **299**

LATHYRUS LATIFOLIUS
♀ **Climber** page **302**

LAVANDULA ANGUSTIFOLIA 'TWICKEL PURPLE'
♀ **Evergreen shrub** page **305**

LAVATERA, SHRUBS
♀ **Shrubs** page **307**

PHLOMIS FRUTICOSA
♀ **Evergreen shrub** page **386**

POTENTILLA FRUTICOSA 'ELIZABETH'
♀ **Shrub** page **410**

ROBINIA HISPIDA
♀ **Shrub** page **446**

SANTOLINA CHAMAECYPARISSUS 'LEMON QUEEN'
Evergreen shrub Silver, feathery foliage, and small, round, pale yellow flowers.
↕↔ 60cm (12in)

THYMUS VULGARIS 'SILVER POSIE'
Evergreen shrub Thyme with variegated leaves and pink flowers.
↕ 15–30cm (6–12in) ↔ 40cm (16in)

PLANTS FOR WET SOIL

Waterlogged, badly-drained soils are inhospitable places for most plants. Roots need to breathe, and if all the air spaces within the soil are filled with water most roots rot and only marginal or aquatic plants can survive. However, if the soil is permanently moist there are many beautiful plants that will thrive, and may survive flooding if it is only for a few days.

ALNUS GLUTINOSA 'IMPERIALIS'
♀ **Tree** page **53**

ASTILBE x *CRISPA* 'PERKEO'
♀ **Perennial** page **75**

ASTILBE 'SPRITE'
♀ **Perennial** page **75**

ASTILBE 'STRAUSSENFEDER'
♀ **Perennial** page **76**

CARDAMINE PRATENSE 'FLORE PLENO'
♀ **Perennial** Double-flowered lady's smock with lilac-pink flowers in spring.
↕ ↔ 20cm (8in)

CHAMAECYPARIS PISIFERA 'BOULEVARD'
♀ **Conifer** page **119**

CIMICIFUGA RACEMOSA
♀ **Perennial** page **126**

CORNUS ALBA 'SIBIRICA'
♀ **Shrub** page **137**

DARMERA PELTATA
♀ **Perennial** page **165**

EUPHORBIA GRIFFITHII 'DIXTER'
♀ **Perennial** Orange bracts in autumn.
↕ 75cm (30in) ↔ 1m (3ft)

FILIPENDULA RUBRA 'VENUSTA'
♀ **Perennial** page **208**

FRITILLARIA MELEAGRIS
♀ **Bulb** page **211**

IRIS FORRESTII
♀ **Perennial** page **282**

IRIS PSEUDACORUS
♀ **Perennial** page **278**

IRIS VERSICOLOR
♀ **Perennial** page **279**

LIGULARIA 'GREGYNOG GOLD'
♀ **Perennial** page **313**

LOBELIA 'QUEEN VICTORIA'
♀ **Perennial** page **322**

LYSICHITON AMERICANUS
♀ **Perennial** page **327**

LYSICHITON CAMTSCHATCENSIS
♀ **Perennial** page **328**

LYSIMACHIA CLETHROIDES
♀ **Perennial** page **328**

LYSIMACHIA NUMMULARIA 'AUREA'
♀ **Perennial** page **329**

LYTHRUM SALICARIA 'FEUERKERZE'
♀ **Perennial** page **329**

MATTEUCCIA STRUTHIOPTERIS
♀ **Hardy fern** page **338**

MIMULUS CARDINALIS
♀ **Perennial** page **341**

OSMUNDA REGALIS
♀ **Hardy fern** page **360**

PRIMULA CANDELABRA TYPES
♀ **Perennials** page **418**

PRIMULA DENTICULATA
♀ **Perennial** page **412**

PRIMULA ROSEA
♀ **Perennial** page **419**

RODGERSIA PINNATA 'SUPERBA'
♀ **Perennial** page **447**

PLANTS FOR HOT, DRY SITES

As problems of water supply become more acute, consider plants that have low water requirements. Often these have silvery or small leaves, and some are fragrant, so the garden can still be interesting through the year. They look attractive growing through gravel, an effective mulch to retain soil moisture. In addition, many establish quickly, and are evergreen and low maintenance.

ABUTILON VITIFOLIUM 'VERONICA TENNANT'
♀ **Shrub** page 30

ACHILLEA TOMENTOSA
♀ **Perennial** page 41

AGAVE VICTORIAE-REGINA
♀ **Perennial** page 48

ALLIUM x *HOLLANDICUM*
Bulb Tall stems bear round heads of small purple flowers in early summer.
↕ 1m (3ft)

ALSTROEMERIA LIGTU HYBRIDS
♀ **Perennial** page 55

ARMERIA JUNIPERIFOLIA 'BEVAN'S VARIETY'
♀ **Perennial** page 68

ARTEMISIA 'POWIS CASTLE'
♀ **Perennial** page 70

ARTEMISIA SCHMIDTIANA 'NANA'
♀ **Perennial** An evergreen that forms a feathery, silver carpet with small yellow flowerheads in summer.
↕ 8cm (3in) ↔ 30cm (12in)

BALLOTA PSEUDODICTAMNUS
♀ **Evergreen shrub** page 79

CEANOTHUS ARBOREUS 'TREWITHEN BLUE'
♀ **Shrub** page 111

CEANOTHUS 'CASCADE'
♀ **Evergreen shrub** The arching branches bear masses of powder-blue flowers in late spring.
↕↔ 4m (12ft)

CEDRONELLA CANARIENSIS
Perennial, not hardy The slightly sticky leaves are aromatic; produces small clusters of mauve flowers in late summer.
↕ 1.2m (4ft) ↔ 60cm (24in)

CERATOSTIGMA PLUMBAGINOIDES
♀ **Perennial** page 114

CISTUS x *AGUILARII* 'MACULATUS'
♀ **Evergreen shrub** page 126

CONVOLVULUS CNEORUM
♀ **Evergreen shrub** page 136

CYTISUS BATTANDIERI
♀ **Shrub** page 159

CYTISUS x *KEWENSIS*
♀ **Shrub** Arching stems are covered with cream flowers in spring.
↕ 30cm (12in) ↔ 1.5m (5ft)

CYTISUS MULTIFLORUS
♀ **Shrub** An upright shrub at first, then spreading, with masses of small white flowers.
↕ 3m (10ft) ↔ 2.5m (8ft)

DIANTHUS 'HAYTOR WHITE'
♀ **Perennial** page 172

DICTAMNUS ALBUS VAR. *PURPUREUS*
♀ **Perennial** page 177

DIERAMA PULCHERRIMUM
Perennial Clumps of grass-like leaves produce arching stems of pendent, pink, bell-shaped flowers.
↕ 1–1.5m (3–5ft) ↔ 60cm (24in)

ERYNGIUM BOURGATII
Perennial Silver-veined spiny leaves form compact clumps with branching stems of steely-blue prickly flowerheads.
↕15–45cm (6–18in) ↔ 30cm (12in)

ERYSIMUM 'BOWLES' MAUVE'
♈ **Evergreen shrub** page **192**

ERYSIMUM 'WENLOCK BEAUTY'
♈ **Perennial** page **193**

ESCHSCHOLZIA CAESPITOSA
♈ **Annual** page **196**

EUPHORBIA CHARACIAS
♈ **Perennial** page **201**

EUPHORBIA CHARACIAS SUBSP. *WULFENII* 'JOHN TOMLINSON'
♈ **Perennial** page **201**

EUPHORBIA MYRSINITES
♈ **Perennial** page **202**

GAURA LINDHEIMERI
♈ **Perennial** page **221**

GERANIUM MADERENSE
♈ **Perennial** A slightly tender, large plant with divided leaves and a colourful display of mauve flowers in summer.
↕↔ 60cm (24in)

GYMNOCALYCIUM ANDRAEA
♈ **Cactus** page **234**

x *HALIMIOCISTUS SAHUCII*
♈ **Evergreen shrub** page **236**

x *HALIMIOCISTUS WINTONENSIS*
♈ **Evergreen shrub** Pale cream flowers with maroon centres are produced in clusters above woolly leaves.
↕60cm (24in) ↔ 90cm (36in)

HALIMIUM LASIANTHUM
♈ **Evergreen shrub** page **237**

HEBE MACRANTHA
♈ **Evergreen shrub** page **242**

HEBE RAKAIENSIS
♈ **Evergreen shrub** page **243**

HIBISCUS SYRIACUS 'WOODBRIDGE'
♈ **Shrub** page **257**

LAVANDULA STOECHAS SUBSP. *PEDUNCULATA*
♈ **Evergreen shrub** page **306**

PARAHEBE PERFOLIATA
♈ **Evergreen shrub** page **368**

ROSMARINUS OFFICINALIS 'BENENDEN BLUE'
Evergreen shrub Upright branches with deep green leaves and vivid blue flowers.
↕↔ 1.5m (5ft)

SPARTIUM JUNCEUM
♈ **Shrub** page **500**

TULIPA LINIFOLIA
♈ **Bulb** page **522**

TULIPA TARDA
♈ **Bulb** Each bulb produces a cluster of yellow and white flowers on short stems.
↕15cm (6in)

VERBASCUM 'GAINSBOROUGH'
♈ **Perennial** page **530**

VERBASCUM DUMULOSUM
♈ **Perennial** page **530**

VERBASCUM 'HELEN JOHNSON'
♈ **Perennial** page **531**

Plants for Damp Shade

A border that has damp soil and is in shade for much of the day is a useful place to grow woodland plants without having to plant a wood. Without the drying effect of the trees the soil will support a greater range of plants, and exotic plants from the Himalayas and South America should thrive, especially if the soil is acidic. It is also a good site for hellebores, lilies and ferns.

Anemone nemorosa 'Robinsoniana'
♀ **Perennial** page **60**

Anemone ranunculoides
♀ **Perennial** page **60**

Athyrium filix-femina
♀ **Hardy fern** page **77**

Cardiocrinum giganteum
♀ **Perennial** page **108**

Digitalis x *mertonensis*
♀ **Perennial** page **178**

Dodecatheon meadia f. *album*
♀ **Perennial** page **179**

Epimedium x *youngianum* 'Niveum'
♀ **Perennial** page **185**

Galanthus (snowdrops)
♀ **Bulbs** pages **217–219**

Gentiana asclepiadea
♀ **Perennial** page **225**

Hacquetia epipactis
♀ **Perennial** A woodland plant; small ruffs of leaves surround clusters of yellow flowers.
↕ 5cm (2in) ↔ 15–30cm (6–12in)

Hepatica transsilvanica
♀ **Perennial** Spring flowers in pastel shades.
↕ 15cm (6in) ↔ 20cm (8in)

Hydrangea aspera Villosa Group
♀ **Shrub** page **265**

Iris graminea
♀ **Perennial** page **283**

Mahonia japonica
♀ **Shrub** page **334**

Onoclea sensibilis
♀ **Hardy fern** page **356**

Polygonatum x *hybridum*
♀ **Perennial** page **408**

Polystichum aculeatum
♀ **Fern** page **409**

Primula sieboldii
♀ **Perennial** Pink or white.
↕ 60cm (24in) ↔ 1m (3ft)

Pulmonaria 'Lewis Palmer'
♀ **Perennial** page **426**

Sarcococca confusa
♀ **Evergreen shrub** page **482**

Thalictrum flavum subsp. *glaucium*
♀ **Perennial** page **511**

Tiarella cordifolia
♀ **Perennial** page **516**

Tolmeia 'Taff's Gold'
♀ **Perennial** page **516**

Trillium luteum
♀ **Perennial** page **519**

Trollius x *cultorum* 'Orange Princess'
♀ **Perennial** page **520**

Uvularia grandiflora
♀ **Perennial** page **526**

Viburnum x *carlcephalum*
♀ **Shrub** page **534**

PLANTS FOR DRY SHADE

The dry shade under trees is not a hospitable place for most plants. Those that survive best are spring-flowering bulbs that disappear underground in summer before the soil dries out. Some evergreens and winter-flowering plants also survive. Even these require good soil preparation and careful watering and feeding for several seasons until they are well-established.

ASPLENIUM SCOLOPENDRIUM
♀ **Hardy fern** page **71**

AUCUBA JAPONICA 'ROZANNIE'
Evergreen shrub Cultivar with green leaves; the flowers are bisexual and self-fertile, resulting in many red berries.
↕↔ 1m (3ft)

BERGENIA CORDIFOLIA
Perennial The large, rounded, evergreen leaves make good ground cover and the pink flowers in spring are showy.'
↕ 60cm (24in) ↔ 75cm (30in)

DRYOPTERIS FILIX-MAS
♀ **Hardy fern** page **179**

EUONYMUS FORTUNEI 'EMERALD 'N' GOLD'
♀ **Evergreen shrub** page **199**

GALANTHUS NIVALIS 'FLORE PLENO'
♀ **Bulb** page **218**

GERANIUM MACRORRHIZUM 'INGWERSEN'S VARIETY'
♀ **Perennial** page **228**

HEDERA HELIX 'LITTLE DIAMOND'
♀ **Evergreen climber** page **246**

HEDERA HIBERNICA 'SULPHUREA'
Evergreen climber An ivy with gold-edged leaves that forms good ground cover.
↕ 3m (10ft)

HELLEBORUS FOETIDUS 'WESTER FLISK'
Perennial One of the best cultivars, with deep red stems, dark green, finely-divided leaves and red-edged green flowers.
↕ 60cm (24in) ↔ 1m (3ft)

IRIS FOETIDISSIMA
♀ **Perennial** Evergreen leaves, purple and brown flowers, and orange seeds in winter.
↕ 30–90cm (12–36in)

IRIS FOETIDISSIMA 'VARIEGATA'
♀ **Perennial** page **282**

KERRIA JAPONICA 'GOLDEN GUINEA'
♀ **Shrub** page **296**

LAMIUM GALEOBDOLON 'HERMANN'S PRIDE'
Perennial Compact, with arching stems marked silver, and pale yellow flowers.
↕ 45cm (18in) ↔ 1m (3ft)

RHODODENDRON 'CECILE'
♀ **Shrub** page **438**

RUBUS TRICOLOR
Evergreen shrub The creeping stems are covered with decorative red bristles, and the leaves are deep green.
↕ 60cm (24in) ↔ 2m (6ft)

RUSCUS ACULEATUS
Perennial No leaves, but the flattened, green and spiny stems are evergreen, and red berries are sometimes produced.
↕ 75cm (30in) ↔ 1m (3ft)

RUSCUS HYPOGLOSSUM
Perennial Arching stems are glossy and evergreen, without spines.
↕ 45cm (18in) ↔ 1m (3ft)

VINCA MINOR 'ATROPURPUREA'
♀ **Perennial** page **538**

PLANTS FOR EXPOSED SITUATIONS

Gardens exposed to strong winds, especially cold ones, make gardening difficult. Choose compact varieties that require less staking, and avoid large-leaved plants that may be damaged. Late-flowering cultivars will not be caught by spring frosts. Protect plants with netting when planting, and in winter to reduce damage, and plant hedges and screens as windbreaks.

ASTER ALPINUS
🏆 **Perennial** page **71**

BERBERIS THUNBERGII 'GOLDEN RING'
Shrub A spiny shrub with purple leaves edged with a thin gold band.
↕1.5m (5ft) ↔ 2m (6ft)

CALLUNA VULGARIS 'KINLOCHRUEL'
🏆 **Evergreen shrub** page **98**

CHAENOMELES × SUPERBA 'PINK LADY'
🏆 **Shrub** page **116**

DEUTZIA × HYBRIDA 'MONT ROSE'
🏆 **Shrub** page **170**

DIANTHUS GRATIANOPOLITANUS
🏆 **Perennial** Low-growing mats of grey foliage and solitary pink flowers.
↕15cm (6in) ↔ 40cm (16in)

ERICA CARNEA 'VIVELLI'
🏆 **Evergreen shrub** page **187**

ERICA × WATSONII 'DAWN'
🏆 **Evergreen shrub** page **189**

HYPERICUM OLYMPICUM
🏆 **Shrub** This compact, deciduous shrub has grey leaves and large yellow flowers.
↕25cm (10in) ↔ 30cm (12in)

PHILADELPHUS 'VIRGINAL'
🏆 **Shrub** Scented, double white flowers are carried in clusters on upright branches.
↕3m (10ft) ↔ 2.5m (8ft)

POTENTILLA FRUTICOSA 'ABBOTSWOOD'
🏆 **Shrub** page **410**

SALIX GRACILISTYLA 'MELANOSTACHYS'
🏆 **Shrub** An upright shrub with grey leaves, and black catkins with red anthers in early spring.
↕3m (10ft) ↔ 4m (12ft)

SALIX RETICULATA
🏆 **Shrub** A low-growing shrub with glossy leaves and erect, yellow and pink catkins in spring.
↕8cm (3in) ↔ 30cm (12in)

SAMBUCUS NIGRA 'GUINCHO PURPLE'
🏆 **Shrub** page **480**

SPIRAEA JAPONICA 'SHIROBANA'
🏆 **Shrub** Mounds of foliage are dotted with both pink and white flowerheads throughout the summer.
↕60cm (24in) ↔ 90cm (36in)

SPIRAEA NIPPONICA 'SNOWMOUND'
🏆 **Shrub** page **502**

TAMARIX TETRANDRA
🏆 **Shrub** page **508**

VIBURNUM OPULUS 'ROSEUM'
🏆 **Shrub** Erect, fast-growing shrub with globular heads of white flowers in summer.
↕↔ 4m (12ft)

PLANTS FOR COASTAL GARDENS

Coastal gardens are windy (see facing page) and, nearer the shore, plants may also have to cope with winds laden with salt. However, as an advantage, they are often free of frost. Silver plants and evergreens often thrive, and if the winds can be ameliorated a wide range of plants should do well. Hydrangeas in particular are often the pride of coastal gardeners.

AGAPANTHUS CAULESCENS
♀ **Perennial** page **47**

ARBUTUS UNEDO
♀ **Evergreen tree** page **66**

BERBERIS DARWINII
♀ **Evergreen shrub** page **84**

BUDDLEJA GLOBOSA
♀ **Shrub** page **94**

CHOISYA TERNATA
♀ **Evergreen shrub** page **121**

CRAMBE MARITIMA
Perennial Silver-blue leaves, and branching heads of white flowers followed by seed pods.
↕↔ 60cm (24in)

ELAEAGNUS x *EBBINGEI* 'GILT EDGE'
♀ **Evergreen shrub** page **182**

ERICA x *WILLIAMSII* 'P. D. WILLIAMS'
♀ **Shrub** page **189**

ESCALLONIA 'DONARD RADIANCE'
♀ **Evergreen shrub** Pink summer flowers.
↕↔ 2.5m (8ft)

ESCALLONIA 'IVEYI'
♀ **Evergreen shrub** page **195**

x *FATSHEDERA LIZEI*
♀ **Shrub** page **205**

FUCHSIA 'MRS POPPLE'
♀ **Shrub** page **212**

FUCHSIA 'TOM THUMB'
♀ **Shrub** page **212**

x *HALIMIOCISTUS WINTONENSIS* 'MERRIST WOOD CREAM'
♀ **Evergreen shrub** page **236**

HEBE 'ALICIA AMHERST'
♀ **Evergreen shrub** Strong-growing, mid-green leaves and dark violet flowers.
↕↔ 1.2m (4ft)

HEBE CUPRESSOIDES 'BOUGHTON DOME'
♀ **Shrub** page **240**

HEBE x *FRANCISCANA* 'VARIEGATA'
♀ **Evergreen shrub** page **241**

HEBE 'LA SÉDUISANTE'
♀ **Evergreen shrub** Large leaves tinted with purple; purple-red flowers in late summer.
↕↔ 1m (3ft)

HYDRANGEA MACROPHYLLA 'ALTONA'
♀ **Shrub** page **266**

HYDRANGEA MACROPHYLLA 'GEOFFREY CHADBUND'
♀ **Shrub** Lacecap with deep red flowers.
↕ 1m (3ft) ↔ 1.5m (5ft)

LEPTOSPERMUM RUPESTRE
♀ **Shrub** page **309**

LUPINUS ARBOREUS
♀ **Shrub** page **326**

MATTHIOLA CINDERELLA SERIES
♀ **Biennial** page **339**

OLEARIA MACRODONTA
Evergreen shrub page **355**

SENECIO CINERARIA 'SILVER DUST'
♀ **Biennial** page **493**

RECOMMENDED TREE AND SOFT FRUIT VARIETIES

The following varieties, if properly cared for, should prove relatively trouble-free and crop reliably.
Apples and pears: cultivars flower at different times, and for effective pollination, and thus good crops, another in the same flowering group (1, 2, 3 or 4, as below) must be nearby.
Plums, cherries: A = needs a pollinator (another cultivar in the same flowering group – 1, 2, 3, 4 or 5); B = crops best with a pollinator. Self-compatible (C) and self-fertile trees crop well alone (and will pollinate other cultivars).

APPLES, COOKING

'ARTHUR TURNER' ♀ 3 Mid-season
'BRAMLEY'S SEEDLING' ♀ 3 Late season
'DUMELOW'S SEEDLING' ♀ 4 Late season
'EDWARD VII' ♀ 6 Late season
'EMNETH EARLY' syn. 'EARLY VICTORIA' ♀ 4 Late season
'GEORGE NEAL' ♀ 6 Late season
'GOLDEN NOBLE' ♀ 4 Mid-season
'GRENADIER' ♀ 3 Early season
'LANE'S PRINCE ALBERT' ♀ 3 Late season
'NEWTON WONDER' ♀ 5 Late season
'PEASGOOD'S NONSUCH' ♀ 3 Mid-season
'WARNER'S KING' ♀ 2 Mid-season

APPLES, DESSERT

'ASHMEAD'S KERNEL' ♀ 4 Late season
'BELLE DE BOSKOOP' ♀ 3 Late season
'BLENHEIM ORANGE' ♀ 3 Mid-season
'CHARLES ROSS' ♀ 3 Mid-season
'CLAYGATE PEARMAIN' ♀ 4 Late season
'DISCOVERY' ♀ 3 Early season
'EGREMONT RUSSET' ♀ 2 Mid-season
'ELLISON'S ORANGE' ♀ 4 Mid-season
'ELSTAR' ♀ 3 Mid-season
'EPICURE' ♀ 3 Early season
'FALSTAFF' ♀ 3 Mid-season
'FIESTA' ♀ 3 Late season
'FORTUNE' SYN. 'LAXTON'S FORTUNE' ♀ 3 Mid-season
'GOLDEN DELICIOUS' ♀ 4 Mid-season
'GREENSLEEVES' ♀ 3 Mid-season
'IDARED' ♀ 2 Late season
'JAMES GRIEVE' ♀ 3 Mid-season
'JONAGOLD' ♀ 3 Late season
'JUPITER' ♀ 3 Late season
'KIDD'S ORANGE RED' ♀ 3 Late season
'KING OF THE PIPPINS' ♀ 5 Mid-season
'KING RUSSET' ♀ 3 Mid-season
'LORD LAMBOURNE' ♀ 2 Mid-season
'MERTON CHARM' ♀ 2 Early season
'MOTHER' syn. 'AMERICAN MOTHER' ♀ 5 Mid season
'PIXIE' ♀ 4 Late season
'RIBSTON PIPPIN' ♀ 2 Mid-season
'ROSEMARY RUSSET' ♀ 3 Late season
'SAINT EDMUND'S PIPPIN' ♀ 2 Mid-season
'SUNSET' ♀ 3 Mid-season
'SUNTAN' ♀ 5 Late season
'WINSTON' ♀ 4 Late season
'WORCESTER PEARMAIN' ♀ 3 Mid-season

BLACKBERRIES

'FANTASIA' ♀ Mid-/late season
'LOCH NESS' ♀ Thornless; mid-/late season

BLACKCURRANTS

'BEN CONNAN' ♀ Early/mid-season
'BEN LOMOND' ♀ Mid-season
'BEN SAREK' ♀ Mid-/late season
'BLACK REWARD' Late season
'JET' Late season

CHERRIES

'MORELLO' ♀ Self-fertile; acid
'STELLA' ♀ Self-fertile; sweet

GOOSEBERRIES

'CARELESS' ♀ Green fruits; mid-season
'GREENFINCH' ♀ Green; mid-season

'INVICTA' ♀ Green; mid-season
'LEVELLER' ♀ Yellow-green; mid-season
'WHINMAN'S INDUSTRY' ♀ Red; mid-season

FIG

'BROWN TURKEY' ♀

HYBRID BERRIES

'SILVANBERRY' ♀
'TAYBERRY' ♀

LOGANBERRIES

'LY 59' ♀ Mid-season
'LY 654' ♀ Thornless; mid-season

MELONS

'AMBER NECTAR' ♀
'EARLIDAWN' ♀
'EARLIQUEEN' ♀
'GALOUBET' ♀
'OGEN' ♀
'SWEETHEART' ♀

PEARS, COOKING

'CATILLAC' ♀ 4 Late season
'PITMASTON DUCHESS' ♀ 4 Mid-season
'WILLIAMS' BON CHRÉTIEN' ♀ 3 Early season

PEARS, DESSERT

'BETH' ♀ 4 Early season
'BEURRÉ HARDY' ♀ 3 Mid-season
'CONCORDE' ♀ 3 Mid-season
'CONFERENCE' ♀ 3 Mid-season
'DOYENNÉ DU COMICE' ♀ 4 Mid-season
'JOSÉPHINE DE MALINES' ♀ 3 Late season
'ONWARD' ♀ 4 Early season
'PITMASTON DUCHESS' ♀ 4 Mid-season
'WILLIAMS' BON CHRÉTIEN' ♀ 3 Early season

PLUMS, COOKING

'CZAR' ♀ Self-fertile; 3; C; Early season
'EARLY LAXTON' ♀ 3; B; Early season
'EDWARDS' ♀ 2; A; Late season
'MARJORIE'S SEEDLING' ♀ 5; C; Late season
'PERSHORE' syn. 'PERSHORE YELLOW EGG' ♀ Self-fertile; 3; C; Early season

PLUMS, DESSERT

'BLUE TIT' ♀ Self-fertile; 5; C
'IMPERIAL GAGE' syn. 'DENNISTON'S SUPERB' ♀ Self-fertile; 2; C; Mid-season
'JEFFERSON' ♀ 1; A; Mid-season
'LAXTON'S DELIGHT' ♀ 3; B; Late season
'OPAL' ♀ 3; C; Early season
'OUILLAN'S GAGE' ♀ Self-fertile; 4; C; Mid-season
'REEVES' ♀ 3; A; Mid-season
'SANCTUS HUBERTUS' ♀ Self-fertile; 3; B; Early season
'VALOR' ♀ A; Mid-season
'VICTORIA' ♀ Self-fertile; 3; C; Mid-season

RASPBERRIES

'AUTUMN BLISS' ♀ Aphid resistant; autumn-fruiting
'GLEN MOY' ♀ Aphid resistant; early season
'GLEN PROSEN' ♀ Mid-season
'LEO' ♀ Some aphid resistance; late season
'MALLING ADMIRAL' ♀ Virus tolerant; mid-/late season
'MALLING DELIGHT' ♀ Some aphid resistance; Early/mid-season
'MALLING JEWEL' ♀ mid-season

REDCURRANTS

'JONKHEER VAN TETS' ♀ Early season
'RED LAKE' ♀ Mid-season
'REDSTART' Late season
'STANZA' ♀ Mid-/late season
'WHITE GRAPE' ♀ Mid-season

STRAWBERRIES

'AROMEL' ♀
'CAMBRIDGE FAVOURITE' ♀ Mid-season
'HAPIL' ♀ Mid-season
'HONEOYE' ♀ Early season
'PEGASUS' ♀ Mid-/Late season
'RHAPSODY' ♀ Late season
'SYMPHONY' ♀ Late season

RECOMMENDED VEGETABLE VARIETIES

New and improved vegetable cultivars are constantly being introduced, both increasing the range of crops that can be grown and ensuring ever-healthier plants. However, there is also a revival of older, more traditional varieties: these will often be listed separately in catalogues under titles such as 'Heritage'. The following lists include not only AGM varieties, but also some relatively new cultivars and others that have proved themselves over time, as well as varieties that grow well in the RHS's vegetable garden at Wisley.

ASPARAGUS

'CONNOVER'S COLOSSAL' 🏆
'FRANKLIM EARLY'
'LIMBRAS' 🏆
'LUCULLAS' 🏆

AUBERGINES

'BONICA' 🏆
'GALINE' 🏆
'MOHICAN' 🏆 White with grey striations
'OVA' 🏆 White
'SLICE RITE'

BEANS, BROAD

'AQUADULCE' 🏆 Early
'AQUADULCE CLAUDIA' 🏆 Early
'EXPRESS' 🏆 Early
'IMPERIAL GREEN LONGPOD' 🏆 Maincrop
'METEOR' 🏆 Maincrop
'RELON' 🏆 Maincrop
'THE SUTTON' 🏆 Early, dwarf
'TOPIC' 🏆 Maincrop
'WITKIEM VROMA' 🏆 Early

BEANS, DWARF FRENCH

'ANNABEL' 🏆
'ARAMIS' 🏆
'CROPPER TEEPEE' 🏆
'DELINEL' 🏆
'PURPLE QUEEN'
'THE PRINCE' 🏆

BEANS, CLIMBING FRENCH

'GOLDMARIE YELLOW'
'KENTUCKY BLUE'
'HUNTER' 🏆
'MASTERPIECE'
'MUSICA' 🏆
'PICKWICK'
'VIOLET PODDED'

BEANS, RUNNER

'ACHIEVEMENT' 🏆
'DESIREE'
'ENORMA' 🏆
'GULLIVER'
'KELVEDON MARVEL' 🏆
'LIBERTY' 🏆

BEETROOT

'BIKORES' 🏆 Round, red
'BOLTARDY' 🏆 Round, red
'BURPEE'S GOLDEN' Round, golden
'CHELTENHAM GREEN TOP' 🏆 Long, red
'FORONO' 🏆 Long, red
'MOBILE' 🏆 Round, red
'MONOGRAM' 🏆 Round, red
'PABLO' 🏆 Round, red
'REGALA' 🏆 Round, red

BROCCOLI

'CLARET' 🏆
'EARLY PURPLE SPROUTING IMPROVED' 🏆
'IMPROVED WHITE'
'LATE PURPLE SPROUTING' 🏆
'RED ARROW' 🏆 Purple sprouting
'RED SPEAR' 🏆 Purple sprouting

BRUSSELS SPROUTS
'CITADEL'
'EDMUND' ♀
'EXTENT' ♀
'ICARUS' ♀ Late
'PEER GYNT'
'STAN' ♀
'TROIKA' Purple

CABBAGES, SPRING
'DUNCAN' ♀
'FIRST EARLY MARKET 218' ♀
'OFFENHAM 1 MYATT'S OFFENHAM COMPACTA' ♀
'PIXIE' ♀ Spring
'SPRING HERO'

CABBAGES, SUMMER/AUTUMN
'DERBY DAY' ♀ Summer
'FIRST OF JUNE' ♀ Summer
'GRENADIER' ♀ Summer
'HISPI' ♀ Early summer
'MINICOLE' Summer/early autumn
'QUICKSTEP' ♀ Late summer/autumn
'STONEHEAD' ♀ Autumn

CABBAGES, WINTER
'FAMOSA' ♀ Savoy
'JULIUS' ♀ Savoy
'MARABEL' ♀ January King Hybrid
'MELISSA' ♀ Savoy
'PROTOVOY' ♀ Savoy
'TRAFALGA' ♀ Savoy
'TUNDRA' Storable, crisphead

CABBAGES, RED
'RODEO' ♀ Early
'ROOKIE' ♀ Early
'RUBY BALL' ♀ Early
'SAPPORO' ♀ Winter, storable

CALABRESE
'ARCADIA' ♀
'CORVET' ♀
'JEWEL' ♀
'SHOGUN' ♀

CARROTS, EARLY
'BANGOR' ♀
'INGOT' ♀
'NANCO' ♀
'NANTES EXPRESS' ♀
'NANTUCKET' ♀
'PARMEX' ♀ Round
'SYTAN' ♀ Some resistance to carrot root fly

CARROTS, MAINCROP
'AUTUMN KING 2' ♀
'BERJO' ♀
'BERTAN' ♀
'FLYAWAY' ♀ Some resistance to carrot root fly
'MOKUM' ♀
'NEW RED' Very long, for exhibition

CAULIFLOWERS, EARLY
'DOK ELGON' ♀
'NAUTILUS'
'PERFECTION' ♀

CAULIFLOWERS, AUTUMN
'AUBADE' ♀
'LIMELIGHT' ♀ Green
'MARMALADE' ♀ Orange
'PLANA' ♀
'RED LION' ♀ Red

CELERIAC
'BRILLIANT'
'MONARCH' ♀

CELERY
'CELEBRITY' ♀
'GIANT PINK-MAMMOTH PINK' ♀
'GREEN CATHEDRAL'
'IVORY TOWER' ♀
'LATHOM SELF BLANCHING' ♀
'STARLIGHT' ♀
'VICTORIA' ♀

CHICORY AND RADICCHIO
'ALLOUETTE' Red
'ANGUSTO' Deep red

'BELGIUM WITLOOF' For forcing chicons
'CRYSTAL HEAD' Sugarloaf

CHINESE CABBAGE
'DASUMI'
'NAGAOKA'
'PAK CHOI-JOYCHOI'

CORN SALAD (LAMB'S LETTUCE)
'VIOLET' Purple
'CAVALLO' ♀
'ENGLISH BROAD-LEAVED'
'JADE'

COURGETTES
'DEFENDER' ♀
'EARLY GEM' ♀
'GOLD RUSH' Yellow
'GREENBUSH'
'SUPREMO' ♀

CUCUMBERS, RIDGE AND GHERKINS
'BUSH CHAMPION' ♀
'FANFARE' Gherkin ♀
'FORTOS' ♀
'MARKETMORE' ♀

CUCUMBERS, UNDER COVER
'ATHENE' ♀
'CARMEN' Resistant to mildew
'DANIMUS' ♀ Small fruit
'FEMDAN' ♀

ENDIVES
'JETI' ♀
'PANCALIERI' ♀
'SALLY' ♀

FLORENCE FENNEL
'ATOS' ♀
'HERACLES' ♀
'ZEFA FINO' ♀

KALE
'BORNICK' ♀
'FRIBOR' ♀
'WINTERBOR' ♀

KOHL RABI
'ADRIANA'
'KONGO GREEN'
'QUICKSTART'
'PURPLE DANUBE' Purple
'RAPID START'
'ROWEL' ♀

LEEKS
'AUTUMN GIANT' Cobra
'CARLTON'
'GIANT WINTER-CATALINA' ♀
'LONGBOW' ♀
'TOLEDO' ♀
'VERINA' ♀
'WINTERREUZEN-GRANADA' ♀

LETTUCE, BUTTERHEAD
'CYNTHIA' Under glass or frame
'DEBBY' ♀
'FORTUNE' Very early
'LILIAN' ♀
'SUMMIT' ♀
'TOM THUMB'

LETTUCE, COS
'BUBBLES' ♀
'CORSAIR' ♀
'JEWEL' ♀
'LITTLE GEM' ♀ Small and compact
'LOBJOIT'S GREEN COS' ♀ Green

LETTUCE, CRISPHEAD
'LAKELAND' ♀
'MALIKA' ♀
'MINIGREEN' ♀
'RED ICEBERG SIOUX'
'WINDERMERE' ♀

LETTUCE, LOOSE-LEAF
'BLACK-SEEDED SIMPSON IMPROVED' ♀
'CATALOGNA' ♀
'COCARDE' ♀

'Frisby' 🏆
'Green Salad Bowl' 🏆
'Lollo Rosso' 🏆
'New Red Fire' 🏆
'Petite Rouge' 🏆
'Raisa' 🏆
'Red Salad Bowl' 🏆
'Rossimo' 🏆

Marrows
'Badger Cross'
'Clarita'
'Minipak'
'Tiger Cross'

Onions, Globe
'Buffalo' 🏆
'Imai Early Yellow' 🏆
'Kelsae'
'Mammoth Red'
'Red Barron'
'Rijnsburger 5 Balstora' 🏆
'Rijnsburger Sito' 🏆
'Royal Oak' 🏆
'Unwin's Exhibition' 🏆

Onions, from Sets
'Centurion' 🏆
'Jagro' 🏆
'Jetset' 🏆
'Sturon' 🏆
'Super Ailsa Craig' 🏆
'Turbo' 🏆

Onions, Salad
'Ishikura' 🏆
'Ramrod' 🏆
'Savel' 🏆
'White Lisbon' 🏆
'White Spear' 🏆
'Winter White Bunching' 🏆

Parsley
'Bravour'
'Favorite'
'Moss Curled 4 - Afro' 🏆
'Plain Leaved'

Parsnips
'Cobham Improved Marrow' 🏆
'Gladiator'
'Javelin' 🏆
'Tender and True' 🏆

Peas, Garden
'Bayard' Leafless
'Cavalier' 🏆 Maincrop
'Early Onward' 🏆 Early
'Feltham First' Early
'Hurst Greenshaft' 🏆 Maincrop
'Kelvedon Wonder' Early

Peas, Sugarpods/Mangetout
'Edula' 🏆
'Honeypod' 🏆
'Oregon Sugarpod' 🏆
'Sugar Snap' 🏆

Potatoes, Early
'Accent'
'Catriona'
'Concorde' 🏆
'Rocket'
'Swift'
'Vanessa' Red

Potatoes, Maincrop
'Avalanche' 🏆
'Cara'
'Croft'
'Famosa' 🏆
'Maxine' 🏆
'Romano' Red
'Stroma' 🏆 Red

Potatoes, Salad
'Charlotte'
'Linzer Delicatess'
'Pink Fir Apple'
'Ratte'

Pumpkins
'Baby Beet' Small, orange
'Jack O'Lantern' Large, orange

RADISHES, SALAD

'CHERRY BELLE' ♀
'FRENCH BREAKFAST' ♀
'SCARLET GLOBE' ♀
'SPARKLER' ♀ White
'SHORT TOP FORCING' ♀
'SUMMER CRUNCH' ♀

RADISH, MOOLI

'APRIL CROSS'
'MINO EARLY'

RADISH, WINTER

'BLACK SPANISH ROUND'
'CHINA ROSE'

SHALLOTS

'ATLANTIC' ♀
'GIANT YELLOW IMPROVED' ♀
'HÂTIVE DE NIORT'
'PIKANT' ♀
'SANTÉ' ♀
'SUCCESS' ♀

SPINACH

'DOMINANT' ♀
'LONG STANDING' ♀
'MAZURKA' ♀
'MONNOPA' ♀
'SIGMALEAF' ♀
'SPACE' ♀
'TRIADE' ♀
'TRIATHLON' ♀

SPINACH BEET AND CHARD

'PERPETUAL SPINACH' Spinach beet
'RAINBOW CHARD' Orange, yellow and pink striped stems
'RHUBARB CHARD' Red stems
SWISS CHARD White stems

SQUASH

'BUTTERNUT' Cream skin, club-shaped
'CUSTARD WHITE' White, flat
'CROWN PRINCE' Steel blue skin
'TIVOLI' Whitish skin, vegetable spaghetti

SWEDES

'MARIAN'
'RUBY'

SWEET CORN

'CANDLE' ♀ Supersweet, early
'FIESTA' ♀ Supersweet
'GOLDEN SWEET' ♀ Supersweet
'GOURMET' ♀ Supersweet
'NORTHERN EXTRA SWEET' ♀ Supersweet
'OVATION' ♀ Supersweet
'PINNACLE' ♀
'SUGAR BOY' ♀ Supersweet
'SUNDANCE' ♀

SWEET PEPPERS

'BELLBOY' ♀ Green to red
'CANAPE' ♀ Green to red
'GYPSY' ♀ Orange
'MAVRAS' ♀ Green to black
'NEW ACE' ♀ Dark green

TOMATOES, OUTDOOR

'GARDENER'S DELIGHT' ♀ Tall, cherry
'GOLD NUGGET' ♀ Tall, cherry, yellow
'OUTDOOR GIRL' ♀ Tall
'SUNGOLD' Yellow
'SWEET 100' Tall, cherry
'TUMBLE' Bush, cherry
'YELLOW PERFECTION' ♀ Tall, yellow

TOMATOES, UNDER COVER

'ALICANTE' Tall
'BLIZZARD' ♀ Tall
'DUMBITO' Beefsteak
'GOLDEN SUNBURST' ♀ Tall, yellow
'SHIRLEY' ♀ Tall
'TIGERELLA' ♀ Tall, striped red, yellow and green

TURNIPS

'MILAN RED' ♀
'MILAN WHITE FORCING' ♀
'PURPLE TOP MILAN'
'TOKYO CROSS' ♀

INDEX

A

B

D

F

G

H

I

M

N

Y

Z

ACKNOWLEDGEMENTS

The publisher would like to thank the following for their kind permission to reproduce the photographs:

l=left, r=right, t=top, c=centre, a=above, b=below.

A-Z Botanical 252tl, Anthony Cooper 312tl, Geoff Kidd 212br, 228br, Malcolm Richards 538tl; **Gillian Beckett** 31tr; **Neil Campbell-Sharp** 72tl, 144tl, 330bl, 336tr; **Garden Picture Library** Brian Carter 47br, 144bl, 320tl, 422tl, 480bl, 493br, John Glover 6, 56bl, 150tl, 219br, 338tl, 342tl, 409br, 511br, 515tr, Sunniva Harte 72bl, Neil Holmes 94bl, Lamontagne 119br, 181br, Jerry Pavia 264tl, Howard Rice 32tl, 77br, 365br, David Russel 508tl, 509tr, JS Sira 402tl; **Derek Gould** 148br; **Diana Grenfell** 259cl; **Photos Horticultural** 55br, 514bl, 531c; **Andrew Lawson** 26, 205tr, 337br; **Clive Nichols** 113br, 193br, 263tr, 271tr, 289br, 325tr, 466tl; **RHS Garden, Wisley** 12; **Eric Sawford** 92tr; **Harry Smith Collection** 47tr, 49tr, 63tr, 68bl, 69tr, 95tr, 110bl, 117l, 118bl, 126tl, 248tl, 300tl, 304tl, 310bl, 319br, 334tl, 344bl, 383tr, 420bl, 481tr, 520bl, 529br.

Dorling Kindersley would also like to thank:

Text contributors and editorial assistance Geoff Stebbings, Candida Frith-Macdonald, Simon Maughan, Andrew Mikolajski, Sarah Wilde, Tanis Smith and James Nugent; at the Royal Horticultural Society, Vincent Square: Susanne Mitchell, Karen Wilson and Barbara Haynes

Design assistance Wendy Bartlet, Ann Thompson

DTP design assistance Louise Paddick

Additional picture research Charlotte Oster, Sean Hunter; special thanks also to Diana Miller, Keeper of the Herbarium at RHS Wisley

Index Ella Skene